A Course in Econometrics

A Course in Econometrics

Arthur S. Goldberger

Harvard University Press
Cambridge, Massachusetts
London, England
1991

Printed in the United States of America
10 9 8 7 6 5 4 3 2 1

This book is printed on acid-free paper, and its binding materials have been chosen for strength and durability.

Library of Congress Cataloging-in-Publication Data

Goldberger, Arthur Stanley, 1930–
A course in econometrics / Arthur S. Goldberger.
p. cm.
Includes bibliographical references and index.
ISBN 0–674–17544–1 (alk. paper)
1. Econometrics. I. Title.
HB139.G634 1991
330′.01′5195—dc20 90–42284
CIP

For Guy H. Orcutt

Contents

Preface

The primary objective of this book is to prepare students for empirical research. But it also serves those who will go on to advanced study in econometric theory. Recognizing that readers will have diverse backgrounds and interests, I appeal to intuition as well as to rigor, and draw on a general acquaintance with empirical economics. I encourage readers to develop a critical sense: students ought to evaluate, rather than simply accept, what they read in journals and textbooks.

The book derives from lecture notes that I have used in the first-year graduate econometrics course at the University of Wisconsin. Students enroll from a variety of departments, including agricultural economics, finance, accounting, industrial relations, and sociology, as well as economics. All have had a year of calculus, a semester of linear algebra, and a semester of statistical inference. Some have had much more course work, including probability theory, mathematical statistics, and econometrics. Others have had substantial empirical research experience.

All of the material can be covered—indeed has been covered—in two semesters. To make that possible, I focus on a few underlying principles, rather than cataloging many potential methods. To accommodate students with varied preparation, the book begins with a review of elementary statistical concepts and methods, before proceeding to the regression model and its variants.

Although the models covered are quite standard, the approach taken is somewhat distinctive. The *conditional expectation function* (CEF) is introduced as the key feature of a multivariate population for economists who are interested in relations among economic variables. The CEF describes how the average value of one variable varies with values of the other variables in the population—a very simple concept. Another key feature of a multivariate population is the linear projection, or best linear predictor (BLP): it provides the best linear approximation to the CEF. Alternative regression models arise according to the sampling scheme used to get drawings from the population.

The focus on CEF's and BLP's is useful. For example, whether a regression specification is "right" or "wrong," least-squares linear regression will typically estimate something in the population, namely, the BLP. Instead of emphasizing the bias (or inconsistency) of least squares, one can consider whether or not the population feature that it does consistently estimate is an interesting one. This approach also avoids visualizing empirical relations as disturbed versions of exact functions. For the most part, "disturbances" are just deviations from a mean, rather than objects that must be added to theoretical relations to produce empirical relations.

The *analogy principle* is relied on to suggest estimators, which are then evaluated according to conventional criteria. Thus least-squares, instrumental-variable, and maximum-likelihood estimators are made plausible by analogy before their sampling properties are studied.

A pedagogical feature of the book is the introduction of technical ideas in simple settings. Many advanced items are covered in the context of simple regression. These include asymptotics, the effect of alternative sampling schemes, and heteroskedasticity-corrected standard errors. The asymptotic theory for the ratio of sample means in sampling from a bivariate population, derived in Chapter 10, serves as a prototype for much more elaborate problems.

From Chapter 16 on, the exercises include real micro-data analyses. These are keyed to the GAUSS programming language, but can readily be adapted to other languages or packages. Virtually all of the exercises have been used as homework assignments or exam questions.

I thank three cohorts of students at Wisconsin, and one class at Stanford (where a portion of the material was used in 1990), for pressing me on details as well as on exposition. Over the years, I have had the benefit of guidance and instruction by several past and present colleagues at Wisconsin, including Guy Orcutt, Harold Watts, Glen Cain, Laurits Christensen, Gary Chamberlain, Charles Manski, and James Powell. I am particularly grateful to Gary Chamberlain for his close critical reading of an early version of the manuscript. Frank Wolak of Stanford provided helpful comments on a later version. They all will recognize their ideas here despite my attempts at camouflage.

I am fortunate to have had the expert editorial advice of Elizabeth Gretz at Harvard University Press, and the proofreading assistance of Donghul Cho and Sangyong Joo. For permission to quote or reproduce their work, I thank Thad W. Mirer, John J. Johnston, and Aptech Systems, Inc. Passages from *Econometric Methods*, 3d ed., by John J. Johnston, copyright © 1984 by McGraw-Hill, are reproduced with permission of McGraw-Hill, Inc.; Table 1.1 is adapted with permission of the Institute for Social Research from *Consumer Behavior of Individual Families over Two and Three Years*, edited by R. Kosobud and J. N. Morgan (Ann Arbor: Institute for Social Research, The University of Michigan, 1964); Table A.6 is reprinted by permission of John Wiley & Sons, Inc., from *Principles of Econometrics* by Henri Theil, copyright © 1971 by John Wiley & Sons; Tables A.3 and A.5 are reprinted by permission of Macmillan Publishing Company from *Economic Statistics and Econometrics*, 2d ed., by Thad W. Mirer, copyright © 1988 by Macmillan Publishing Company.

Madison, Wisconsin
November 1990

A Course in Econometrics

1 *Empirical Relations*

1.1. Theoretical and Empirical Relations

Most of economics is concerned with relations among variables. For example, economists might consider how

- the output of a firm is related to its inputs of labor, capital, and raw materials;
- the inflation rate is related to unemployment, change in the money supply, and change in the wage rate;
- the quantity demanded of a product depends on household income, price of the product, and prices of substitute products;
- the proportion of income saved varies with the level of family income;
- the earnings of a worker are related to her age, education, race, region of residence, and years of work experience.

In theoretical economics, the relations are characteristically treated as exact relations, that is, as deterministic relations, that is, as (single-valued) functions. For example, consider the relation between savings and income. Let

Y = savings rate = savings/income = proportion of income saved,
X = income.

In theoretical economics, one might consider $Y = g(X)$, where $g(\cdot)$ is a function in the mathematical sense, that is, a single-valued function. Henceforth we will always use the word "function" in this strict sense. Corresponding to each value of X, there is a unique value of Y. An economist might ask such questions as: Is $g(X)$ constant with respect to X? Is $g(X)$ increasing in X? Is $g(X)$ linear in X?

The same applies when there are several explanatory variables $X_1, \ldots, X_k$, as when a firm's output is related to its inputs of labor, capital, and raw materials. In theory one considers $Y = g(X_1, \ldots, X_k)$, where corresponding to each set of values for $X_1, \ldots, X_k$, there is a unique value of Y. So g is again a (single-valued) function. The relation of Y to the X's is an exact one, that is, a deterministic one.

This is what relations look like in theory. What happens when we look at *empirical relations*, that is, at real-world data on economic variables?

Table 1.1 refers to 1027 U.S. "consumer units" (roughly, families) interviewed by the University of Michigan's Survey Research Center in 1960, 1961, and 1962. Income is averaged over the two years 1960 and 1961; the savings rate is the ratio of two-year savings to two-year income. In the source, the data were presented in grouped form, with ten brackets for income and nine brackets for the savings rate. For convenience, we have assumed that all observations in a bracket were located at a single point (the approximate midpoint of the bracket) and have labeled the values of X and Y accordingly. Across the top of the table are the ten distinct values of X = income (in thousands of dollars), which we refer to as x_i ($i = 1, \ldots, 10$). Down the left-hand side of the table are the nine distinct values of Y = savings rate, which we refer to as y_j

Table 1.1 Joint frequency distribution of X = income and Y = savings rate.

	X_i									
Y_j	0.5	1.5	2.5	3.5	4.5	5.5	6.7	8.8	12.5	17.5
.50	.001	.011	.007	.006	.005	.005	.008	.009	.014	.004
.40	.001	.002	.006	.007	.010	.007	.008	.009	.008	.007
.25	.002	.006	.004	.007	.010	.011	.020	.019	.013	.006
.15	.002	.009	.009	.012	.016	.020	.042	.054	.024	.020
.05	.010	.023	.033	.031	.041	.029	.047	.039	.042	.007
0	.013	.013	.000	.002	.001	.000	.000	.000	.000	.000
−.05	.001	.012	.011	.005	.012	.016	.017	.014	.004	.003
−.18	.002	.008	.013	.006	.009	.008	.008	.008	.006	.002
−.25	.009	.009	.010	.006	.009	.007	.005	.003	.002	.003
$p(x)$	.041	.093	.093	.082	.113	.103	.155	.155	.113	.052

Source: Adapted from R. Kosobud and J. N. Morgan, eds., *Consumer Behavior of Individual Families over Two and Three Years* (Ann Arbor: Institute for Social Research, The University of Michigan, 1964), Table 5–5.

($j = 1, \ldots, 9$). So there are $90 = 10 \times 9$ cells in the cross-tabulation. In the i, j cell one finds

$$p(x_i, y_j) = \text{the proportion of the 1027 families who reported the combination } (X = x_i \text{ and } Y = y_j).$$

This table gives the *joint frequency distribution* of Y and X for this data set.

Here is some general notation for a joint frequency distribution of variables X and Y, where X takes on distinct values x_i ($i = 1, \ldots, I$) and Y takes on distinct values y_j ($j = 1, \ldots, J$). The joint frequencies $p(x_i, y_j)$ are defined for each of the $I \times J$ cells. Clearly $\Sigma_i \Sigma_j p(x_i, y_j) = 1$, where $\Sigma_i = \Sigma_{i=1}^{I}$, $\Sigma_j = \Sigma_{j=1}^{J}$. From the joint frequency distribution it is easy to calculate the *marginal frequency distribution* of X:

$$\begin{aligned} p(x_i) &= \sum_j p(x_i, y_j) \\ &= \text{proportion of observations having } X = x_i \\ &\quad (i = 1, \ldots, I). \end{aligned}$$

Then $\Sigma_i p(x_i) = \Sigma_i[\Sigma_j p(x_i, y_j)] = 1$.

Return to the joint frequency distribution of Table 1.1. For each of the ten columns, add down the rows to get the marginal frequency distribution $p(x)$ in the last row—the bottom margin—and observe that the entries in the last row do add up to 1.

Evidently, the empirical relation between Y and X is not a deterministic one. For if it were, then in any column of the body of the table, there would be only a single nonzero entry. But in every column, there are several nonzero entries. Indeed, in most columns, all nine entries are nonzero. Corresponding to each value of X, there is a whole set of values of Y rather than a single value of Y. What we see is a distribution rather than a function. This is characteristic of the real world: empirical relations are not deterministic, not exact, not functional relations.

Now focus attention on the distribution of Y corresponding to a particular value of X. Take $X = x_i$, say, and ask what proportion of the observations that have $X = x_i$, also have the values $Y = y_1, \ldots, y_J$. The answers give the *conditional frequency distribution* of Y given $X = x_i$:

$$p(y_j | x_i) = \frac{p(x_i, y_j)}{p(x_i)} \qquad (j = 1, \ldots, J).$$

It follows for each $i = 1, \ldots, I$ that

$$\sum_j p(y_j|x_i) = \sum_j \frac{p(x_i, y_j)}{p(x_i)} = \frac{\Sigma_j p(x_i, y_j)}{p(x_i)} = \frac{p(x_i)}{p(x_i)} = 1.$$

Divide the entries in each column of Table 1.1 by the column sum. The resulting Table 1.2 gives the conditional frequency distributions of Y given X, one such distribution for each distinct value of X. Observe that each column sum in this table is equal to 1. The nondeterministic character of empirical relations is again apparent. If $Y = g(X)$ as in theoretical economics, each column in the body of Table 1.2 would have a single unity, all other entries being zero. But Table 1.2 does not look like that.

So we face a dilemma. We would like to use economic theory to guide our analysis of data, and to use data to implement the theory. But the savings-income relation in economic theory is deterministic, while in the empirical data it is not deterministic. How shall we resolve the dilemma?

The theory seems to say that all families with the same value of X should have the same Y. If so, the data seem to indicate that these families did not do what they should. Perhaps they tried to, but made mistakes? If so, the conditional distributions are all due to error—there

Table 1.2 Conditional frequency distributions of Y = savings rate for given values of X = income.

	X									
Y	0.5	1.5	2.5	3.5	4.5	5.5	6.7	8.8	12.5	17.5
.50	.024	.118	.075	.073	.044	.049	.052	.058	.124	.077
.40	.024	.022	.064	.086	.088	.068	.052	.058	.071	.135
.25	.049	.064	.043	.086	.088	.107	.129	.123	.115	.115
.15	.049	.097	.097	.146	.142	.194	.271	.348	.212	.384
.05	.244	.247	.355	.378	.363	.281	.303	.252	.372	.135
0	.317	.140	.000	.024	.009	.000	.000	.000	.000	.000
−.05	.024	.129	.118	.061	.106	.155	.109	.090	.035	.058
−.18	.049	.086	.140	.073	.080	.078	.052	.052	.053	.038
−.25	.220	.097	.108	.073	.080	.068	.032	.019	.018	.058
Total	1.000	1.000	1.000	1.000	1.000	1.000	1.000	1.000	1.000	1.000
$m_{Y\|X}$	−.012	.065	.048	.099	.079	.083	.112	.129	.154	.161

is a true value of Y for each value of X, but the families erred in their savings behavior, or perhaps in reporting savings to the interviewer. This is surely possible, but to rely on errors alone is unappealing.

We know that the families differ in characteristics other than income that may be relevant to their savings behavior. The gap between theory and reality might diminish if the theory introduced more explanatory variables, $X_2, \ldots, X_k$. Then instead of looking at $p(y_j|x_i)$ we would be looking at $p(y_j|x_{1i}, \ldots, x_{kh})$. Presumably there will be less dispersion of Y within those narrowly defined cells than there is in the coarsely defined cells of our tables. But even then the empirical relation would not be deterministic. For example, consider all households who have the same income, family size, and race. We would still see differences in their Y values. Because a gap would remain in any case, for present purposes we may as well continue with the single-X case.

1.2. Sample Means and Population Means

To resolve the dilemma, we first reinterpret the economic theory. When the theorist speaks of Y being a function of X, let us say that she means that the *average* value of Y is a function of X. If so, when she says that $g(X)$ increases with X, she means that on average, the value of Y increases with X. Or, when she says that $g(X)$ is constant, she means that the average value of Y is the same for all values of X. With that interpretation in mind, let us re-examine our data set, seeking the empirical counterpart of the theorist's average value.

Here is some algebra that shows how to calculate the average of a frequency distribution. First, for the variable X = income: if the marginal frequency distribution of X is given by $p(x_i)$ ($i = 1, \ldots, I$), then the *marginal mean* of X is

$$m_X = \sum_i x_i p(x_i).$$

Similarly, the marginal mean of Y is $m_Y = \Sigma_j y_j p(y_j)$. Further, if the conditional frequency distribution of Y given $X = x_i$ is $p(y_j|x_i)$ ($j = 1, \ldots, J$), then the *conditional mean* of Y given $X = x_i$ is

$$m_{Y|x_i} = \sum_j y_j p(y_j|x_i).$$

There are I such conditional means, one for each distinct value of X.

Observe that the average of the conditional means equals the marginal mean:

$$\begin{aligned}\sum_i m_{Y|x_i} p(x_i) &= \sum_i \left[\sum_j y_j p(y_j \mid x_i)\right] p(x_i) \\ &= \sum_i \sum_j y_j p(x_i, y_j) \\ &= \sum_j y_j \left[\sum_i p(x_i, y_j)\right] = \sum_j y_j p(y_j) = m_Y.\end{aligned}$$

?

Return to Table 1.2. The conditional means of Y have been calculated, one for each of the ten values of X, and are presented in the last row of the table. If we extract the top row (the x_i) and the bottom row (the $m_{Y|x_i}$), we have the *conditional mean function*, or cmf, for Y given X, which we will refer to as $m_{Y|X}$.

The cmf is a deterministic relation—that is, a function—in our data. The cmf specifies how the average value of Y is functionally related to X in the data set. For an economist who is concerned with the relation of the savings rate to income, this cmf $m_{Y|X}$ is the most interesting feature of the joint frequency distribution. We can plot it, and study it in terms of the economic theorist's concerns: Does $m_{Y|X}$ vary with X? Does it vary linearly with X, that is, is $\Delta m/\Delta X$ constant?

In Figure 1.1, the ten points that make up the cmf are plotted and, for convenience, are connected by line segments. Looking at the plot, we see a cmf that is too ragged and erratic to be taken seriously by a theorist. So a gap remains between theory and reality.

To proceed, we recognize that the theorist who discussed the relation between the savings rate and income was not talking about $m_{Y|X}$ for these particular 1027 families in 1960–1961. If the Survey Research Center had happened to interview a different 1027 families, or even a 1028th family, or even the same 1027 families in a different year, we would have had a different $p(x, y)$ table, different $p(y \mid x)$ columns, and no doubt a different $m_{Y|X}$ function.

The next step is obvious. We suppose that what we observe is only a *sample* from a *population*. Our cmf displays sample means, not population means. Presumably the theorist was referring to population means, not sample means. It will be adequate for present purposes to think of the population itself as represented by a joint frequency distribution, one that refers to millions of families rather than to our 1027. (For conve-

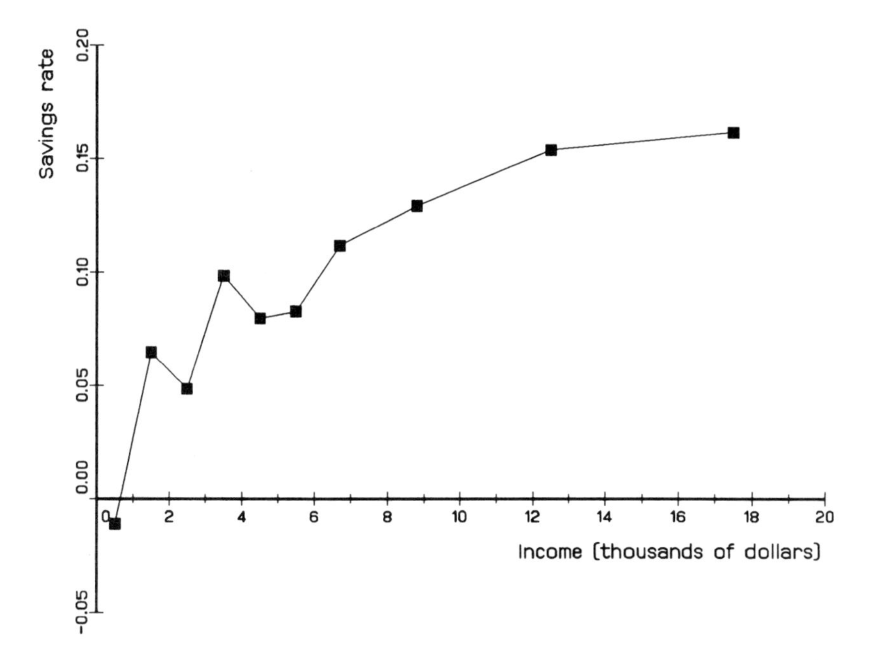

Figure 1.1 Conditional mean function: savings rate on income.

nience, we continue to suppose that the X, Y pairs are confined to the same 90 combinations.) In the population the joint frequencies are given by $\pi(x_i, y_j)$, say, with $\Sigma_i \Sigma_j \pi(x_i, y_j) = 1$. So in the population, the marginal frequencies of X are

$$\pi(x_i) = \sum_j \pi(x_i, y_j),$$

and the conditional frequencies of Y given X are

$$\pi(y_j \mid x_i) = \pi(x_i, y_j)/\pi(x_i).$$

Further, the population mean of X is

$$\mu_X = \sum_i x_i \pi(x_i),$$

and the population conditional means of Y given X are

$$\mu_{Y \mid x_i} = \sum_j y_j \pi(y_j \mid x_i).$$

We have arrived at the following position. When a theorist talks about $Y = g(X)$, she is really referring to the population conditional mean function $\mu_{Y|X} = g(X)$, which is indeed a function of X. We now have the theorist referring to the population features $\pi(x, y)$, $\pi(y \mid x)$, $\mu_{Y|X}$, while the empirical material refers to the sample features $p(x, y)$, $p(y \mid x)$, $m_{Y|X}$. This leaves us with the gap between the hypothetical population π's and μ's and the observed sample p's and m's.

1.3. Sampling

Imagine this physical representation of the population and sample. Each family in the population is represented by a chip on which its (X, Y) pair is printed. The millions of chips are in a barrel. The joint frequency distribution in the barrel is $\pi(x, y)$. We draw 1027 chips with replacement, record the x, y combinations, and tabulate the (relative) frequencies as $p(x, y)$. Our $p(x, y)$ table is just one of the possible $p(x, y)$ tables that might have been obtained in this manner. Our data set is just one sample from the population. In general none of the possible sample $p(x, y)$ tables will be identical to the population $\pi(x, y)$ table, and none of the possible sample $m_{Y|X}$ functions will coincide with the population $\mu_{Y|X}$ function.

The dilemma has been substantially resolved. The questions that remain include: What sort of samples come from a population? How do sample joint frequency distributions, conditional frequency distributions, and cmf's depart from the population joint frequency distribution, conditional frequency distribution, and cmf? How can we best use a sample to learn about the population from which it came? How confident can we be in our conclusions? These are precisely the questions that are addressed in classical statistical theory.

1.4. Estimation

A large part of empirical econometrics is concerned with estimating population conditional mean functions from a sample. That is, economists very often want to learn how the average value of one variable varies in a population with one, or several, other variables.

If so, what remains to be discussed? After all, in introductory statistics courses, we have learned all about estimating population means. In particular we have learned that the sample mean is an attractive estimator of a population mean—perhaps even that it is the best estimator. That attractiveness should carry over to the present situation, where we are concerned with a population conditional mean *function*. A population cmf is just a set of population means, and our joint sample can be viewed as a collection of conditional subsamples. So it is natural to use the sample conditional means as estimates of the population conditional means. That is, it is natural to take $m_{Y|x_i}$ as the estimate of $\mu_{Y|x_i}$, thus taking $m_{Y|X}$ as the estimate of $\mu_{Y|X}$, bearing in mind that $m \neq \mu$.

But is that always the right way to proceed? Suppose that an economic theory says that the population cmf for the savings rate on income is linear: $\mu_{Y|X} = \alpha + \beta X$, with α and β unknown. As Figure 1.1 shows, our sample $m_{Y|X}$ is not linear in income—the ten $m_{Y|x_i}$'s do not fall on a straight line. As empirical economists who wish to be guided by economic theory, shall we retain the ten sample $m_{Y|x_i}$'s as they stand? Or shall we smooth the sample $m_{Y|x_i}$'s by fitting a straight line to the ten points, obtaining $m^*_{Y|X} = a + bX$, and use those $m^*_{Y|x_i} = a + bx_i$ $(i = 1, \ldots, 10)$ as the estimates of the $\mu_{Y|x_i}$, thus using a and b as the estimates of α and β? If we decide to smooth, how shall we fit the line? And do we know that the smoothed estimates are better than the sample means as estimates of the population means? Or suppose a theory said that the population cmf is exponential: $\mu_{Y|X} = \alpha X^{\beta}$. How should we fit that *curve*? And does the smoothed sample line $m^*_{Y|X}$ still tell us anything about the population curve $\mu_{Y|X}$?

These are typical of the issues that arise in this book. To address them seriously, we turn to a review of the framework provided by the random variable–probability distribution model of classical statistics.

Exercises

The following all refer to the empirical joint frequency distribution of Tables 1.1 and 1.2.

1.1 Calculate the marginal frequency distribution of Y. Then calculate the mean of the conditional means of Y, verifying that it equals the marginal mean of Y (up to round-off error).

1.2 Calculate the conditional frequency distributions of X given Y, and the conditional mean function of X given Y.

1.3 Plot the two conditional mean functions $m_{Y|X}$ and $m_{X|Y}$ on a single diagram, using the horizontal axis for x-values and the vertical axis for y-values. Comment on the differences between those two functions.

1.4 Let Z = savings (in thousands of dollars), so $Z = XY$. The savings of a family with income x_i and savings rate y_j is $z_{ij} = x_i y_j$, so that the mean savings for the families in our sample is given by

$$m_Z = \sum_i \sum_j x_i y_j p(x_i, y_j).$$

Will this equal $m_X m_Y$? That is, can mean savings be obtained by multiplying mean savings rate by mean income? Explain.

2 *Univariate Probability Distributions*

2.1. Introduction

The general framework of probability theory involves an experiment that has various possible outcomes. Each distinct outcome is represented as a point in a set, the sample space. Probabilities are assigned to certain outcomes in accordance with certain axioms, and then the probabilities of other events, which are subsets of the sample space, are deduced. Let S denote the sample space, A denote an event, and $\Pr(\cdot)$ the probability assignment. Then the axioms are $0 \leq \Pr(A) \leq 1$, $\Pr(S) = 1$, and, where $A_1, A_2, \ldots$ are disjoint events, $\Pr(\cup_j A_j) = \Sigma_j \Pr(A_j)$.

We proceed somewhat more concretely. The distinct possible outcomes are identified, that is, distinguished, by the value of a single variable X. Each trial of the experiment produces one and only one value of X. Here X is called a *random variable*, a label that merely indicates that X is a variable whose value is determined by the outcome of an experiment. The values that X takes on are denoted by x. So we may refer to events such as $\{X = x\}$ and $\{X \leq x\}$. We distinguish two cases of probability distributions: discrete and continuous.

2.2. Discrete Case

In the *discrete case*, the number of distinct possible outcomes is either finite or countably infinite, so one can compile a list of them: x_1, x_2, ... The convention is to list these *mass points* in increasing order: $x_1 < x_2 < \ldots$. The assignment of probabilities is done via a function $f(x)$, with these properties:

$$f(x) \geq 0 \quad \text{everywhere,}$$

$$f(x) = 0 \quad \text{except at the mass points } x_1, x_2, \ldots,$$

$$\sum_i f(x_i) = 1,$$

where Σ_i denotes summation over all the mass points. The function $f(\cdot)$ is called a *probability mass function*, or pmf.

The initial assignment of probabilities is $\Pr(X = x) = f(x)$. That is, the probability that the random variable capital X takes on the value lowercase x is $f(x)$. Then the probabilities of various events are deducible by rules of probability theory. For example, supposing that the list is in increasing order, $\Pr(X \leq x_5) = \Sigma_{i=1}^{5} f(x_i)$. For another example, if x_0 is not a mass point, then $\Pr(X = x_0) = f(x_0) = 0$.

Observe that the pmf $f(\cdot)$ has exactly the formal properties that $p(\cdot)$ had in univariate *frequency* distributions. (And observe the perverse notation: we used $p(\cdot)$ for *f*requency distributions, and now use $f(\cdot)$ for *p*robability distributions.) Because of the formal resemblance, it may be helpful to interpret the pmf $f(\cdot)$ as the $\pi(\cdot)$ of Chapter 1, namely the frequency distribution in a population.

Here are several examples of discrete univariate probability distributions:

(1) *Bernoulli* with parameter p $(0 \leq p \leq 1)$. Here

$$f(x) = p^x(1 - p)^{(1-x)} \quad \text{for } x = 0, 1,$$

with $f(x) = 0$ elsewhere. So $\Pr(X = 0) = f(0) = p^0(1 - p)^1 = 1 - p$, $\Pr(X = 1) = f(1) = p^1(1 - p)^0 = p$, and $\Pr(X = x) = f(x) = 0$ for all other values of x. Observe that $f(x) \geq 0$ everywhere, that $f(x) = 0$ except at the two mass points $x = 0$ and $x = 1$, and that $\Sigma_i f(x_i) = f(0) + f(1) = 1$, as required. So this is a legitimate pmf.

In what contexts might the Bernoulli distribution be relevant? That is, for what experiments might it be appropriate? A familiar example is a coin toss: $X = 1$ if heads, $X = 0$ if tails. The Bernoulli pmf says that $\Pr(X = 1) = p$, $\Pr(X = 0) = 1 - p$. Special cases are $p = 0.5$ (fair coin), and $p = 0.7$ (loaded coin). A more interesting example concerns unemployment. Let $X = 1$ if unemployed, $X = 0$ otherwise, the experiment being drawing an adult at random from the U.S. population. The Bernoulli pmf says that the probability of being unemployed is p.

(2) *Discrete Uniform* with parameter N (N positive integer). Here

$$f(x) = 1/N \quad \text{for } x = 1, 2, \ldots, N,$$

with $f(x) = 0$ elsewhere. Observe that $f(x) \geq 0$ everywhere, that $f(x) = 0$ except at the N mass points, and that $\Sigma_i f(x_i) = 1/N + \cdots + 1/N = 1$. So this is a legitimate pmf.

In what contexts might a discrete uniform distribution be relevant? A very familiar example is the roll of a fair die: X = the number on the face that comes up, and $N = 6$. This discrete uniform pmf says that $\Pr(X = 1) = \Pr(X = 2) = \cdots = \Pr(X = 6) = 1/6$.

(3) *Binomial* with parameters n, p (n positive integer, $0 \leq p \leq 1$). Here

$$f(x) = \frac{n!}{x!(n-x)!} p^x(1-p)^{(n-x)} \quad \text{for } x = 0, 1, 2, \ldots, n,$$

with $f(x) = 0$ elsewhere. (Recall factorial notation: $0! = 1$, $1! = 1$, $2! = 2$, $3! = 6$, ...) Observe that $f(x) \geq 0$ everywhere, that $f(x) = 0$ except at the mass points, and (as can be confirmed by summing from 0 to n) that $\Sigma_i f(x_i) = [p + (1-p)]^n = 1$.

In what contexts might the binomial distribution be relevant? Suppose we toss n identical coins at once, and let X = number of heads. That is, we run the Bernoulli(p) experiment n times, independently, and record the number of 1's. Or if we observe an adult over n months, let X = number of months unemployed. The binomial distribution may be appropriate.

Special cases of the binomial include:

(a) $$n = 1: f(0) = 1 \times p^0(1-p)^1 = (1-p),$$
$$f(1) = 1 \times p^1(1-p)^0 = p.$$

So the binomial distribution with parameters $(1, p)$ is the same as the Bernoulli distribution with parameter p.

(b) $$n = 2: f(0) = 1 \times p^0(1-p)^2 = (1-p)^2,$$
$$f(1) = 2 \times p^1(1-p)^1 = 2p(1-p),$$
$$f(2) = 1 \times p^2(1-p)^0 = p^2.$$

Clearly $f(0) + f(1) + f(2) = [p + (1-p)]^2 = 1$.

(4) *Poisson* with parameter λ ($\lambda > 0$). Here

$$f(x) = e^{-\lambda} \lambda^x/x! \quad \text{for } x = 0, 1, 2, \ldots,$$

with $f(x) = 0$ elsewhere. Observe that $f(x) \geq 0$ everywhere, that $f(x) = 0$ except at the mass points, and (using the series expansion

$$e^{\lambda} = \sum_{x=0}^{\infty} (\lambda^x/x!) = 1 + \lambda + \lambda^2/2 + \lambda^3/6 + \ldots)$$

that $\Sigma_i f(x_i) = 1$. In the Poisson distribution, the number of distinct possible outcomes is countably infinite.

Applications of the Poisson distribution might include the number of phone calls received at a switchboard in an hour, or the number of job offers an individual receives in a year.

2.3. Continuous Case

In the *continuous case*, we again consider an experiment whose outcomes are distinguished by the value of a single real variable X. But now there is a continuum of distinct possible outcomes, so we cannot compile them in a list.

The assignment of probabilities is done via a function $f(x)$ with these properties: $f(x) \geq 0$ everywhere, $\int_{-\infty}^{\infty} f(x)\, dx = 1$. This function $f(\cdot)$ is called a *probability density function*, or pdf. The initial assignment of probabilities via $f(x)$ is as follows. For any pair of numbers a, b with $a \leq b$:

$$\Pr(a \leq X \leq b) = \int_a^b f(x)\, dx.$$

That is, the probability that the random variable X lies in the closed interval $[a, b]$ is given by the area under the $f(x)$ curve between the points a and b.

To see what we are committed to in the continuous case, consider several specific events:

(1) $A = \{-\infty \leq X \leq \infty\}$.

Here $a = -\infty$, $b = \infty$, so $\Pr(A) = \int_{-\infty}^{\infty} f(x)\, dx = 1$, as it should, since A exhausts the sample space.

(2) $A = \{X \leq b\} = \{-\infty \leq X \leq b\}$.

Here $a = -\infty$, so $\Pr(A) = \int_{-\infty}^{b} f(x)\, dx$.

(3) $\qquad A = \{X = a\} = \{a \leq X \leq a\}.$

Here $b = a$, so $\Pr(A) = \int_a^a f(x)\,dx = 0$.

Consider (3), which says that in the continuous case $\Pr(X = x) = 0$ for every x. This means that the probability that X takes on a particular value x is zero, for every such particular value. And yet on every run of the experiment some value of x is taken on. Is that a contradiction? No, not unless one confuses two distinct concepts, zero probability and impossibility. In the continuous case, a zero-probability event is not an impossible event. Although this seems awkward, no other assignment of probabilities to events of the form $\{X = x\}$ is possible when the distinct possible outcomes form a continuum.

Further, the following events all have the same probability, namely $\int_a^b f(x)\,dx$:

$$A_1 = \{a \leq X \leq b\}, \qquad A_2 = \{a \leq X < b\},$$
$$A_3 = \{a < X \leq b\}, \qquad A_4 = \{a < X < b\}.$$

For example, $A_1 = A_2 \cup A_0$ where $A_0 = \{X = b\}$. But A_2 and A_0 are disjoint, and $\Pr(A_0) = 0$, so $\Pr(A_2) = \Pr(A_1)$.

The *cumulative distribution function*, or cdf, is defined as

$$F(x) = \int_{-\infty}^{x} f(t)\,dt,$$

with t being a dummy argument. The cdf gives the area under the pdf from $-\infty$ up to x, so $F(x) = \Pr(X \leq x)$. Some properties of a cdf are immediate:

- $F(-\infty) = 0$, $F(\infty) = 1$.
- $F(\cdot)$ is monotonically nondecreasing (because $f(t) \geq 0$).
- Wherever differentiable, $dF(x)/dx = f(x)$, because $F = \int f(t)\,dt$, and the derivative of an integral with respect to its upper limit is just the argument (the integrand) evaluated at the upper limit.

In the continuous case the cdf is convenient because

$$\Pr(a \leq X \leq b) = \int_a^b f(x)\,dx = \int_{-\infty}^{b} f(x)\,dx - \int_{-\infty}^{a} f(x)\,dx$$
$$= F(b) - F(a).$$

The cdf could have been introduced in the discrete case as $F(x) = \Pr(X \leq x)$, but it is not so crucial there.

Here are several examples of continuous univariate probability distributions:

(1) *Rectangular* (or continuous uniform) on the interval $[a, b]$, with parameters $a < b$. The pdf is

$$f(x) = 1/(b - a) \quad \text{for } a \le x \le b,$$

with $f(x) = 0$ elsewhere. Observe that $f(x) \ge 0$ everywhere, and that

$$\int_{-\infty}^{\infty} f(x)\, dx = \int_{-\infty}^{a} 0\, dx + \int_{a}^{b} [1/(b - a)]\, dx + \int_{b}^{\infty} 0\, dx$$
$$= [1/(b - a)]\, x \|_a^b = 1.$$

(Note: The symbol $\|$ is used to denote an integral to be evaluated.) So this is a legitimate pdf. It plots as a rectangle, with base $b - a$ and height $1/(b - a)$; the area of the rectangle is base $\times$ height $= 1$. The cdf is

$$F(x) = \int_{-\infty}^{x} f(t)\, dt = \begin{cases} 0 & \text{for } x < a, \\ (x - a)/(b - a) & \text{for } a \le x \le b, \\ 1 & \text{for } b < x. \end{cases}$$

(2) *Exponential* with parameter $\lambda > 0$. The pdf is

$$f(x) = \lambda\, e^{-\lambda x} \quad \text{for } x > 0,$$

with $f(x) = 0$ for $x \le 0$. The relevant indefinite integral is

$$\int \lambda\, e^{-\lambda t}\, dt = \lambda \int e^{-\lambda t}\, dt = \lambda\, (e^{-\lambda t})/(-\lambda) = -e^{-\lambda t},$$

so the cdf is

$$F(x) = \begin{cases} 0 & \text{for } x \le 0, \\ 1 - e^{-\lambda x} & \text{for } x > 0. \end{cases}$$

The exponential pdf and cdf for $\lambda = 2$ are plotted in Figure 2.1.

The exponential distribution may be appropriate for the length of time until a light bulb fails. It may also be relevant for the duration of unemployment among those who leave a job, with time being measured continuously.

(3) *Standard Normal.* The standard normal distribution plays a central role in statistical theory. The pdf is

$$f(x) = (2\pi)^{-1/2} \exp(-x^2/2),$$

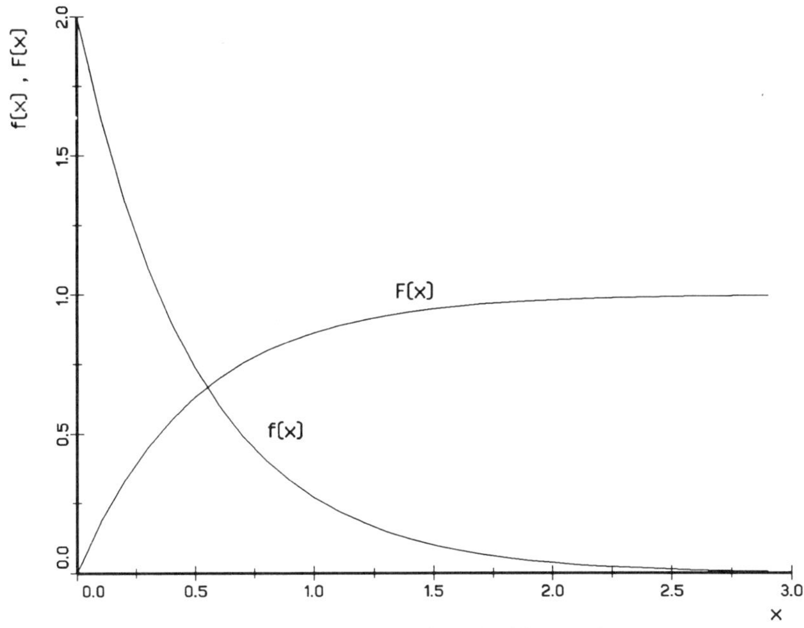

Figure 2.1 Exponential distribution: pdf and cdf, $\lambda = 2$.

which plots as a familiar bell-shaped curve, as in Figure 2.2. Some features of the curve are apparent by inspection of the pdf formula. The curve is symmetric about zero: $f(-x) = f(x)$. The ordinate at zero is $f(0) = 1/\sqrt{(2\pi)} = 0.3989$. The slope is

$$f'(x) = (2\pi)^{-1/2} \exp(-x^2/2)(-x) = -xf(x).$$

So $f'(x) > 0$ for $x < 0$, $f'(x) = 0$ for $x = 0$, $f'(x) < 0$ for $x > 0$. The second derivative is $f''(x) = -[xf'(x) + f(x)] = -[-x^2 f(x) + f(x)] = (x^2 - 1)f(x)$. So the curve has inflection points at $x = 1$ and $x = -1$.

The cdf is

$$F(x) = \Pr(X \leq x) = \int_{-\infty}^{x} f(t)\, dt.$$

No closed form is available, but the standard normal cdf is plotted in Figure 2.2 and tabulated in Table A.1. The tabulation is confined to $x > 0$, which suffices because the symmetry of $f(x)$ about 0 implies that $F(-x) = 1 - F(x)$.

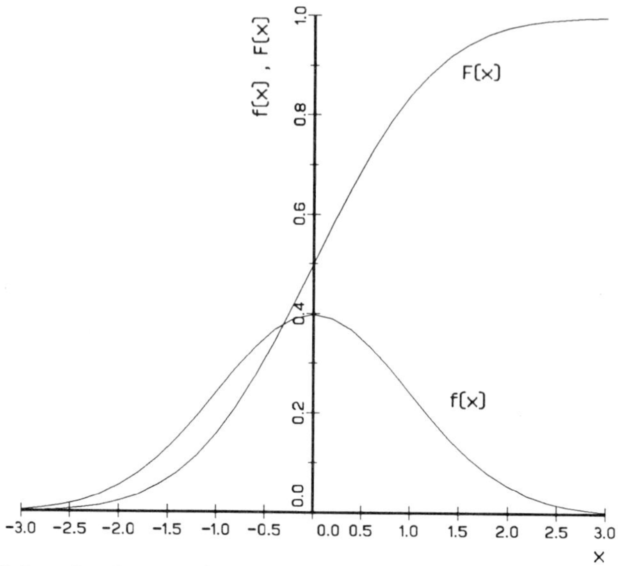

Figure 2.2 Standard normal distribution: pdf and cdf.

(4) *Standard Logistic.* The pdf is

$$f(x) = e^x/(1 + e^x)^2.$$

It is easy to verify that this pdf plots as a symmetric bell-shaped curve, very similar to the standard normal, and that the cdf is

$$F(x) = \int_{-\infty}^{x} f(t)\, dt = e^x/(1 + e^x).$$

(5) *Power* on interval [0, 1] with parameter $\theta > 0$. The pdf is

$$f(x) = \theta\, x^{\theta - 1} \quad \text{for } 0 \leq x \leq 1,$$

with $f(x) = 0$ elsewhere. The relevant indefinite integral is

$$\int \theta\, t^{\theta - 1}\, dt = \theta \int t^{\theta - 1}\, dt = (\theta/\theta)\, t^{\theta} = t^{\theta},$$

so the cdf is

$$F(x) = \begin{cases} 0 & \text{for } x < 0, \\ x^{\theta} & \text{for } 0 \leq x \leq 1, \\ 1 & \text{for } x > 1. \end{cases}$$

The power pdf and cdf for $\theta = 3$ are plotted in Figure 2.3. The power distribution may have no natural application, but we will use it for examples because the integration is so simple.

2.4. Mixed Case

A mixture of the discrete and continuous cases may also be relevant. Let X = dollars spent in a year on car repairs, the experiment being drawing a family at random from the U.S. population. An appropriate model would allow for a mass point at $X = 0$, with a continuous distribution over positive values of X.

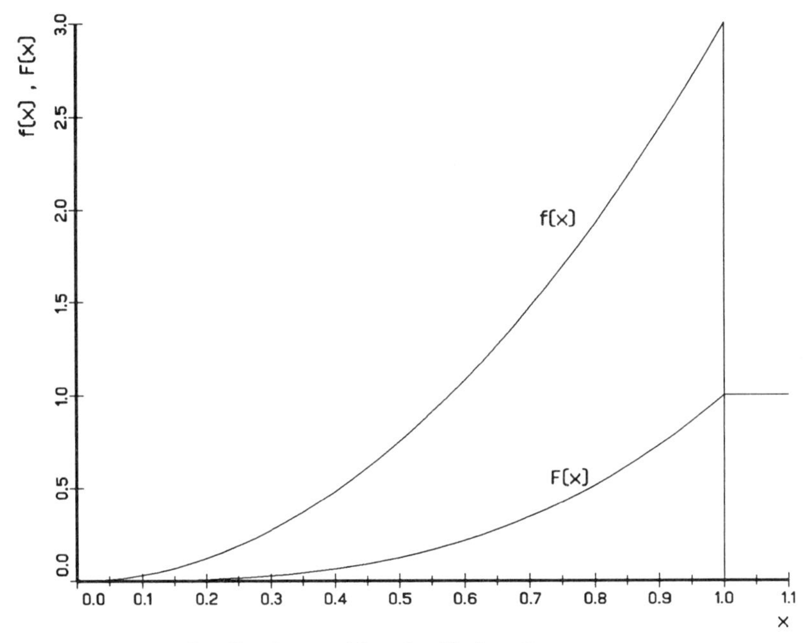

Figure 2.3 Power distribution: pdf and cdf, $\theta = 3$.

2.5. Functions of Random Variables

We now have in hand a stock of univariate distributions to draw upon. Once probabilities are initially assigned via $f(x)$, we are committed to many other probabilities. Here are two trivial examples:

(a) Suppose that $X \sim$ Poisson(1), that is, the random variable X has the Poisson distribution with parameter 1. Let $A = \{2 < X \leq 4\}$. To find Pr(A), write $A = A_1 \cup A_2$, where $A_1 = \{X = 3\}$, $A_2 = \{X = 4\}$. Because $A_1 \cap A_2 = 0$, it follows that $\Pr(A) = \Pr(A_1) + \Pr(A_2) = f(3) + f(4)$. But $f(x) = e^{-1}1^x/x! = 1/(ex!)$, so $f(3) = 0.0613$, $f(4) = 0.0153$, and $\Pr(A) = 0.0766$.

(b) Suppose that $X \sim$ standard normal. Let $A = \{|X| < 2\}$. To find Pr(A), the calculation is

$$\Pr(|X| < 2) = \Pr(-2 < X < 2) = F(2) - F(-2) = F(2) - [1 - F(2)]$$
$$= 2\,F(2) - 1 = 0.954,$$

using Table A.1 to get $F(2) = 0.977$.

Now suppose $Y = h(X)$ is a (single-valued) function of X. Let B be any event that is defined in terms of Y. Then B can be translated into an event defined in terms of X, so we can deduce Pr(B). Indeed we can deduce the probability distribution of the random variable Y.

We illustrate the procedure with a few examples.

(1) Suppose that the pmf of X is

$$f(x) = \begin{cases} 1/8 & \text{for } x = -1, \\ 2/8 & \text{for } x = 0, \\ 5/8 & \text{for } x = 1, \end{cases}$$

with $f(x) = 0$ elsewhere, and suppose that $Y = X^2$. The possible outcomes for Y are 0 (which occurs iff $X = 0$) and 1 (which occurs iff $X = -1$ or $X = 1$). Now $\Pr(Y = 0) = \Pr(X = 0) = 2/8$, and $\Pr(Y = 1) = \Pr(X = -1$ or $X = 1) = \Pr(X = -1) + \Pr(X = 1) = 1/8 + 5/8 = 6/8$. So the pmf of Y is

$$g(y) = \begin{cases} 1/4 & \text{for } y = 0, \\ 3/4 & \text{for } y = 1, \end{cases}$$

with $g(y) = 0$ elsewhere.

(2) Suppose that $X \sim$ standard normal and that

$$Y = \begin{cases} 1 & \text{if } X < -1, \\ 2 & \text{if } -1 \le X \le 2, \\ 3 & \text{if } 2 < X. \end{cases}$$

Then $\Pr(Y = 1) = F(-1)$, $\Pr(Y = 2) = F(2) - F(-1)$, $\Pr(Y = 3) = 1 - F(2)$, where $F(\cdot)$ denotes the standard normal cdf. So, referring to Table A.1, the pmf of Y is

$$g(y) = \begin{cases} 0.159 & \text{for } y = 1, \\ 0.818 & \text{for } y = 2, \\ 0.023 & \text{for } y = 3, \end{cases}$$

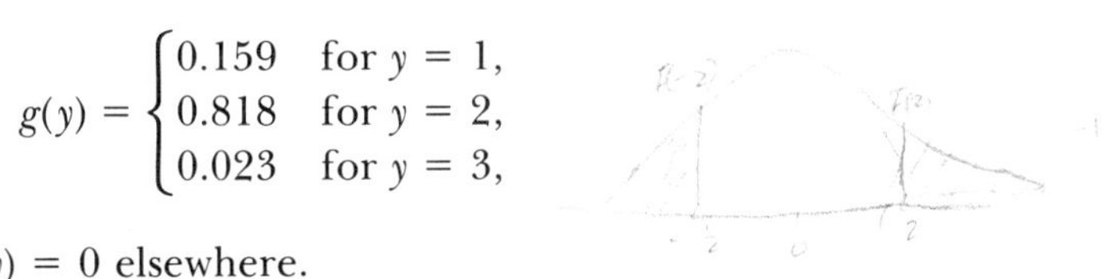

with $g(y) = 0$ elsewhere.

(3) Suppose that $X \sim$ rectangular on the interval $[-1, 1]$ and that $Y = X^2$. Now the pdf of X is $f(x) = 1/2$ for $-1 \le x \le 1$, with $f(x) = 0$ elsewhere. It may be tempting to say that $\Pr(Y = y) = \Pr(X = -\sqrt{y}) + \Pr(X = \sqrt{y})$, but this will not help since all three events have zero probability. Instead, we proceed via cdf's. The cdf of X is

$$F(x) = \begin{cases} 0 & \text{for } x < -1, \\ (1 + x)/2 & \text{for } -1 \le x \le 1, \\ 1 & \text{for } x > 1. \end{cases}$$

Let $G(y) = \Pr(Y \le y)$ be the cdf of Y. Clearly Y is confined to the interval $[0, 1]$, so $G(y) = 0$ for $y < 0$. For $y \ge 0$,

$$\begin{aligned} G(y) &= \Pr(Y \le y) = \Pr(X^2 \le y) = \Pr(-\sqrt{y} \le X \le \sqrt{y}) \\ &= F(\sqrt{y}) - F(-\sqrt{y}). \end{aligned}$$

Now for $0 \le y \le 1$,

$$F(\sqrt{y}) - F(-\sqrt{y}) = (1 + \sqrt{y})/2 - (1 - \sqrt{y})/2 = \sqrt{y},$$

while for $y > 1$, $F(\sqrt{y}) - F(-\sqrt{y}) = 1 - 0 = 1$. So the cdf of Y is

$$G(y) = \begin{cases} 0 & \text{for } y \le 0, \\ \sqrt{y} & \text{for } 0 < y \le 1, \\ 1 & \text{for } y > 1. \end{cases}$$

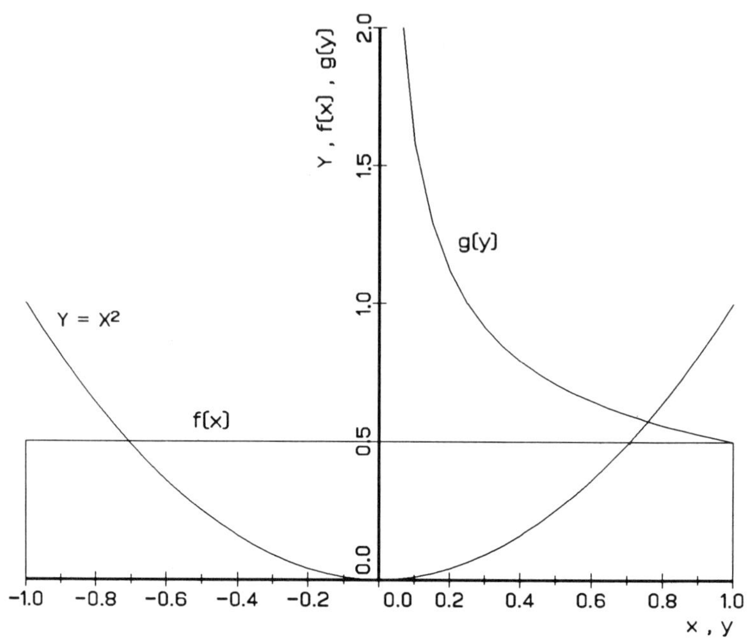

Figure 2.4 Distribution of a function.

Finally, we get the pdf of y by differentiating its cdf:

$$g(y) = \partial G(y)/\partial y = \begin{cases} 0 & \text{for } y \leq 0, \\ 1/(2\sqrt{y}) & \text{for } 0 < y \leq 1, \\ 0 & \text{for } y > 1. \end{cases}$$

Plotting the pdf as in Figure 2.4, we see a curve that slopes downward over the unit interval; it runs along the horizontal axis elsewhere. The shape of the pdf of Y may not have been anticipated from inspecting the rectangular shape of the pdf of X and the parabolic shape of the function $Y = X^2$.

(4) *Linear Functions.* Suppose that the continuous random variable X has pdf $f(x)$ and cdf $F(x)$ and that $Y = a + bX$ is a linear function of X, with a and $b > 0$ being constants. To find the pdf of Y, we follow the approach of (3). The cdf is

$$\begin{aligned} G(y) &= \Pr(Y \leq y) = \Pr(a + bX \leq y) = \Pr[X \leq (y - a)/b] \\ &= F[(y - a)/b] = F(x), \end{aligned}$$

where $x = (y - a)/b$. So the pdf of Y is

$$g(y) = G'(y) = [\partial F(x)/\partial x](\partial x/\partial y) = f(x)/b = (1/b)f[(y - a)/b].$$

If b had been negative, then the $1/b$ would have been replaced by $1/(-b)$.

As a special case, suppose that $X \sim$ standard normal and that $Y = a + bX$, with $b > 0$. The pdf of X is $f(x) = (2\pi)^{-1/2} \exp(-x^2/2)$, so the pdf of Y is

$$g(y) = (1/b)\,(2\pi)^{-1/2} \exp\{-[(y - a)/b]^2/2\}.$$

This specifies the (general) *normal distribution* with parameters a and b^2, which will be discussed in Chapter 7.

Exercises

2.1 One ball will be drawn from a jar containing white, blue, yellow, and green balls. The probability that the ball drawn will be green is 0.25, and the probability that the ball drawn will be yellow is 0.14. What is the probability that the ball drawn will be white or blue?

2.2 The probability that family A will buy a car is 0.7, the probability that family B will buy a car is 0.5, and the probability that both families will buy is 0.35. Find the probability of each of these events:

(a) Neither buys.
(b) Only one buys.
(c) At least one buys.

2.3 In a city, 65% of the families subscribe to the morning paper, 50% subscribe to the afternoon paper, and 80% subscribe to at least one of the two papers. What proportion of the families subscribes to both papers?

2.4 Consider two events A and B such that $\Pr(A) = 1/2$ and $\Pr(B) = 1/3$. Find $\Pr(B \cap \text{not } A)$ for each of these cases:

(a) A and B are disjoint.
(b) $B \subset A$.
(c) $\Pr(B \cap A) = 1/7$.

2.5 Consider two events A and B with $\Pr(A) = 0.5$ and $\Pr(B) = 0.7$. Determine the minimum and maximum values of $\Pr(A \cap B)$ and the conditions under which each is attained.

2.6 Consider an experiment in which a loaded coin (probability of heads is 5/6) is tossed once *and* a fair die is rolled once.

(a) Specify the sample space for the experiment.
(b) What is the probability that the coin will be heads *and* the number that appears on the die will be even?

2.7 Discuss the appropriateness of the Bernoulli distribution as a model for unemployment.

2.8 Consider these seven events:

$A = \{X = 1\}$ $\quad B = \{X = 2\}$ $\quad C = \{X \leq 3\}$
$D = \{X \text{ is even}\}$ $\quad E = \{1 < X < 5\}$ $\quad F = C \cup D$
$G = C \cap D$

Consider also these three discrete probability distributions:

(a) Bernoulli with parameter $p = 0.4$.
(b) Discrete uniform with parameter $N = 9$.
(c) Binomial with parameters $n = 2$, $p = 0.4$.

For each distribution calculate the probability of each event. Treat 0 as an even number.

2.9 Consider these three events:

$$A = \{X = 2\}, \qquad B = \{X = 3\}, \qquad C = \{X = 4\}.$$

Consider also these two discrete probability distributions:

(a) Binomial with parameters $n = 4$, $p = 0.6$.
(b) Poisson with parameter $\lambda = 1.5$.

For each distribution calculate the probability of each event.

2.10 Consider these five events:

$A = \{0 \leq X \leq 1/2\}$ $\quad B = \{X \leq 1/2\}$ $\quad C = \{X = 1/2\}$
$D = \{1/4 \leq X \leq 3/4\}$ $\quad E = \{1 < X \leq 2\}$

Consider also these two continuous probability distributions:

(a) Rectangular on the interval [0, 2].
(b) Power on the interval [0, 1] with parameter $\theta = 2$.

For each distribution calculate the probability of each event.

2.11 Consider these six events:

$A = \{0 \leq X \leq 1\}$ $B = \{0 \leq X \leq 2\}$ $C = \{1 \leq X \leq 3\}$
$D = \{-1 \leq X \leq 1\}$ $E = \{-2 \leq X \leq 2\}$ $F = \{-3 \leq X \leq 3\}$

Consider also these three continuous distributions:

(a) Exponential with parameter $\lambda = 2$.
(b) Standard normal.
(c) Standard logistic.

For each distribution calculate the probability of each event.

2.12 For each of the following, use the cdf approach to obtain the pdf of Y:

(a) X distributed exponential, $Y = 2X$.
(b) X distributed rectangular on (0, 1), $Y = -\log(X)$. (Note: In this book, "log" always denotes natural logarithm.)
(c) X distributed standard normal, $Y = X^2$.

3 Expectations: Univariate Case

3.1. Expectations

Our discussion of empirical frequency distributions in Chapter 1 placed considerable emphasis on the mean, $m_X = \Sigma_i x_i p(x_i)$. In probability distributions, the mean again plays an important role. The name for mean in a probability distribution is *expectation*, or *expected value*.

Suppose that the random variable X has pmf or pdf $f(x)$, a situation that we write as $X \sim f(x)$. Then the expectation of X is defined as

$$E(X) = \begin{cases} \sum_i x_i f(x_i) & \text{in the discrete case,} \\ \int_{-\infty}^{\infty} x f(x)\, dx & \text{in the continuous case.} \end{cases}$$

Let $Z = h(X)$ be a function of X. To obtain the expectation of Z, one might first deduce $g(z)$, the pmf or pdf of Z, and apply the definition: $E(Z) = \Sigma_j z_j g(z_j)$ or $\int_{-\infty}^{\infty} z g(z)\, dz$. Equivalently, one can get the expectation of $Z = h(X)$ as

$$E(Z) = \begin{cases} \sum_i h(x_i) f(x_i) & \text{in the discrete case,} \\ \int_{-\infty}^{\infty} h(x) f(x)\, dx & \text{in the continuous case.} \end{cases}$$

The symbol μ is also used to denote an expectation, so we will write $\mu_X = E(X)$ and $\mu_Z = E(Z)$.

Example. Suppose $X \sim$ Bernoulli(p). This is a discrete distribution with $f(0) = 1 - p$, $f(1) = p$, where $0 \le p \le 1$. We calculate $E(X) = 0(1 - p) + 1(p) = p$. Also, let $Z = X^2$. Then $E(Z) = 0^2(1 - p) + 1^2 p =$

p. Observe that $E(X) = p$, so (unless $p = 0$ or $p = 1$) the expected value of X will be a value of X that never occurs. Observe also that $p = E(X^2) \neq [E(X)]^2 = p^2$, so in general the expectation of a function is not the function of the expectation: $E[h(X)] \neq h[E(X)] = h(\mu_X)$.

Example. Suppose $X \sim$ rectangular on [0, 2]. This is a continuous distribution with $f(x) = 1/2$ for $0 \leq x \leq 2$, $f(x) = 0$ elsewhere. We calculate

$$E(X) = \int_{-\infty}^{\infty} xf(x)\, dx = \int_0^2 (x/2)\, dx = (1/2) \int_0^2 x\, dx.$$

But $\int x\, dx = x^2/2$ and $(x^2/2)\|_0^2 = 2$. So $E(X) = (1/2)2 = 1$. Also, let $Z = X^2$. Then

$$E(Z) = \int_0^2 (x^2/2)\, dx = (1/2) \int_0^2 x^2\, dx.$$

But $\int x^2\, dx = x^3/3$ and $(x^3/3)\|_0^2 = 8/3$. So $E(Z) = (1/2)(8/3) = 4/3$. Observe again that $E(X^2) = 4/3 \neq 1 = [E(X)]^2$.

Caution: There are distributions whose expectations are infinite, or do not exist, but we ignore those possibilities throughout this book.

3.2. Moments

The *moments* of X are the expectations of integer powers of X, or of $X^* = X - \mu_X$. For nonnegative integers r,

$E(X^r)$ is the rth *raw moment*, or moment about zero, of X,

$E(X^{*r})$ is the rth *central moment*, or moment about the mean, of X.

Each of the moments provides some information about the distribution. Taking $r = 1$, we have $E(X) = \mu$ and $E(X^*) = 0$. Taking $r = 2$, we have the second raw moment $E(X^2)$, and the *variance*:

$$E(X^{*2}) = V(X) = \sigma^2.$$

Table 3.1 Expectations and variances of illustrative distributions.

Distributions	$E(X)$	$V(X)$
Discrete		
(1) Bernoulli, parameter p	p	$p(1 - p)$
(2) Discrete uniform, parameter N	$(N + 1)/2$	$(N^2 - 1)/12$
(3) Binomial, parameters n, p	np	$np(1 - p)$
(4) Poisson, parameter λ	λ	λ
Continuous		
(1) Rectangular on the interval $[a, b]$	$(a + b)/2$	$(b - a)^2/12$
(2) Exponential, parameter λ	$1/\lambda$	$1/\lambda^2$
(3) Standard normal	0	1
(4) Standard logistic	0	$\pi^2/3$
(5) Power on $[0, 1]$, parameter θ	$\theta/(1 + \theta)$	$\theta/[(1+\theta)^2(2+\theta)]$

The variance of a random variable X is the expectation of the squared deviation of X from its expectation. It serves as a measure of the spread of the distribution. If $V(X) = 0$, then X is a constant, and conversely.

For the distributions introduced in Chapter 1, Table 3.1 gives the expectations and variances.

3.3. Theorems on Expectations

Several useful theorems on expectations and moments are easy to establish:

T1. LINEAR FUNCTIONS. For linear functions, the expectation of the function is the function of the expectation, and the variance of the function is the slope squared multiplied by the variance. That is, if $Z = a + bX$ where a and b are constants, then

$$E(Z) = a + bE(X), \qquad V(Z) = b^2V(X).$$

Proof (for continuous case).

$$E(Z) = \int (a + bx)f(x)\,dx = \int af(x)\,dx + \int bxf(x)\,dx$$

$$= a\int f(x)\,dx + b\int xf(x)\,dx = a\,1 + bE(X) = a + bE(X).$$

Then $Z^* = Z - E(Z) = a + bX - [a + bE(X)] = bX - bE(X) = bX^*$, so $V(Z) = E(Z^{*2}) = E(bX^*)^2 = b^2E(X^{*2}) = b^2V(X)$. ■

(Note: The symbol $\int$ is used in this chapter as shorthand for $\int_{-\infty}^{\infty}$.) A parallel proof applies to the discrete case.

T2. VARIANCE. The variance of a random variable is equal to the expectation of its square minus the square of its expectation. That is,

$$V(X) = E(X^2) - E^2(X).$$

Proof. Write $V(X) = E(X^{*2})$, where $X^* = X - E(X)$. Now $X^{*2} = X^2 + E^2(X) - 2E(X)X$, so using T1 extended to handle two variables gives $E(X^{*2}) = E(X^2) + E^2(X) - 2E(X)E(X) = E(X^2) - E^2(X)$. ■

(Note: $E^2(X)$ denotes $[E(X)]^2$.) Because $E(X^{*2}) \geq 0$, we can conclude that $E(X^2) \geq E^2(X)$, with equality iff $V(X) = 0$, that is, iff X is a constant.

T3. MEAN SQUARED ERROR. Let c be any constant. Then the mean squared error of a random variable about the point c is

$$E(X - c)^2 = \sigma^2 + (c - \mu)^2.$$

Proof. Write $(X - c) = (X - \mu) - (c - \mu) = X^* - (c - \mu)$. So $(X - c)^2 = X^{*2} + (c - \mu)^2 - 2(c - \mu)X^*$. Then using T1 gives

$$E(X - c)^2 = E(X^{*2}) + (c - \mu)^2 - 2(c - \mu)E(X^*).$$

But $E(X^*) = 0$ and $E(X^{*2}) = \sigma^2$. ■

T4. MINIMUM MEAN SQUARED ERROR. The value of c that minimizes $E(X - c)^2$ is $c = \mu$.

Proof. From T3, $E(X - c)^2 = \sigma^2 + (c - \mu)^2$. But $(c - \mu)^2 \geq 0$ with equality iff $c - \mu = 0$, that is, iff $c = \mu$. ■

3.4. Prediction

Thus far, the expected value of a random variable is simply the mean in its probability distribution. We now offer a practical interpretation.

Suppose that the random variable X has a known pmf or pdf $f(x)$. A single draw will be made from the distribution of X. You are asked to forecast, predict, or guess the outcome, using a constant c as the predictor. What is the best guess, that is, what is the best predictor? Suppose that your criterion for good prediction is minimum mean squared forecast error. Then you will choose c to minimize $E(U^2)$, where $U = X - c$ is the forecast error. By T4, the solution to your problem is $c = \mu$. The best predictor of a random drawing from a known probability distribution is the expected value of the random variable, when the criterion for predictive success is minimum mean squared forecast error.

When you use μ as the predictor, the forecast error is $X - \mu = \epsilon$, say, so the expected forecast error is $E(\epsilon) = 0$, and the expected squared forecast error is $E(\epsilon^2) = E(X - \mu)^2 = \sigma^2$. A predictor for which the expected forecast error is zero is called an *unbiased predictor*. So μ is an unbiased predictor, but there are many unbiased predictors. Let $Z = \mu + W$ where W is any random variable with $E(W) = 0$. Then Z is also an unbiased predictor of X, because $E(X - Z) = E(X - \mu - W) = E(X - \mu) - E(W) = 0 - 0 = 0$. But (unless W is correlated with X), $E(X - Z)^2 = \sigma^2 + V(W) \geq \sigma^2$.

Different criteria for predictive success lead to different choices: It can be shown that to minimize $E(|U|)$, you should choose $c = \text{median}(X)$, and that to maximize $\Pr(U = 0)$, you should (in the discrete case!) choose $c = \text{mode}(X)$. In econometrics, it is customary to adopt the minimum mean squared error criterion; as we have seen, this leads to the expected value as the best predictor. This is true even when, as in a Bernoulli distribution, the expected value is not a possible value of X.

3.5. Expectations and Probabilities

Any probability can be interpreted as an expectation. Define the variable Z which is equal to 1 if event A occurs, and equal to zero if event A does not occur. Then it is easy to see that $\Pr(A) = E(Z)$.

How much information about the probability distribution of a random variable X is provided by the expectation and variance of X? There are three useful theorems here.

MARKOV INEQUALITY. If Y is a nonnegative random variable, that is, if $\Pr(Y < 0) = 0$, and k is any positive constant, then $\Pr(Y \geq k) \leq E(Y)/k$.

Proof (for continuous case). Write

$$E(Y) = \int_0^\infty yf(y)\,dy = \int_0^k yf(y)\,dy + \int_k^\infty yf(y)\,dy = a + b,$$

say. Now $a \geq 0$, so $E(Y) \geq b$. Also $b \geq k \int_k^\infty f(y)\,dy = k \Pr(Y \geq k)$. So $E(Y) \geq k \Pr(Y \geq k)$. ■

CHEBYSHEV INEQUALITY #1. If X is a random variable, c is any constant, and d is any positive constant, then $\Pr(|X - c| \geq d) \leq E(X - c)^2/d^2$.

Proof. Let $Y = (X - c)^2$, so Y is a nonnegative random variable, and $|X - c| \geq d \Leftrightarrow Y \geq d^2$. Let $k = d^2$, and apply the Markov Inequality to get $E(Y) \geq d^2 \Pr(Y \geq d^2)$. ■

CHEBYSHEV INEQUALITY #2. If X is a random variable with expectation $E(X) = \mu$ and variance $V(X) = \sigma^2$, and d is any positive constant, then $\Pr(|X - \mu| \geq d) \leq \sigma^2/d^2$.

Proof. Apply Chebyshev Inequality #1 with $c = \mu$. ■

How much information about the expectation of a function is provided by the expectation of a random variable? As we have seen in T1, for linear functions the expectation of the function is the function of the expectation. But if $Y = h(X)$ is nonlinear, then in general $E(Y) \neq h[E(X)]$: the direction of the inequality may depend on the distribution of X. For certain functions, we can be more definite.

Let $E(X) = \mu$, $Y = h(X)$, $\partial Y/\partial X = h'(X)$. Let Z be the tangent line to

$h(X)$ at the point μ, that is, $Z = h(\mu) + h'(\mu)(X - \mu)$. Since Z is linear in X, while $h(\mu)$ and $h'(\mu)$ are constants, we have

$$E(Z) = h(\mu) + h'(\mu)\,E(X - \mu) = h(\mu).$$

If $Y = h(X) \leq Z$ everywhere, then regardless of the distribution of X, we are assured that $E(Y) \leq E(Z) = h(\mu)$. Now, a concave function lies everywhere below its tangent line, no matter where the tangent line is drawn. Thus we have shown

JENSEN'S INEQUALITY. If $Y = h(X)$ is concave and $E(X) = \mu$, then $E(Y) \leq h(\mu)$.

For example, the logarithmic function is concave, so $E[\log(X)] \leq \log[E(X)]$ regardless of the distribution of X. Similarly, if $Y = h(X)$ is convex, so that it lies everywhere above its tangent line, then $E(Y) \geq h(\mu)$. For example, the square function is convex, so $E(X^2) \geq [E(X)]^2$ regardless of the distribution of X, as we have already seen.

Exercises

3.1 Verify the entries of expectations and variances in Table 3.1, except those for the standard normal and standard logistic. Note: The following definite integral is well known:

$$\int_0^\infty t^n e^{-at}\,dt = n!/a^{n+1} \quad \text{for } a > 0 \text{ and } n \text{ positive integer.}$$

3.2 For each of the following distributions for the random variable X, calculate $E(X)$ and $V(X)$:

(a) Discrete uniform, parameter $N = 9$.
(b) Binomial, parameters $n = 2$, $p = 0.4$.
(c) Binomial, parameters $n = 4$, $p = 0.6$.
(d) Poisson, parameter $\lambda = 3/2$.
(e) Rectangular on the interval $[0, 2]$.
(f) Exponential, parameter $\lambda = 2$.
(g) Power on $[0, 1]$, parameter $\theta = 2$.

3.3 Suppose X has the power distribution on [0, 1] with parameter $\theta = 5$. Let $Z = 1/X^2$. Find $E(Z)$, $E(Z^2)$, $V(Z)$.

3.4 Suppose X has the exponential distribution with parameter $\lambda = 4$. Let $Z = \exp(X)$. Find $E(Z)$, $E(Z^2)$, $V(Z)$.

3.5 Suppose X has the rectangular distribution on the interval [0, 3]. Let $Z = F(X)$ where $F(\cdot)$ is the cdf of X. Find $E(Z)$.

3.6 Suppose $X \sim f(x)$. Let $Z = F(x)$ where $F(\cdot)$ is the cdf of X. Find $E(Z)$.

3.7 Let $A = \{X \geq 1\}$ and $B = \{|X - \mu| \geq 2\sigma\}$. Consider these three distributions: (i) Rectangular on the interval [0, 2], (ii) Exponential with parameter $\lambda = 2$, (iii) Power on [0, 1] with parameter $\theta = 3$.

(a) For each distribution, use the Markov or Chebyshev Inequality to calculate an upper bound on Pr(A) and Pr(B).
(b) For each distribution, use the appropriate cdf to calculate the exact Pr(A) and Pr(B).
(c) Comment on the usefulness of the inequalities.

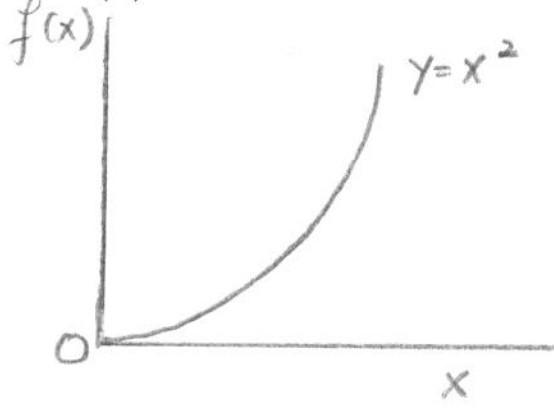

4 *Bivariate Probability Distributions*

4.1. Joint Distributions

The focus in this book is on relations between variables, where the relations are not deterministic ones. So we need more than one variable in our probability distributions. We take up the bivariate case in detail. Consider an experiment that has various distinct possible outcomes. The outcomes are distinguished by the values of a pair of random variables X, Y. The values they take on are labeled x, y. Each trial of the experiment produces one value of the pair (x, y). We refer to the pair (X, Y) as a *random vector*, a name that merely indicates a set of random variables whose joint values are determined by the outcome of an experiment. As in the univariate situation, we distinguish two cases.

Discrete Case

In the discrete case, there are a finite, or countably infinite, number of distinct possible values for X, and also for Y, and thus for the pair (X, Y). So we can list the distinct possible pairs, say as the column and row headings of a two-way array. The points on the list are called *mass points*. There is a function $f(x, y)$, called the *joint probability mass function*, or joint pmf, of the distribution. It must satisfy these requirements: $f(x, y) \geq 0$ everywhere, $f(x, y) > 0$ only at the mass points, and

$$\sum_i \sum_j f(x_i, y_j) = 1.$$

The joint pmf gives the basic assignment of probabilities via:

$$\Pr(X = x, Y = y) = f(x, y).$$

Probabilities of other events then follow in the usual fashion.

If we enter the f's in the cells of the two-way array, we have a table that looks like an empirical joint frequency distribution, such as that for savings rate and income. But now the entries in the table are probabilities rather than frequencies, or if you like, they are frequencies in a population rather than in a sample.

Example: *Trinomial Distribution with parameters n, p, q.* Here n is a positive integer, $0 \leq p \leq 1$, $0 \leq q \leq 1$, and $p + q \leq 1$. The joint pmf is:

$$f(x, y) = \frac{n!}{[x!y!(n - x - y)!]} p^x q^y (1 - p - q)^{(n-x-y)}$$

for $x = 0, 1, \ldots, n$ and $y = 0, 1, \ldots, n - x$, with $f(x, y) = 0$ otherwise. This might be a relevant model for the labor force status of individuals over n months, using a three-way breakdown of status for each month: employed, unemployed, or not in labor force. The variables over an n-month period would be X = number of months employed, Y = number of months unemployed, and $n - X - Y$ = number of months not in labor force.

Continuous Case

In the continuous case, there is a continuum of distinct possible outcomes for X and also for Y, and thus a two-dimensional continuum of possible outcomes for the pair (X, Y). There is a function $f(x, y)$, called the *joint probability density function*, or joint pdf, of the distribution. It must satisfy these requirements: $f(x, y) \geq 0$ everywhere, and

$$\int_{-\infty}^{\infty} \int_{-\infty}^{\infty} f(x, y)\, dy\, dx = 1.$$

The joint pdf gives the basic assignment of probabilities as follows. For any $a \leq b$, $c \leq d$:

$$\Pr(a \leq X \leq b, c \leq Y \leq d) = \int_a^b \int_c^d f(x, y)\, dy\, dx.$$

Probabilities of other events follow in the usual fashion. As in the univariate case, the pdf does not give the probability of being at a point: $\Pr(X = x, Y = y) = 0$ everywhere, even where $f(x, y) \neq 0$.

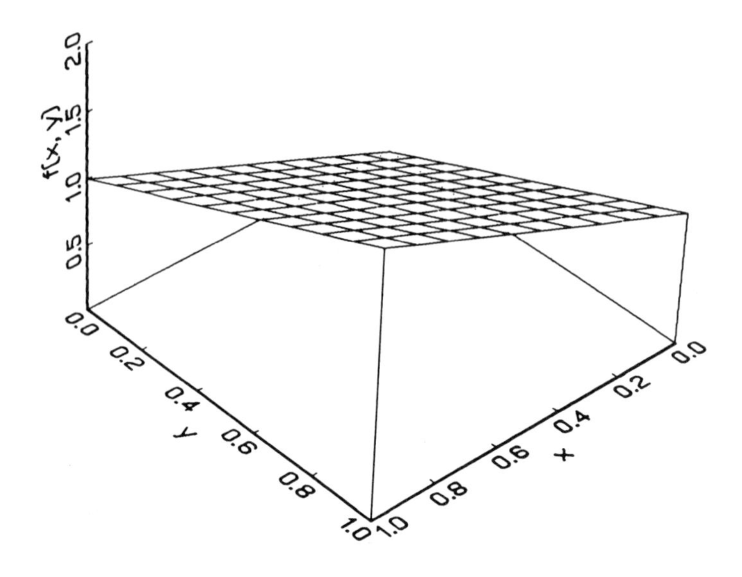

Figure 4.1 Roof distribution: joint pdf.

Example: *Roof Distribution.* The joint pdf is

$$f(x, y) = (x + y) \quad \text{for } 0 \leq x \leq 1 \text{ and } 0 \leq y \leq 1,$$

with $f(x, y) = 0$ elsewhere. Clearly $f(x, y) \geq 0$ everywhere, and

$$\begin{aligned} \int_{-\infty}^{\infty} \int_{-\infty}^{\infty} f(x, y)\, dy\, dx &= \int_0^1 \int_0^1 (x + y)\, dy\, dx \\ &= \int_0^1 (xy + y^2/2) \|_0^1\, dx \\ &= \int_0^1 (x + 1/2)\, dx = (x^2/2 + x/2) \|_0^1 \\ &= 1/2 + 1/2 = 1. \end{aligned}$$

So this is a legitimate joint pdf. The plot in Figure 4.1 accounts for the name "roof distribution." We will use this joint pdf as an example because the integration is simple.

The *joint cumulative distribution function*, or joint cdf, may be defined as $F(x, y) = \int_{-\infty}^{x} \int_{-\infty}^{y} f(s, t)\, dt\, ds = \Pr(X \leq x, Y \leq y)$.

4.2. Marginal Distributions

We proceed to implications of the initial assignment of probabilities in a bivariate probability distribution.

Discrete Case

Let $A = \{X = x\}$ and $A_j = \{X = x, Y = y_j\}$ for $j = 1, 2, \ldots$. Recognizing that $A = \cup_j A_j$ is a union of disjoint events, we calculate

$$\Pr(X = x) = \Pr(A) = \sum_j \Pr(A_j) = \sum_j f(x, y_j) = f_1(x),$$

say. This new function $f_1(x)$ is called the *marginal* pmf of X. Observe that $f_1(x) \geq 0$ everywhere, that $f_1(x) > 0$ only for points in the list, and that

$$\sum_i f_1(x_i) = \sum_i \left[\sum_j f(x_i, y_j) \right] = 1.$$

So $f_1(x)$ is a legitimate univariate probability mass function. Similarly, $f_2(y) = \Sigma_i f(x_i, y) = \Pr(Y = y)$ is the marginal pmf of Y. The subscripts 1 and 2 merely distinguish the two functions.

Example. For the trinomial distribution, we can verify that

$$f_1(x) = \frac{n!}{[x!(n - x)!]} p^x (1 - p)^{(n-x)},$$

for $x = 0, 1, \ldots, n$, with $f_1(x) = 0$ otherwise. This is recognizable as the pmf of a binomial distribution with parameters n, p.

Continuous Case

Let $A = \{a \leq X \leq b\} = \{a \leq X \leq b, -\infty \leq Y \leq \infty\}$. Then

$$\Pr(a \leq X \leq b) = \Pr(A) = \int_a^b \int_{-\infty}^{\infty} f(x, y)\, dy\, dx = \int_a^b f_1(x)\, dx,$$

say, where

$$f_1(x) = \int_{-\infty}^{\infty} f(x, y)\, dy.$$

This new function $f_1(x)$ is called the *marginal* pdf of X. Observe that $f_1(x) \geq 0$ everywhere (since it is the integral of a nonnegative function) and that $\int_{-\infty}^{\infty} f_1(x)\, dx = \int_{-\infty}^{\infty} \int_{-\infty}^{\infty} f(x, y)\, dy\, dx = 1$. So $f_1(x)$ is a legitimate univariate pdf.

Similarly, $f_2(y) = \int_{-\infty}^{\infty} f(x, y)\, dx$ is the marginal pdf of Y.

> ***Example.*** For the roof distribution,
>
> $$f_1(x) = \int_0^1 (x + y)\, dy = (xy + y^2/2)\|_0^1 = x + 1/2 \quad \text{for } 0 \leq x \leq 1,$$

with $f_1(x) = 0$ elsewhere. Figure 4.2 plots this marginal pdf.

4.3. Conditional Distributions

We continue to draw implications of the initial assignment of probabilities in a bivariate distribution.

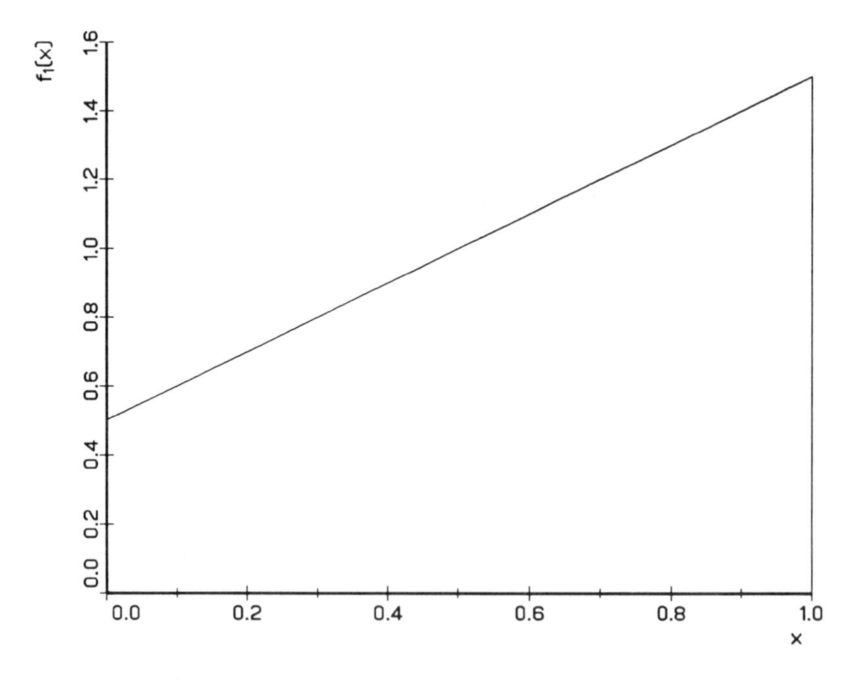

Figure 4.2 Roof distribution: marginal pdf.

Discrete Case

Let $A = \{X = x\}$, $B = \{Y = y\}$. In probability theory, if $\Pr(B) \neq 0$, one defines the probability that A occurs given that B occurs as

$$\Pr(A|B) = \Pr(A \cap B)/\Pr(B).$$

Now in the discrete case, $\Pr(A \cap B) = f(x, y)$ and $\Pr(B) = f_2(y)$, so

$$\Pr(A|B) = f(x, y)/f_2(y) = g_1(x|y),$$

say, a function that is defined only for y such that $f_2(y) \neq 0$.

For any such y value, observe that $g_1(x|y)$ is a function of x alone, $g_1(x|y) \geq 0$ (because $f(x, y) \geq 0$ and $f_2(y) > 0$), and

$$\sum_i g_1(x_i|y) = \sum_i [f(x_i, y)/f_2(y)] = \left[\sum_i f(x_i, y)\right] \Big/ f_2(y)$$

$$= f_2(y)/f_2(y) = 1.$$

So for any y value with positive mass, $g_1(x|y)$ is a legitimate univariate pmf for the random variable X. It is called the *conditional* pmf of X given $Y = y$, and may be used in the ordinary way. For example, the probability that the random variable X takes on the value x given that the random variable Y takes on the value y_j is $\Pr(X = x|Y = y_j) = g_1(x|y_j)$.

Running across j, there is a set of conditional probability distributions of X—one distribution of X corresponding to each distinct possible value of Y. Conditioning on Y may be viewed as partitioning the bivariate population into subpopulations. Within each subpopulation, the value of Y is constant while X varies.

Similarly, $g_2(y|x) = f(x, y)/f_1(x)$, defined for all x such that $f_1(x) \neq 0$, is the conditional pmf of Y given $X = x$. There is one such distribution for each distinct value of X. The pattern here is precisely the same as in the empirical joint frequency distribution of income and savings rate.

Continuous Case

In the continuous case, we proceed rather differently. For each y such that $f_2(y) \neq 0$, define the function

$$g_1(x|y) = f(x, y)/f_2(y),$$

leaving $g_1(\cdot|y)$ undefined elsewhere. This $g_1(x|y)$ is called the *conditional* pdf of X given $Y = y$. It is easy to confirm that $g_1(x|y) \geq 0$ everywhere where defined, and that

$$\int_{-\infty}^{\infty} g_1(x|y)\, dx = \int_{-\infty}^{\infty} [f(x, y)/f_2(y)]\, dx = \left[\int_{-\infty}^{\infty} f(x, y)\, dx\right] \Big/ f_2(y)$$

$$= f_2(y)/f_2(y) = 1.$$

So for any y value with positive density, $g_1(x|y)$ is a legitimate univariate pdf for the random variable X.

There are an infinity of such conditional distributions, one for each value of Y. Each of them can be used in the ordinary way. For example,

$$\Pr(a \leq X \leq b | Y = y) = \int_a^b g_1(x|y)\, dx.$$

Observe that we have succeeded in defining $\Pr(A|B)$ even though $\Pr(B) = 0$. This would be nonsense in the discrete case, but it is quite meaningful in the continuous case, where zero probability events do occur.

For a quite distinct example, suppose we want $\Pr(A|B)$, where $A = \{a \leq X \leq b\}$ and $B = \{c \leq Y \leq d\}$. Provided that $\Pr(B) \neq 0$, we have

$$\Pr(A|B) = \Pr(A \cap B)/\Pr(B) = \left[\int_a^b \int_c^d f(x, y)\, dy\, dx\right] \Big/ \int_c^d f_2(y)\, dy.$$

Similarly the conditional pdf of Y given x, defined for all x such that $f_1(x) \neq 0$, is $g_2(y|x) = f(x, y)/f_1(x)$.

Example. For the roof distribution, $g_2(y|x)$ is defined only for $0 \leq x \leq 1$. There

$$g_2(y|x) = f(x, y)/f_1(x) = (x + y)/(x + 1/2) \quad \text{for } 0 \leq y \leq 1,$$

with $g_2(y|x) = 0$ elsewhere. Figure 4.3 plots this function for $x = 0$, 0.5, 1.

Mixed cases may arise in a bivariate population. For example, if Y = family income and X = number of persons in family, then a natural model would have X discrete and Y continuous. In such situations, the joint distribution is most conveniently specified in terms of the marginal pmf of the discrete variable and the conditional pdf of the continuous variable.

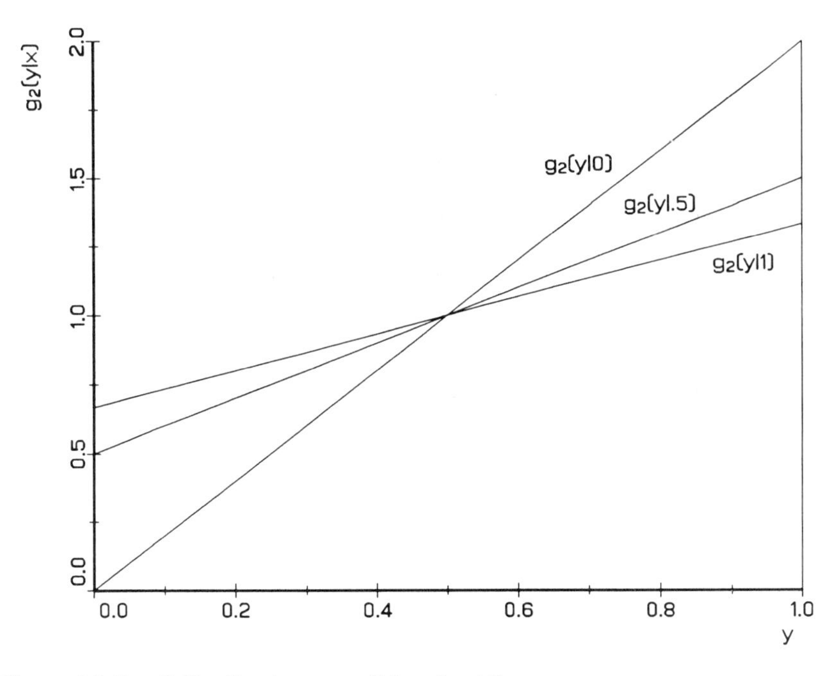

Figure 4.3 Roof distribution: conditional pdf's.

Exercises

4.1 *Curved-roof Distribution.* Consider the continuous bivariate probability distribution whose joint probability density function is

$$f(x, y) = 3(x^2 + y)/11 \quad \text{for } 0 \leq x \leq 2,\ 0 \leq y \leq 1,$$

with $f(x, y) = 0$ elsewhere. The plot of this pdf in Figure 4.4 accounts for its name.

(a) Show that the marginal pdf of X, plotted in Figure 4.5, is

$$f_1(x) = 3(2x^2 + 1)/22 \quad \text{for } 0 \leq x \leq 2,$$

with $f_1(x) = 0$ elsewhere.

(b) Derive $f_2(y)$, the marginal pdf of Y.

(c) For $0 \leq x \leq 2$, derive $g_2(y|x)$, the conditional pdf of Y given X, plotted in Figure 4.6 for $x = 0, 1, 2$.

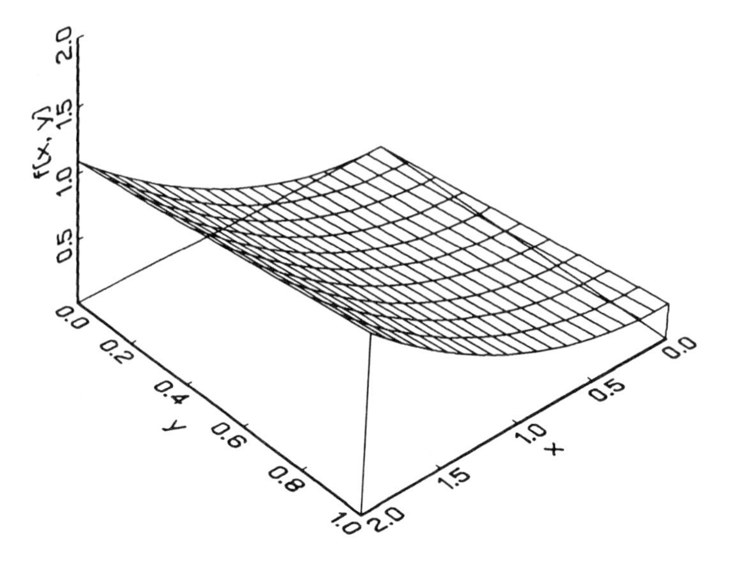

Figure 4.4 Curved-roof distribution: joint pdf.

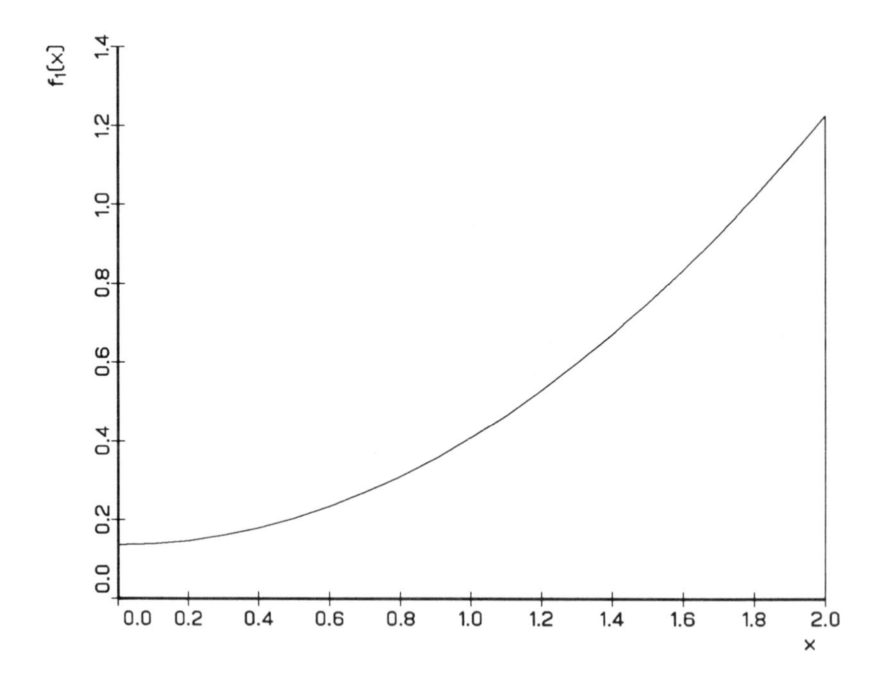

Figure 4.5 Curved-roof distribution: marginal pdf.

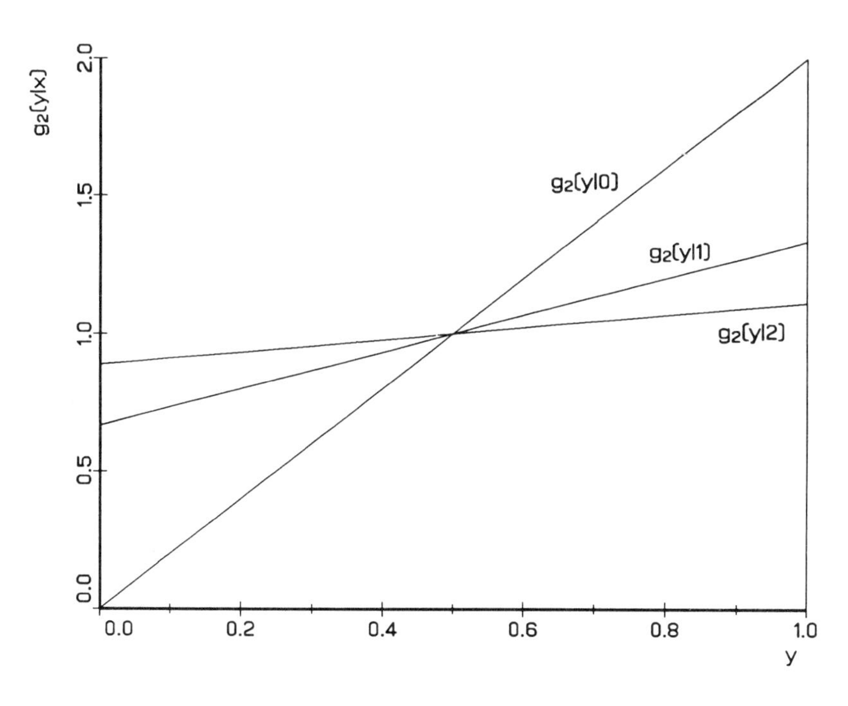

Figure 4.6 Curved-roof distribution: conditional pdf's.

4.2 For the curved-roof distribution, let $A = \{0 \leq Y \leq 1/2\}$. Calculate $\Pr(A)$, and calculate $\Pr(A|x)$ for $x = 0, 1, 2$.

4.3 For the curved-roof distribution, derive $g_1(x|y)$, the conditional pdf of X given Y.

5 Expectations: Bivariate Case

5.1. Expectations

Suppose that the random vector (X, Y) has joint pdf or pmf $f(x, y)$. Let $Z = h(X, Y)$ be a scalar function of (X, Y). Then the expectation of the random variable Z is defined as

$$E(Z) = \begin{cases} \int_{-\infty}^{\infty} \int_{-\infty}^{\infty} h(x, y) f(x, y)\, dy\, dx & \text{in the continuous case,} \\ \sum_i \sum_j h(x_i, y_j) f(x_i, y_j) & \text{in the discrete case.} \end{cases}$$

If in fact Z is a function of only one of the two variables, then its expectation is also computable from the marginal distribution of that variable. That is, if $h(x, y) = h_1(x)$, then

$$\begin{aligned} E(Z) &= \sum_i \sum_j h_1(x_i) f(x_i, y_j) = \sum_i h_1(x_i) \left[\sum_j f(x_i, y_j) \right] \\ &= \sum_i h_1(x_i) f_1(x_i). \end{aligned}$$

(Note: Here and subsequently, we will usually report derivations for only one of the two cases, discrete and continuous. The understanding is that a parallel derivation applies to the other case.)

The moments of the joint distribution are the expectations of certain functions of (X, Y), or of (X^*, Y^*), where $X^* = X - E(X)$, $Y^* = Y - E(Y)$. For nonnegative integers r, s:

$E(X^r Y^s)$ is the (r, s)th raw moment, or moment about zero,
$E(X^{*r} Y^{*s})$ is the (r, s)th central moment, or moment about the mean.

In particular,

$$\begin{aligned}
r = 1, s = 0&: E(X^1Y^0) &&= E(X) = \mu_X, && && \\
r - 2, s = 0&: E(X^{*2}Y^{*0}) &&= E[X - E(X)]^2 && = V(X) && = \sigma_X^2, \\
r = 1, s = 1&: E(X^*Y^*) &&= E\{[X - E(X)][Y - E(Y)]\} && = C(X, Y) && = \sigma_{XY}.
\end{aligned}$$

The last of these is called the *covariance* of X and Y. Thus the covariance of a pair of random variables is the expected cross-product of their deviations from their respective expectations.

The *standard deviation* of a random variable is the square root of its variance. So the standard deviation of X is $\sigma_X = \sqrt{\sigma_X^2}$, and similarly for Y. The *correlation coefficient*, or simply *correlation*, of a pair of random variables is the ratio of their covariance to the product of their standard deviations. So the correlation coefficient of X and Y is

$$\rho = C(X, Y)/[\sqrt{V(X)}\sqrt{V(Y)}] = \sigma_{XY}/(\sigma_X\sigma_Y).$$

Some useful theorems are readily established:

T5. LINEAR FUNCTION. Suppose that $Z = a + bX + cY$, where a, b, c are constants. Then

$$E(Z) = a + bE(X) + cE(Y),$$

$$V(Z) = b^2V(X) + c^2V(Y) + 2bcC(X, Y).$$

Proof. For the expectation,

$$\begin{aligned}
E(Z) &= \sum_i \sum_j (a + bx_i + cy_j)f(x_i, y_j) \\
&= a\sum_i \sum_j f(x_i, y_j) + b\sum_i x_i \left[\sum_j f(x_i, y_j)\right] + c\sum_j y_j \left[\sum_i f(x_i, y_j)\right] \\
&= a \quad 1 \quad + b\sum_i x_i f_1(x_i) \quad + c\sum_j y_j f_2(y_j).
\end{aligned}$$

For the variance, $V(Z) = E(Z^{*2})$, where

$$Z^* = Z - E(Z) = b[X - E(X)] + c[Y - E(Y)] = bX^* + cY^*.$$

Expanding the square gives Z^{*2} as a linear function of X^{*2}, Y^{*2}, and X^*Y^*. Use the rule for expectation of a linear function, extended to handle three variables. ■

T6. PAIR OF LINEAR FUNCTIONS. Suppose that

$$Z_1 = a_1 + b_1X + c_1Y, \quad \text{where } a_1, b_1, c_1 \text{ are constants,}$$
$$Z_2 = a_2 + b_2X + c_2Y, \quad \text{where } a_2, b_2, c_2 \text{ are constants.}$$

Then $C(Z_1, Z_2) = b_1b_2V(X) + c_1c_2V(Y) + (b_1c_2 + b_2c_1)C(X, Y)$.

Proof. Extend the approach used for $V(Z)$ in the proof of T5. ■

T7. COVARIANCE AND VARIANCE.

$$C(X, Y) = E(XY) - E(X)E(Y) = C(Y, X).$$
$$V(X) = E(X^2) - E^2(X) = C(X, X).$$

Proof. $C(X, Y) = E(X^*Y^*) = E\{[X - E(X)]Y^*\} = E(XY^*) - E(X)E(Y^*)$

$$= E(XY^*) = E\{X[Y - E(Y)]\}$$
$$= E(XY) - E(X)E(Y). \quad ■$$

Example. For the roof distribution introduced in Section 4.1, these moments are calculated by integration in the joint pdf:

$$E(X) = E(Y) = 7/12, \qquad E(X^2) = E(Y^2) = 5/12, \qquad E(XY) = 1/3.$$

Then

$$V(X) = E(X^2) - E^2(X) = 5/12 - 49/144 = 11/144 = V(Y),$$
$$C(X, Y) = E(XY) - E(X)E(Y) = 1/3 - 49/144 = -1/144.$$

5.2. Conditional Expectations

In a bivariate probability distribution, the *conditional expectation of Y given* $X = x$ is the counterpart of the sample conditional mean $m_{Y|x}$ that was introduced in Chapter 1.

DEFINITION. Let the random vector (X, Y) have joint pdf $f(x, y) = g_2(y|x)f_1(x)$, and let $Z = h(X, Y)$ be a function of (X, Y). Then the conditional expectation of the random variable Z, given $X = x$, is

$$E(Z|x) = \int_{-\infty}^{\infty} h(x, y)g_2(y|x)\, dy$$

in the continuous case, and similarly in the discrete case.

The symbol $E(\cdot|x)$ denotes an expectation taken in the distribution $g_2(y|x)$, so $E(Z|x) = \mu_{Z|x}$ is just the expected value of $h_2(Y) = h(x, Y)$ in a particular univariate distribution. If we then allow X to vary, we get a set of conditional expectations, denoted collectively as $E(Z|X) = \mu_{Z|X}$.

To illustrate the concepts, consider some special cases. Here a, b, c are constants, X is a random variable, and x is a particular value of that variable. Given $X = x$, then any function of X alone is constant. With that in mind, the following results are immediate:

(i) Let $Z = h(X)$. Then $E(Z|x) = h(x)$.
(ii) Let $Z = h_1(X)Y$. Then $E(Z|x) = h_1(x)E(Y|x)$.
(iii) Let $Z = a + bX + cY$. Then $E(Z|x) = a + bx + cE(Y|x)$.
(iv) Let $Z = Y$. Then $E(Z|x) = E(Y|x) = \mu_{Y|x}$, the conditional expectation of Y given that $X = x$.
(v) Let $Z = (Y - \mu_{Y|X})$. Then $E(Z|x) = E(Y|x) - \mu_{Y|x} = 0$.
(vi) Let $Z = (Y - \mu_{Y|X})^2$. Then $E(Z|x) = V(Y|x) = \sigma^2_{Y|x}$, the *conditional variance* of Y given that $X = x$.
(vii) Let $Z = (Y - \mu_Y)$. Then $E(Z|x) = E(Y|x) - \mu_Y = \mu_{Y|x} - \mu_Y$.
(viii) Let $Z = (Y - \mu_Y)^2$. Then $E(Z|x) = \sigma^2_{Y|x} + (\mu_{Y|x} - \mu_Y)^2$.

Proof of (viii). Write $Y - \mu_Y = (Y - \mu_{Y|X}) + (\mu_{Y|X} - \mu_Y)$, so

$$(Y - \mu_Y)^2 = (Y - \mu_{Y|X})^2 + (\mu_{Y|X} - \mu_Y)^2 + 2(\mu_{Y|X} - \mu_Y)(Y - \mu_{Y|X}).$$

Take expectations conditional on $X = x$, using (v), (vi), and the conditional constancy of $(\mu_{Y|X} - \mu_Y)$. ■

Now allow X to vary, so that $E(Z|X)$ is itself a random variable, taking on the values $E(Z|x)$. Several key theorems are easily established:

T8. LAW OF ITERATED EXPECTATIONS. The (marginal) expectation of $Z = h(X, Y)$ is the expectation of its conditional expectations:

$$E(Z) = E_X[E(Z|X)].$$

(Note: The symbol E_X, read as "the expectation over X," is the expectation taken in the marginal distribution of X. The subscript may be omitted if there is no risk of confusion.)

Proof.

$$\begin{aligned}
E(Z) &= \int_{-\infty}^{\infty} \int_{-\infty}^{\infty} h(x, y) f(x, y)\, dy\, dx \\
&= \int_{-\infty}^{\infty} \int_{-\infty}^{\infty} h(x, y)[g_2(y|x) f_1(x)]\, dy\, dx \\
&= \int_{-\infty}^{\infty} \left[\int_{-\infty}^{\infty} h(x, y) g_2(y|x)\, dy \right] f_1(x)\, dx \\
&= \int_{-\infty}^{\infty} E(Z|x) f_1(x)\, dx. \quad \blacksquare
\end{aligned}$$

T9. MARGINAL AND CONDITIONAL MEANS. The (marginal) expectation of Y is equal to the expectation of its conditional expectations:

$$\mu_Y = E(Y) = E_X[E(Y|X)] = E(\mu_{Y|X}).$$

T10. ANALYSIS OF VARIANCE. The (marginal) variance of Y is equal to the expectation of its conditional variances plus the variance of its conditional expectations:

$$\sigma_Y^2 = V(Y) = E_X[V(Y|X)] + V_X[E(Y|X)] = E(\sigma_{Y|X}^2) + V(\mu_{Y|X}).$$

Proof. Write $V(Y) = E(Z)$ where $Z = (Y - \mu_Y)^2$, and apply T8 to item (viii) in the list above.

T11. EXPECTED PRODUCT. The expected product of X and Y is equal to the expected product of X and the conditional expectation of Y given X:

$$E(XY) = E_X[XE(Y|X)] = E(X\mu_{Y|X}).$$

T12. COVARIANCE. The covariance of X and Y is equal to the covariance of X and the conditional expectation of Y given X:

$$C(X, \mu_{Y|X}) = E(X\mu_{Y|X}) - E(X)E(\mu_{Y|X}) = E(XY) - E(X)E(Y)$$
$$= C(X, Y).$$

5.3. Conditional Expectation Function

As we have seen, the conditional expectation of Y given that $X = x$ is

$$E(Y|x) = \mu_{Y|x} = \int_{-\infty}^{\infty} y g_2(y|x)\, dy.$$

As we change x, that is, allow X to vary, we get $E(Y|X) = \mu_{Y|X}$, a function of X, known as the *conditional expectation function*, or CEF, or "population regression function" of Y given X. Similarly, $V(Y|x)$ is the conditional variance of Y given that $X = x$, and $V(Y|X)$ is the *conditional variance function*, or CVF, of Y given X. The shapes of the CEF and CVF are determined ultimately by the joint pmf or pdf. (Note the confusing language: the CEF and CVF of Y given X are, mathematically speaking, functions of X.)

Example. For the roof distribution, the CEF of Y given X is defined only for $0 \leq x \leq 1$. There

$$E(Y|x) = \int_0^1 [y(x + y)/(x + 1/2)]\, dy$$
$$= [1/(x + 1/2)](xy^2/2 + y^3/3)\|_0^1 = [1/(x + 1/2)](x/2 + 1/3)$$
$$= (3x + 2)/(6x + 3).$$

This function is plotted in Figure 5.1.

The deviation of Y from its CEF has certain characteristic properties. Let $\epsilon = Y - E(Y|X)$. Because ϵ is just the deviation from a (conditional) expected value, we have

$$E(\epsilon|X) = 0, \tag{5.1}$$

$$V(\epsilon|X) = \sigma^2_{Y|X}. \tag{5.2}$$

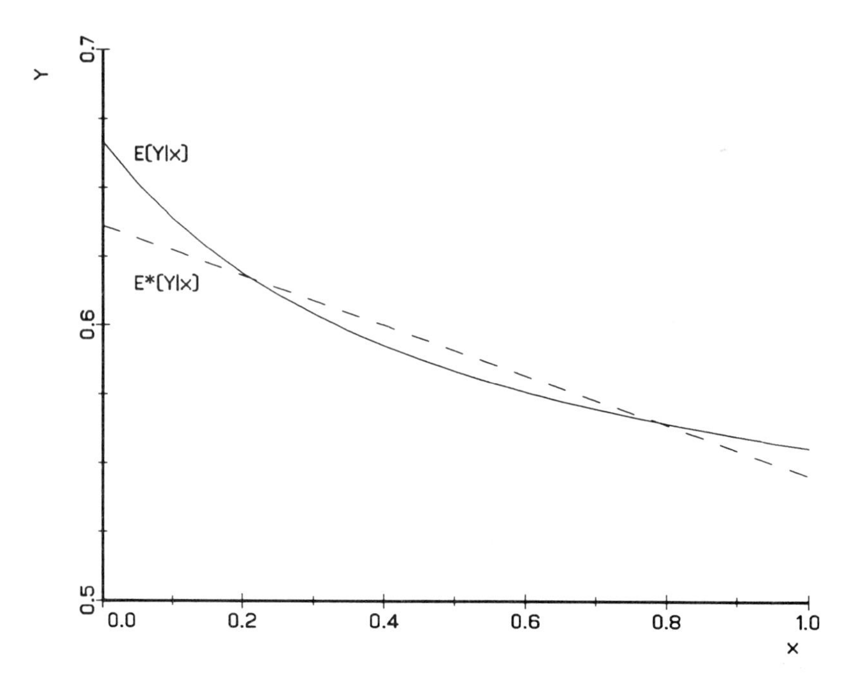

Figure 5.1 Roof distribution: CEF and BLP.

From these it follows that

(5.3) $E(\epsilon) = 0,$

(5.4) $C(X, \epsilon) = 0,$

(5.5) $V(\epsilon) = E(\sigma^2_{Y|X}),$

(5.6) If $Z = h(X)$, then $C(Z, \epsilon) = 0.$

Proofs. By T8 and Eq. (5.1), $E(\epsilon) = E_X[E(\epsilon|X)] = E_X(0) = 0$. By T12 and Eq. (5.1), $C(X, \epsilon) = C[X, E(\epsilon|X)] = C(X, 0) = 0$. By T10, Eq. (5.2), and Eq. (5.1), $V(\epsilon) = E(\sigma^2_{\epsilon|X}) + V(\mu_{\epsilon|X}) = E(\sigma^2_{Y|X})$. Finally,

$$C(Z, \epsilon) = E(Z\epsilon) - E(Z)E(\epsilon) = E(Z\epsilon) = E_X[ZE(\epsilon|X)] = E(Z0) = 0. \quad \blacksquare$$

We conclude that the deviation of Y from the function $E(Y|X)$ is a random variable that has zero expectation, and zero covariance with every function of the conditioning variable X. No other function of X yields deviations with the latter property.

5.4. Prediction

Recall the univariate prediction problem introduced in Section 3.4: The random variable Y has known pmf or pdf $f(y)$. A single draw is made and you are asked to predict the value of Y. The best constant predictor c, in the sense of minimizing $E(U^2)$ where $U = Y - c$, is $c = \mu_Y$. For that optimal choice of c, we have $U = Y - \mu_Y$, with $E(U) = 0$ and $E(U^2) = E(Y - \mu_Y)^2 = \sigma_Y^2$.

Now consider this prediction problem for the bivariate case: The random vector (X, Y) has known joint pmf or pdf $f(x, y)$. A single draw is made. You are told the value of X that was drawn and asked to predict the value of Y that accompanied it. You are free to use any function of X, say $h(X)$, as your predictor. What is the best predictor, in the sense of minimizing $E(U^2)$ where $U = Y - h(X)$? The answer is $h(X) = E(Y|X)$. That is, the best predictor of Y given X is the CEF.

Proof. Let $h(X)$ be any function of X, let $U = Y - h(X)$, $\epsilon = Y - E(Y|X)$, and $W = E(Y|X) - h(X)$, so that $U = \epsilon + W$, with W being a function of X alone. For a particular $X = x$, we have $W = E(Y|x) - h(x) = w$, say. So at $X = x$, we have $U = \epsilon + w$, so $U^2 = \epsilon^2 + w^2 + 2w\epsilon$, so

$$E(U^2|x) = E(\epsilon^2|x) + w^2 + 2wE(\epsilon|x) = \sigma_{Y|x}^2 + w^2,$$

using Eqs. (5.1) and (5.2). Across all X,

$$E(U^2) = E_X[E(U^2|X)] = E(\sigma_{Y|X}^2) + E(W^2).$$

The last term is nonnegative and vanishes iff $W \equiv 0$, that is, iff $h(X) = E(Y|X)$. ■

For the optimal choice of $h(X)$, the prediction error is $U = Y - E(Y|X) = \epsilon$, with $E(\text{U}) = E(\epsilon) = 0$, $E(U^2) = E(\epsilon^2) = E(\sigma_{Y|X}^2)$. From T10 we see that $E(Y|X)$ is a better (strictly, no worse) predictor than $E(Y)$. Both predictors are unbiased, but in general, the additional information provided by knowledge of the X-value improves the prediction of Y.

Continuing with the bivariate setting of the prediction problem, suppose that we confine the choice of predictors to *linear* functions of X:

$h(X) = a + bX$. The best such function, in the sense of minimizing $E(U^2)$, where $U = Y - h(X)$, is the line

$$E^*(Y|X) = \alpha + \beta X,$$

where

$$\beta = \sigma_{XY}/\sigma_X^2, \qquad \alpha = \mu_Y - \beta\mu_X.$$

This line is called the *linear projection* (or LP) of Y on X, or *best linear predictor* (BLP) of Y given X.

Proof. Write $U = Y - (a + bX)$, and use the linearity of the expectation operator to calculate

$$\partial E(U^2)/\partial a = E(\partial U^2/\partial a) = 2E(U\partial U/\partial a) = -2E(U)$$
$$\partial E(U^2)/\partial b = E(\partial U^2/\partial b) = 2E(U\partial U/\partial b) = -2E(XU).$$

The first-order conditions are $E(U) = 0$ and $E(XU) = 0$, which together are equivalent to $E(U) = 0$ and $C(X, U) = 0$. Substituting for U, we get

$$E(Y) = a + bE(X),$$
$$C(X, Y) = bV(X).$$

The solution values are denoted as α and β, and it can be confirmed that they locate a minimum. ■

The minimized value of the criterion is

$$E(U^2) = V[Y - (\alpha + \beta X)] = V(Y) + \beta^2 V(X) - 2\beta C(X, Y)$$
$$= V(Y) - \beta^2 V(X).$$

Example. For the roof distribution, use the moments previously obtained to calculate

$$\beta = (-1/144)/(11/144) = -1/11, \qquad \alpha = 7/12 - (-1/11)7/12 = 7/11.$$

So $E^*(Y|X) = 7/11 - (1/11)X$. This BLP is plotted along with the CEF in Figure 5.1. ■

The deviation of Y from its BLP has several characteristic properties. Let $U = Y - E^*(Y|X) = Y - (\alpha + \beta X)$. Then we have

(5.7) $E(U) = 0,$

(5.8) $C(X, U) = 0,$

(5.9) $V(U) = V(Y) - \beta^2 V(X).$

Proofs. The first-order conditions are equivalent to $E(U) = 0$ and $C(X, U) = 0$. With $E(U) = 0$, we have $V(U) = E(U^2)$. ■

We conclude that the deviation of Y from the function $E^*(Y|X)$ is a random variable that has zero expectation, and zero covariance with the conditioning variable X.

5.5. Conditional Expectations and Linear Predictors

We have just developed two predictors of Y given X: the CEF, which is the best predictor, and the BLP, which is the best linear predictor. We were already familiar with the marginal expectation $E(Y)$, which is the best constant predictor. It will be useful to recapitulate the connections among these concepts.

Because $E(Y|X)$, $E^*(Y|X)$, and $E(Y)$ solve successively more constrained minimization problems, it is clear that, as a predictor of Y, the BLP is worse (no better) than the CEF, and better (no worse) than the marginal expectation.

A sharp distinction between the CEF and the BLP refers to prediction errors. Let $U = Y - E^*(Y|X)$ and $\epsilon = Y - E(Y|X)$; then U has zero covariance with X, while ϵ has zero covariance with every function of X.

A pair of theorems relates the BLP to the CEF:

T13. LINEAR APPROXIMATION TO CEF. The best linear approximation to the CEF, in the sense of minimizing $E(W^2)$ where $W = E(Y|X) - (a + bX)$, is the BLP, namely $E^*(Y|X) = \alpha + \beta X$, with $\beta = C(X, Y)/V(X)$, $\alpha = E(Y) - \beta E(X)$.

Proof. This is formally the same linear prediction problem as was solved in Section 5.4, except that W plays the role of U and $E(Y|X)$ plays the role of Y. So the solution values must be

$$a = E(\mu_{Y|X}) - bE(X), \qquad b = C(X, \mu_{Y|X})/V(X).$$

But $E(\mu_{Y|X}) = E(Y)$ by T8, and $C(X, \mu_{Y|X}) = C(X, Y)$ by T12. ■

T14. LINEAR CEF. If the CEF is linear, then it coincides with the BLP: if $E(Y|X) = a + bX$, then $b = C(X, Y)/V(X) = \beta$ and $a = E(Y) - \beta E(X) = \alpha$.

With this as background, we may be able to clarify the concept of *linear relation* as used in empirical econometrics. One often reads that a dependent variable Y is assumed to be a linear function of X plus an error (or disturbance). Some care in interpreting such statements is needed. Taken by itself, $Y = a + bX + U$ is a vacuous statement. When supplemented by $E(U) = 0$, it amounts only to stating that $E(Y) = a + bE(X)$. When further supplemented by $C(X, U) = 0$, it amounts only to announcing that the BLP is being labeled as $a + bX$. But to say that $Y = a + bX + U$, with $E(U|X) = 0$ for all X, is to *assume* something, namely that the CEF is linear.

Finally, referring back to Chapter 1, we see that the CEF is the population counterpart of the sample conditional mean function $m_{Y|X}$, while the BLP may be viewed as the population counterpart of a certain smoothed sample line $m^*_{Y|X}$.

Exercises

5.1 *Curved-roof Distribution.* Recall the bivariate distribution introduced in Exercise 4.1, whose joint pdf is

$$f(x, y) = 3(x^2 + y)/11 \quad \text{for } 0 \le x \le 2,\ 0 \le y \le 1,$$

with $f(x, y) = 0$ elsewhere. Figures 4.4, 4.5, and 4.6 plotted $f(x, y)$, the marginal pdf $f_1(x)$, and selected conditional pdf's $g_2(y \mid x)$.

(a) For $0 \le x \le 2$, find the CEF of Y given X.
(b) Calculate $E(X)$, $E(Y)$, $E(X^2)$, $E(Y^2)$, $E(XY)$, $V(X)$, $V(Y)$, $C(X, Y)$.
(c) Find the BLP of Y given X.

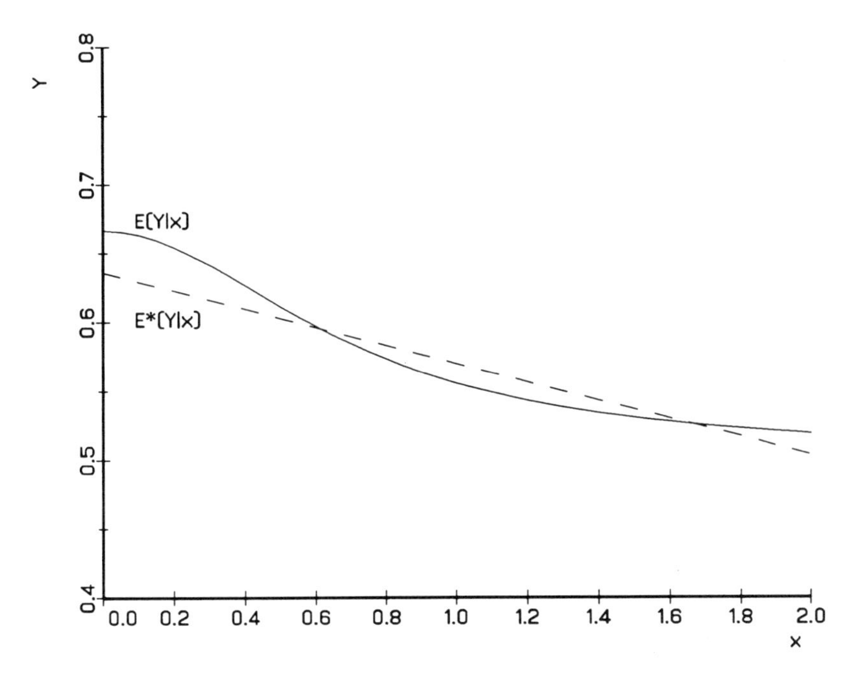

Figure 5.2 Curved-roof distribution: CEF and BLP.

(d) Comment on Figure 5.2, where the CEF and BLP are plotted.
(e) In the figure, the BLP appears closer to the CEF at high, rather than low, values of x. Why does that happen? Hint: See Figure 4.5.

5.2 For the joint pmf in the table below:

(a) Find the conditional expectation function $E(Y \mid X)$.
(b) Find the best linear predictor $E^*(Y \mid X)$.
(c) Prepare a table that gives $E(Y \mid x)$ and $E^*(Y \mid x)$ for $x = 1, 2, 3$.

	$x = 1$	$x = 2$	$x = 3$
$y = 0$	0.15	0.10	0.15
$y = 1$	0.15	0.30	0.15

5.3 Suppose that the random variables Z (= permanent income in thousands of dollars) and W (= transitory income in thousands of dollars) have zero covariance, with $E(Z) = 42$, $E(W) = 0$, $V(Z) = 2500$, $V(W) = 500$. Further, X (= current income in thousands of dollars) is determined as $X = Z + W$.

(a) Calculate $E(X)$, $C(Z, X)$, $C(W, X)$, and $V(X)$.
(b) Find the BLP of current income given permanent income.
(c) Predict as best you can the current income of a person whose permanent income is $z = 54$.
(d) Find the BLP of permanent income given current income.
(e) Predict as best you can the permanent income of a person whose current income is $x = 54$.
(f) Comment on the relation between the answers to (c) and (e).

5.4 For the setup of Exercise 5.3, suppose also that Y (= consumption in thousands of dollars) is determined as $Y = (6/7)Z + U$ where U (= transitory consumption in thousands of dollars) has $E(U) = 0$, $V(U) = 250$, $C(Z, U) = 0$, $C(W, U) = 0$. Find $E^*(Y|Z)$, the BLP of consumption given permanent income, and also find $E^*(Y|X)$, the BLP of consumption given current income. Comment on the distinctions between these two BLP's.

5.5 Provide a counterexample to this proposition: If V_1, V_2, V_3 are three random variables with $V_1 = V_2 + V_3$, then V_1 and V_3 must have nonzero covariance. Hint: Use the setup of Exercise 5.3.

5.6 The random variables X and Y are jointly distributed. Let $\epsilon = Y - E(Y|X)$ and $U = Y - E^*(Y|X)$, where $E(Y|X)$ is the CEF and $E^*(Y|X)$ is the BLP. Determine whether the following is true or false: $C(\epsilon, U) = V(\epsilon)$.

5.7 Suppose that the criterion for successful prediction were changed from minimizing $E(U^2)$ to minimizing $E(|U|)$, where $U = Y - h(X)$.

(a) Show that the best predictor would change from $E(Y|X)$ to $M(Y|X)$, the *conditional median function* of Y given X, defined as the curve that gives the conditional medians of Y as a function of X.
(b) Comment on the attractiveness of the conditional median function as a description of the relation of Y to X in a bivariate population.

5.8 In a bivariate population, let us define the *best proportional predictor* of Y given X as the ray through the origin, $E^{**}(Y|X) = \gamma X$, with γ being the value for c that minimizes $E(U^2)$, where now $U = Y - cX$.

(a) Show that $\gamma = E(XY)/E(X^2)$.
(b) Is $E^{**}(Y|X)$ an unbiased predictor? Explain.
(c) Let $U = Y - \gamma X$. Does $C(X, U) = 0$?
(d) Find the minimized value of $E(U^2)$.
(e) Compare this $E(U^2)$ with those that result when the marginal expectation is used, and when the BLP is used.

6 *Independence in a Bivariate Distribution*

6.1. Introduction

We can recognize at least three possible responses to the question, How is Y related to X in a bivariate population?: the conditional pdf's (or pmf's) $g_2(y|x)$, the conditional expectation function $E(Y|X)$, and the best linear predictor $E^*(Y|X)$. Correspondingly, we can recognize three possible responses to the question, What does it mean to say that Y is *not* related to X in the bivariate population?

6.2. Stochastic Independence

In any bivariate probability distribution we can write the joint pdf (or pmf) as the product of a conditional and a marginal pdf (or pmf):

$$f(x, y) = g_2(y|x)f_1(x) \quad \text{for all } (x, y) \text{ such that } f_1(x) \neq 0.$$

To start, we say that Y is *stochastically independent* of X iff

$$g_2(y|x) = f_2^*(y) \qquad \text{for all } (x, y) \text{ such that } f_1(x) \neq 0,$$

where $f_2^*(y)$ does not depend on x. In other words, the conditional probability distribution of Y given x is the same for all x-values; that is, the conditional probability distribution does not vary with—"is independent of"—X. One implication of stochastic independence is immediate: the marginal pdf of Y is

$$f_2(y) = \int g_2(y|x)f_1(x)\,dx = \int f_2^*(y)f_1(x)\,dx = f_2^*(y)\int f_1(x)\,dx$$

$$= f_2^*(y).$$

(Note: In this chapter the symbol $\int$ is used as shorthand for $\int_{-\infty}^{\infty}$.) That is, if the conditional distribution of Y is the same for all values of X, then the marginal distribution of Y coincides with that common conditional distribution. So

$$Y \text{ is stochastically independent of } X \text{ iff } f(x, y) = f_1(x)f_2(y).$$

That is, Y is stochastically independent of X if and only if the joint pdf (or pmf) factors into the product of the marginal pdf's (or pmf's). In that event, the conditional pdf (or pmf) of X given Y is

$$g_1(x|y) = f(x, y)/f_2(y) = f_1(x)f_2(y)/f_2(y) = f_1(x).$$

We conclude that Y is stochastically independent of X if and only if X is stochastically independent of Y. So stochastic independence is a symmetric relation, which leads us to the equivalent, more traditional

DEFINITION. X and Y are stochastically independent iff

$$f(x, y) = f_1(x)f_2(y) \quad \text{for all } (x, y).$$

For brevity, the unqualified term *independent* is often used instead of stochastically independent.

The following implications are straightforward.

If X and Y are independent, then:

I1. If A is an event defined in terms of X alone, and B is an event defined in terms of Y alone, then $\Pr(A \cap B) = \Pr(A)\Pr(B)$; that is, A and B are independent events.

I2. If $Z = h(X)$ is a function of X alone, then Z and Y are independent random variables.

Proof. Recall from Section 2.5 how, in a univariate distribution, one goes from $f(x)$, the pdf or pmf of X, to $g(z)$, the pdf (or pmf) of the function $Z = h(X)$. Apply the same method here to go from each $g_1(x|y)$ to a $g_1^*(z|y)$. If $g_1(x|y)$ is the same for all y, then $g_1^*(z|y)$ must also be the same for all y. ■

I3. Let $Z_1 = h_1(X)$ be a function of X alone, and $Z_2 = h_2(Y)$ be a function of Y alone. Then Z_1 and Z_2 are independent.

Proof. By I2, Z_1 and Y are independent. Apply I2 again with Z_1 taking the role of Y, and Y taking the role of X. ■

6.3. Roles of Stochastic Independence

Stochastic independence will play several roles in this book.

• Independence serves as a stringent baseline for discussing relations among variables. If X and Y are independent random variables, then it is certainly appropriate to say that there is "no relation" between them in the population.

• Independence serves as a device for building a joint distribution from a pair of marginal distributions. If $X \sim f_1(x)$, $Y \sim f_2(y)$, and X and Y are independent, then $f(x, y) = f_1(x)f_2(y)$.

Example: *Tossing Two Coins.* Suppose $X \sim$ Bernoulli(p) and $Y \sim$ Bernoulli(p), so for all (x, y) with x and y being 0 or 1,

$$f_1(x) = p^x(1 - p)^{1-x}, \qquad f_2(y) = p^y(1 - p)^{1-y}.$$

What is the joint distribution of (X, Y)? That is, how does one fill in the probabilities of the four possible paired outcomes? Without an assumption of independence (or some other information), we cannot fill them in; with it, we can.

A couple of remarks on this example:

(1) If X and Y are independent, then we will have a random vector (X, Y) in which the two random variables are *independent and identically distributed.* We will then say that (X, Y) is a *random sample* of size 2 from the Bernoulli(p) population.

(2) "Independent" and "identically distributed" are distinct concepts. We might have had $X \sim$ Bernoulli(p) and $Y \sim$ Bernoulli(p^*), with $p^* \neq p$, in which case the variables would not be identically distributed. Or we might have had them both Bernoulli(p) but not independent.

Independence serves as a device for building up a behavioral model. For example, suppose $Y = 1$ if a household purchases a car, $Y = 0$ if not, for the population of households in 1990. Suppose further that Y is determined by X = income and U = taste for cars, according to this rule:

$$Y = \begin{cases} 1 & \text{if } a + bX + U > 0, \\ 0 & \text{if } a + bX + U \leq 0. \end{cases}$$

Here a and b are constants, X and Y are observable, while U is an unobserved variable. How does the probability of car purchase vary, if at all, with income? Suppose that $U \sim$ standard normal, with X and U independent. The question is, what is $\Pr(Y = 1|x)$ at income level x? Now

$$Y = 1 \quad \Leftrightarrow \quad (a + bx + U) > 0 \quad \Leftrightarrow \quad U > -(a + bx).$$

So

$$\begin{aligned} \Pr(Y = 1|x) &= \Pr[U > -(a + bx)|x] = 1 - F[-(a + bx)] \\ &= F(a + bx), \end{aligned}$$

where $F(\cdot)$ denotes the standard normal cdf. Here $F(a + bX)$ is not the cdf of income, but rather the CEF of the binary variable Y, conditional on the continuous variable X. Observe how the assumption of independence was used to assert that $U|x \sim$ standard normal for every x.

6.4. Mean-Independence and Uncorrelatedness

We turn to less stringent concepts of the absence of a relation between the variables in a bivariate distribution.

Mean-Independence

In any bivariate probability distribution, we have

$$E(Y|x) = \int y g_2(y|x)\, dy.$$

We say that Y is *mean-independent* of X iff

$$E(Y|x) = \mu_Y^* \quad \text{for all } x \text{ such that } f_1(x) \neq 0,$$

where μ_Y^* does not depend on x. In other words, the conditional expectation of Y given x is the same for all x-values; that is, the population conditional mean does not vary with—"is independent of"—X. One implication of mean-independence is immediate: the marginal expectation of Y is

$$\mu_Y = E_X[E(Y|X)] = E(\mu_Y^*) = \mu_Y^*,$$

by the Law of Iterated Expectations (T8, Section 5.2). That is, if the conditional expectation of Y is the same for all values of X, then the marginal expectation of Y coincides with that common conditional expectation.

Another implication is:

M1. If Y is mean-independent of X, and $Z = h(X)$ is a function of X alone, then Y is mean-independent of Z.

Proof. If $h(\cdot)$ is one-to-one, then the implication is immediate, because $E(\cdot|z)$ is equivalent to $E(\cdot|x)$ with $x = h^{-1}(z)$. Otherwise, let $i(k)$ denote the set of all i such that $z(x_i) = z_k$, and let $\Sigma_{i(k)}$ denote summation over all i in that set. Then the joint probability mass at the point (z_k, y_j) is $f^*(z_k, y_j) = \Sigma_{i(k)} f(x_i, y_j)$, so that the marginal probability mass at z_k is

$$f_1^*(z_k) = \sum_j f^*(z_k, y_j) = \sum_{i(k)} \left[\sum_j f(x_i, y_j)\right] = \sum_{i(k)} f_1(x_i).$$

So the conditional probability mass for y_j given z_k is

$$g_2^*(y_j|z_k) = f^*(z_k, y_j)/f_1^*(z_k) = \left[\sum_{i(k)} g_2(y_j|x_i)f_1(x_i)\right] \Big/ \left[\sum_{i(k)} f_1(x_i)\right]$$

$$= \sum_{i(k)} g_2(y_j|x_i)w_{ik},$$

say, with $\Sigma_{i(k)} w_{ik} = 1$. Then

$$E(Y|z_k) = \sum_j y_j g_2^*(y_j|z_k) = \sum_j y_j \left[\sum_{i(k)} g_2(y_j|x_i)w_{ik}\right]$$

$$= \sum_{i(k)} w_{ik} \left[\sum_j y_j g_2(y_j|x_i)\right] = \sum_{i(k)} w_{ik} E(Y|x_i).$$

If $E(Y|x_i)$ is constant across all i, then $E(Y|z_k)$ will be constant at that same value across all k. ■

What is the connection between independence and mean-independence? If two distributions are the same, they must have the same mean, so independence implies mean-independence. But two distributions may have the same mean, and yet have different variances, or third moments, or medians. So mean-independence is weaker than stochastic independence, because it refers only to the expectation rather than to the entire distribution. Nevertheless, to the extent that economists do interpret "the relation" between Y and X to refer to the population conditional mean function, then it is mean-independence rather than stochastic independence that should serve as the natural baseline for "no relation."

Mean-independence is not a symmetric relation: if Y is mean-independent of X, then X may or may not be mean-independent of Y.

Example: *Three-point Distribution.* Suppose that (X, Y) is discrete, with $f(x, y) = 1/3$ at each of three mass points, namely $(1, -1)$, $(0, 0)$, and $(1, 1)$. Then

$$E(Y|x = 1) = 0 = E(Y|x = 0),$$

but

$$E(X|y = -1) = 1, \qquad E(X|y = 0) = 0, \qquad E(X|y = 1) = 1.$$

Uncorrelatedness

Recall the definition of the covariance in a bivariate probability distribution:

$$C(X, Y) = E\{[X - E(X)][Y - E(Y)]\}.$$

We say that Y is *uncorrelated* with X iff $C(X, Y) = 0$. Clearly, uncorrelatedness is a symmetric relation.

What is the connection between uncorrelatedness and mean-independence? Two results will shed some light:

M2. If Y is mean-independent of X, then Y is uncorrelated with X.

Proof. By T12 (Section 5.2), $C(X, Y) = C[X, E(Y|X)]$. If $E(Y|X) = \mu_Y$ for all X, then $C(X, Y) = C(X, \mu_Y) = 0$, because μ_Y is a constant. ■

M3. If Y is mean-independent of X, and $Z = h(X)$ is a function of X alone, then Y is uncorrelated with Z.

Proof. By M1, Y is mean-independent of Z. Use M2 with Z playing the role of X. ■

Clearly, mean-independence is stronger than uncorrelatedness. The three-point distribution example illustrates this. With $f(x, y) = 1/3$ at each of the three mass points, namely $(1, -1)$, $(0, 0)$, and $(1, 1)$, we calculate $C(X, Y) = 0$. But X is not mean-independent of Y in that example.

Indeed, Y can be uncorrelated with many functions of X without being mean-independent of X. What is true is that:

M4. If Y is uncorrelated with $E(Y|X)$, then Y is mean-independent of X.

Proof. Let $Z = E(Y|X)$, so $Y = Z + \epsilon$ with $C(Z, \epsilon) = 0$. Then $C(Y, Z) = C(Z + \epsilon, Z) = C(Z, Z) + C(Z, \epsilon) = V(Z) \geq 0$, with equality iff Z is constant. ■

6.5. Types of Independence

One useful way to distinguish among uncorrelatedness, mean-independence, and stochastic independence is in terms of the joint moments of the probability distribution, namely the $E(X^r Y^s)$, where r and s are positive integers. We see that:

If Y is uncorrelated with X, then $E(XY) = E(X)E(Y)$.

Proof. $C(X, Y) = E(XY) - E(X)E(Y)$. ■

If Y is mean-independent of X, then $E(X^r Y) = E(X^r)E(Y)$ for all r.

Proof. Let $Z = X^r = h(X)$. Then by M3, $C(Z, Y) = 0$. ■

If Y is independent of X, then $E(X^r Y^s) = E(X^r)E(Y^s)$ for all r, s.

Proof. By I3, X^r and Y^s are independent; hence they are uncorrelated. ■

Another informative distinction between uncorrelatedness and mean-independence concerns prediction:

If Y is mean-independent of X, then $E(Y|X) = E(Y)$ for all X, the CEF of Y given X is a horizontal line, and the best predictor of Y given X is $E(Y)$. If Y is uncorrelated with X, then $E^*(Y|X) = E(Y)$ for all X, the BLP of Y given X is a horizontal line, and the best linear predictor of Y given X is $E(Y)$.

Recalling the discussion of deviations in Sections 5.3 and 5.4, we may now say that:

If $\epsilon = Y - E(Y|X)$, then ϵ is mean-independent of X.
If $U = Y - E^*(Y|X)$, then U is uncorrelated with X.

In applied econometrics, one sometimes reads that "X and Y are uncorrelated, so there is no relation between them." This statement is ambiguous. Only if "the relation" of Y to X refers to the BLP rather than to the CEF would such a statement be appropriate. One also reads that "there is no linear relation" between Y and X. That statement too is ambiguous, and should not be interpreted to say that "there *is* a nonlinear relation" between Y and X. Properly speaking, "no linear relation" means that $C(X, Y) = 0$, so the best-fitting linear relation between Y and X is a horizontal line. And that leaves two possibilities open: perhaps the CEF is also a horizontal line (Y is mean-independent of X), or perhaps the CEF is a curve, the best linear approximation to which happens to be a horizontal line.

Example. For the three-point distribution introduced in Section 6.4, the CEF $E(X|Y)$ is V-shaped (nonlinear), while the BLP $E^*(X|Y)$ is horizontal because $C(X, Y) = 0$.

Example. Let Y = earnings and X = age. Because of natural life-cycle development, it is plausible that $E(Y|X)$ plots as an inverted U. If so, it is quite possible that $E^*(Y|X)$ is horizontal.

6.6. Strength of a Relation

In some contexts, it is interesting to measure the extent of dependence between Y and X in a bivariate population. It seems natural to base such measures on the analysis of prediction.

Recalling the definition of the correlation coefficient (Section 5.1), we show:

CAUCHY-SCHWARTZ INEQUALITY.

$$\text{If } \rho = C(X, Y)/[\sqrt{V(X)}\sqrt{V(Y)}], \text{ then } 0 \leq \rho^2 \leq 1.$$

Proof. From $\beta = C(X, Y)/V(X)$ and Eq. (5.9), it follows that

$$\rho^2 = C^2(X, Y)/[V(X)V(Y)] = \beta^2 V(X)/V(Y) = [V(Y) - V(U)]/V(Y).$$

So $\rho^2 = 1 - V(U)/V(Y)$. But $0 \leq V(U) \leq V(Y)$. ■

If $\rho^2 = 1$, we say that X and Y are perfectly correlated. This ρ^2, the population *coefficient of determination*, measures the proportional reduction in expected squared prediction error that is attributable to using the BLP $E^*(Y|X)$ rather than the marginal expectation $E(Y)$ for predicting Y given X. It is commonly used as an indicator of the strength of "the linear relation" between Y and X in a population.

Example. For the roof distribution,

$$\rho^2 = (-1/144)^2/[(11/144)(11/144)] = 1/121.$$

A related measure relies on the CEF rather than on the BLP. Referring to the Analysis of Variance formula (T10, Section 5.2), define the *correlation ratio* for Y on X as

$$\eta^2 = V_X[E(Y|X)]/V(Y) = 1 - E_X[V(Y|X)]/V(Y).$$

Clearly $0 \leq \eta^2 \leq 1$. This η^2 measures the proportional reduction in expected squared prediction error that is attributable to using the CEF $E(Y|X)$ rather than the marginal expectation $E(Y)$ for predicting Y given X. It may be used as an indicator of the strength of the relation of Y to X, when "the relation" is interpreted to be the CEF. Because the BLP solves a constrained version of the prediction problem solved by the CEF, it follows that $\rho^2 \leq \eta^2$, with equality iff the CEF is linear.

One should not confuse either of these measures of strength with measures of steepness such as the slope of the BLP, $\partial E^*(Y|X)/\partial X = \beta$, or the slope of the CEF, $\partial E(Y|X)/\partial X$. In most economic contexts, slope, rather than strength, will be of primary interest.

Exercises

6.1 Suppose that X_1 and X_2 are stochastically independent Bernoulli variables, with parameters p_1 and p_2 respectively. Let $Y = X_1X_2$ and $W = X_1 + X_2$. Determine whether each of the following statements is true or false.

(a) The random variable Y is distributed Bernoulli with parameter p_1p_2.
(b) The expectation of W^2 is equal to $p_1^2 + p_2^2$.

6.2 Two economists know the joint distribution of X = price and Y = quantity. One decides to predict quantity given price, using the BLP $E^*(Y|X)$; his prediction error is $U = Y - E^*(Y|X)$. The other decides to predict price given quantity; her prediction error is $V = X - E^*(X|Y)$. Let $\sigma_{XY} = C(X, Y)$, $\sigma_{UV} = C(U, V)$, and ρ = correlation of X and Y.

(a) Show that $\sigma_{UV} = -(1 - \rho^2)\, \sigma_{XY}$.
(b) What does that result imply about the sign and magnitude of σ_{UV} as compared with the sign and magnitude of σ_{XY}?

6.3 Suppose that $Z = XY$, where X and Y are independent random variables. Show that $V(Z) = V(X)V(Y) + E^2(X)V(Y) + E^2(Y)V(X)$.

6.4 Suppose that $Z = XY$, where Y is mean-independent of X and the conditional variance of Y given X is constant. Show that the conclusion in Exercise 6.3 is still correct.

6.5 Suppose that $Y = Z - X$ is independent of Z and of X. Show that Y is a constant.

6.6 Suppose that $Y = Z/X$ is independent of Z and of X, where X and Z are positive random variables. Show that Y is a constant.

6.7 Suppose that X and W are independent random variables with $E(X) = 0$, $E(X^2) = 1$, $E(X^3) = 0$, $E(W) = 1$, $E(W^2) = 2$. Let $Y = W + WX^2$.

(a) Find the CEF $E(Y|X)$ and the BLP $E^*(Y|X)$.
(b) Change the assumption $E(X^3) = 0$ to $E(X^3) = 1$. Find the CEF $E(Y|X)$ and the BLP $E^*(Y|X)$.
(c) Which relation remained the same in going from (a) to (b)? Which changed? Why?

7 Normal Distributions

7.1. Univariate Normal Distribution

Recall from Section 2.3 that the random variable Z has the standard normal distribution iff its pdf is

$$f(z) = \exp(-z^2/2)/\sqrt{(2\pi)}.$$

It can be shown by integration that

$$E(Z) = 0, \quad E(Z^2) = 1, \quad E(Z^3) = 0, \quad E(Z^4) = 3.$$

Suppose $Z \sim$ standard normal, and let $X = a + bZ$, where a and b are constants with $b > 0$. By the linear function rule (T1, Section 3.3),

$$E(X) = a + bE(Z) = a, \qquad V(X) = b^2V(Z) = b^2,$$

so we may as well rewrite the linear function as

$$X = \mu + \sigma Z, \quad \text{with } \sigma > 0.$$

As in Section 2.5, to find the pdf of X, first find its cdf:

$$G(x) = \Pr(X \leq x) = \Pr(\mu + \sigma Z \leq x) = \Pr[Z \leq (x - \mu)/\sigma]$$
$$= F(z),$$

with $z = (x - \mu)/\sigma$; here $F(\cdot)$ denotes the standard normal cdf. So the pdf of X is

$$g(x) = \partial G(x)/\partial x = \partial F(z)/\partial x = [\partial F(z)/\partial z](\partial z/\partial x) = f(z)/\sigma,$$

with $z = (x - \mu)/\sigma$. That is,

$$g(x) = \sigma^{-1}(2\pi)^{-1/2}\exp(-z^2/2) = \exp\{-[(x - \mu)/\sigma]^2/2\}/\sqrt{(2\pi\sigma^2)}.$$

We write this as $X \sim \mathcal{N}(\mu, \sigma^2)$. This is a two-parameter family, the (general) *univariate normal distribution* with parameters μ and σ^2. The standard normal distribution is the special case $X \sim \mathcal{N}(0, 1)$. Observe that $E(X) = \mu$ and $V(X) = \sigma^2$, as the notation suggested.

What we have shown is that if $Z \sim \mathcal{N}(0, 1)$ and $X = \mu + \sigma Z$ with $\sigma > 0$, then $X \sim \mathcal{N}(\mu, \sigma^2)$. But, as can be verified, the argument reverses: if $X \sim \mathcal{N}(\mu, \sigma^2)$ and $Z = (X - \mu)/\sigma$, then $Z \sim \mathcal{N}(0, 1)$. The conclusion is that

$$X \sim \mathcal{N}(\mu, \sigma^2) \quad \text{iff} \quad (X - \mu)/\sigma \sim \mathcal{N}(0, 1).$$

It follows that there is no need to tabulate cdf's for general univariate normal distributions: the $\mathcal{N}(0, 1)$ cdf table suffices to provide the probabilities of events for any $\mathcal{N}(\mu, \sigma^2)$ distribution. Just translate the event in terms of X into an event in terms of Z. (Remark: For $b < 0$, use the fact that $-Z \sim \mathcal{N}(0, 1)$.)

An immediate implication is that *a* (nontrivial) *linear function of a normal variable is itself normal.* That is:

If $X \sim \mathcal{N}(\mu, \sigma^2)$ and $Y = a + bX$ with $b \neq 0$, then $Y \sim \mathcal{N}(a + b\mu, b^2\sigma^2)$.

Proof. It suffices to show that $Y = a^* + b^*Z$ where $Z \sim \mathcal{N}(0, 1)$. Let $Z = (X - \mu)/\sigma$, so $X = \mu + \sigma Z$. Then $Y = a + bX = a + b(\mu + \sigma Z) = (a + b\mu) + b\sigma Z = a^* + b^*Z$. ■

The trivial case $b = 0$ is ruled out because it would make Y be a constant. Some writers allow $b = 0$, and would say that $Y = a$ has a "degenerate normal distribution."

7.2. Standard Bivariate Normal Distribution

Suppose that U_1, U_2 are independent $\mathcal{N}(0, 1)$ variables. Let ρ be any constant with $|\rho| < 1$, and let

$$Z_1 = U_1, \qquad Z_2 = \rho U_1 + \sqrt{(1 - \rho^2)}U_2.$$

We show that the joint pdf of (Z_1, Z_2) is

$$g(z_1, z_2) = (2\pi)^{-1}(1 - \rho^2)^{-1/2}\exp(-w/2), \tag{7.1}$$

with

(7.2) $$w = (z_1^2 + z_2^2 - 2\rho z_1 z_2)/(1 - \rho^2).$$

Proof. The joint pdf will be the product of conditional and marginal pdf's: $g(z_1, z_2) = h(z_2|z_1) f_1(z_1)$. Clearly $Z_1 \sim \mathcal{N}(0, 1)$ so

$$f_1(z_1) = f(u_1),$$

with $u_1 = z_1$, where $f(\cdot)$ denotes the $\mathcal{N}(0, 1)$ pdf. Next, for given $Z_1 = z_1$, we see that

$$Z_2 = \rho z_1 + \sqrt{(1 - \rho^2)}U_2$$

is a linear function of U_2, with $U_2 \sim \mathcal{N}(0, 1)$ independently of Z_1. So by the linear-function result in Section 7.1, we know that $Z_2|z_1$ is normally distributed, with $E(Z_2|z_1) = \rho z_1$ and $V(Z_2|z_1) = (1 - \rho^2)$. That is, $Z_2|z_1 \sim \mathcal{N}(\rho z_1, 1 - \rho^2)$. So the conditional pdf is

$$h(z_2|z_1) = f(u_2)/\sqrt{(1 - \rho^2)},$$

with $u_2 = (z_2 - \rho z_1)/\sqrt{(1 - \rho^2)}$. So the joint pdf is

$$\begin{aligned} g(z_1, z_2) &= f(u_1) f(u_2)/\sqrt{(1 - \rho^2)} \\ &= (2\pi)^{-1}(1 - \rho^2)^{-1/2} \exp[-(u_1^2 + u_2^2)/2]. \end{aligned}$$

But

$$\begin{aligned} u_1^2 + u_2^2 &= z_1^2 + (z_2 - \rho z_1)^2/(1 - \rho^2) \\ &= [(1 - \rho^2)z_1^2 + z_2^2 + \rho^2 z_1^2 - 2\rho z_1 z_2]/(1 - \rho^2) \\ &= (z_1^2 + z_2^2 - 2\rho z_1 z_2)/(1 - \rho^2). \quad \blacksquare \end{aligned}$$

The pdf in Eqs. (7.1)–(7.2) defines a one-parameter family, the *standard bivariate normal,* or SBVN, *distribution* with parameter ρ. We write this as $(Z_1, Z_2) \sim \text{SBVN}(\rho)$. Figures 7.1, 7.2, and 7.3 plot the SBVN(ρ) distribution for three values of ρ; the variables are labeled x and y.

It is easy to verify that the derivation reverses, so that

$$(Z_1, Z_2) \sim \text{SBVN}(\rho) \quad \text{iff} \quad U_1, U_2 \text{ are independent } \mathcal{N}(0, 1) \text{ variables},$$

where $U_1 = Z_1$, $U_2 = (Z_2 - \rho Z_1)/\sqrt{(1 - \rho^2)}$. Consequently, we can deduce

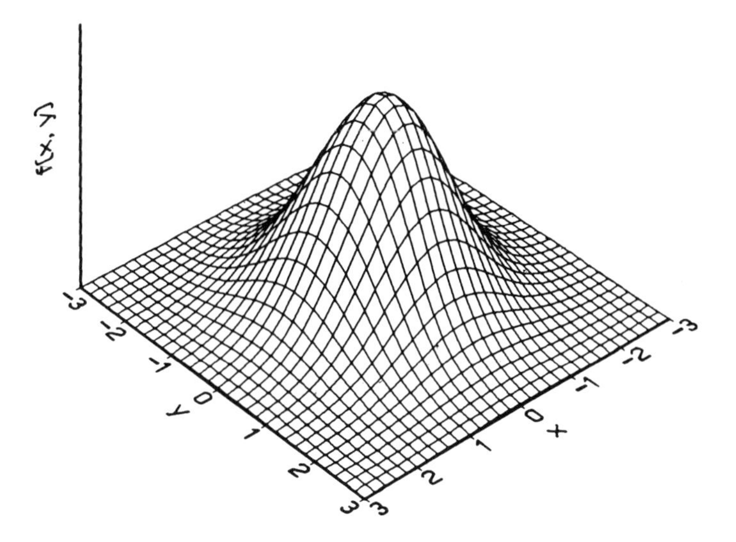

Figure 7.1 SBVN distribution: joint pdf, ρ = 0.

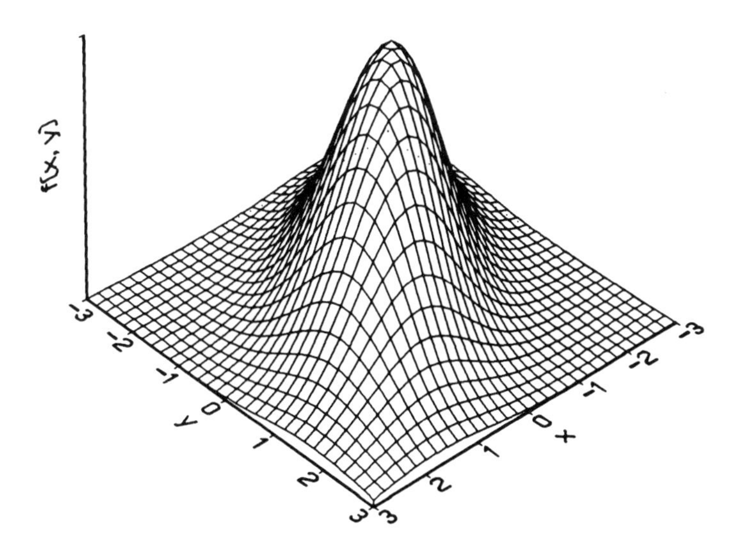

Figure 7.2 SBVN distribution: joint pdf, ρ = 0.6.

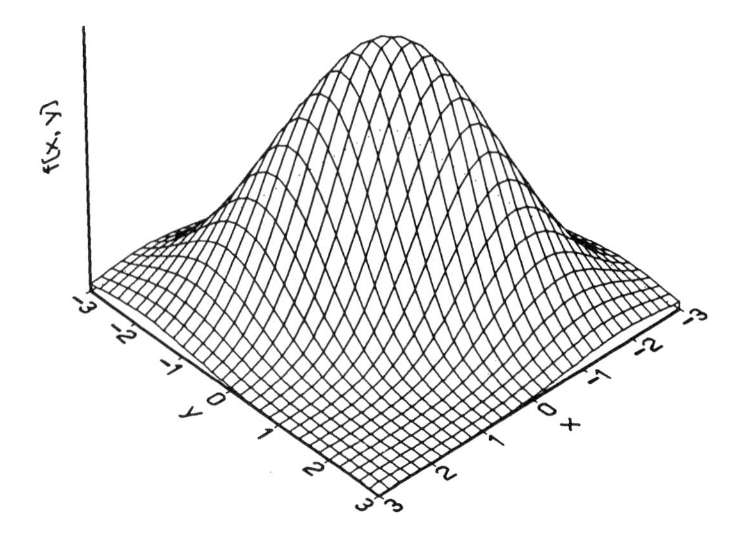

Figure 7.3 SBVN distribution: joint pdf, $\rho = -0.6$.

properties of the SBVN(ρ) distribution directly by relying on the representation in terms of U_1, U_2. If $(Z_1, Z_2) \sim$ SBVN(ρ), then

$$E(Z_1) = E(U_1) = 0, \qquad V(Z_1) = V(U_1) = 1,$$

$$\begin{aligned} C(Z_1, Z_2) &= C[U_1, \rho U_1 + \sqrt{(1-\rho^2)}U_2] \\ &= \rho C(U_1, U_1) + \sqrt{(1-\rho^2)}C(U_1, U_2) = \rho. \end{aligned}$$

So, as the notation suggested, ρ is the correlation of Z_1 and Z_2. Further,

$$Z_1 \sim \mathcal{N}(0, 1), \qquad Z_2|Z_1 \sim \mathcal{N}(\rho Z_1, 1-\rho^2),$$

and

$$\begin{aligned} \rho = 0 \quad &\Leftrightarrow \quad Z_2|Z_1 \sim \mathcal{N}(0, 1) \text{ for all } Z_1 \\ &\Rightarrow \quad Z_1 \text{ and } Z_2 \text{ are independent.} \end{aligned}$$

Of course the roles of Z_1 and Z_2 can be reversed.

So in the SBVN distribution, the marginals are normal, the conditionals are normal, the conditional expectation functions are linear, the conditional variance functions are constant, and uncorrelatedness is equivalent to stochastic independence. Figure 7.4 plots selected contours

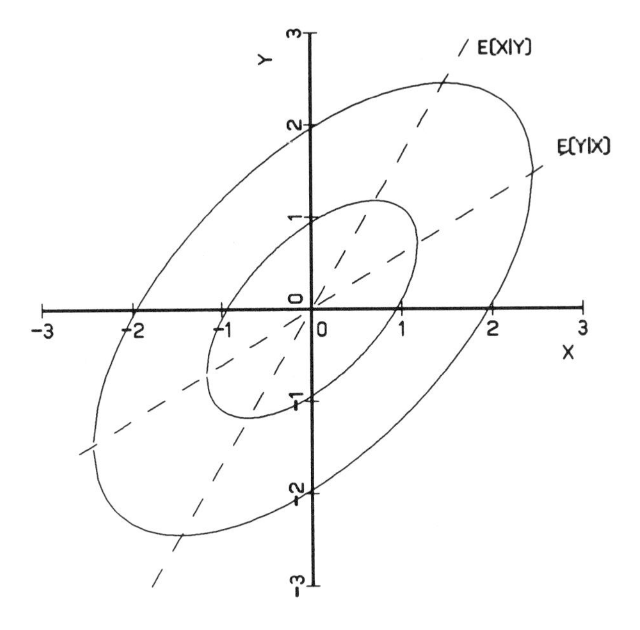

Figure 7.4 SBVN distribution: contours and CEF's, $\rho = 0.6$.

of the SBVN(0.6) distribution along with the two CEF's; the variables are labeled X and Y.

We can calculate the probabilities of various events in an SBVN distribution by translating into a standard normal event.

Example. Suppose $(X, Y) \sim \text{SBVN}(0.6)$. Find $\Pr(Y \leq 2|x = 2)$. We know that $U = (Y - \rho x)/\sqrt{(1 - \rho^2)} \sim \mathcal{N}(0, 1)$, so at $x = 2$, we have $U = (Y - 1.2)/0.8$, and $Y \leq 2 \Leftrightarrow U \leq 1$. So $\Pr(Y \leq 2|x = 2) = F(1) = 0.841$.

7.3. Bivariate Normal Distribution

Suppose that $(Z_1, Z_2) \sim \text{SBVN}(\rho)$. Let $\mu_1, \mu_2, \sigma_1 > 0, \sigma_2 > 0$ be constants, and let

$$X_1 = \mu_1 + \sigma_1 Z_1, \qquad X_2 = \mu_2 + \sigma_2 Z_2.$$

We show that the joint pdf of X_1, X_2 is

$$(7.3)\qquad \phi(x_1, x_2) = \exp(-w/2)/(2\pi\psi),$$

with

$$(7.4)\qquad \begin{aligned} w &= (z_1^2 + z_2^2 - 2\rho z_1 z_2)/(1 - \rho^2), \\ z_1 &= (x_1 - \mu_1)/\sigma_1, \\ z_2 &= (x_2 - \mu_2)/\sigma_2, \\ \psi^2 &= \sigma_1^2\sigma_2^2\,(1 - \rho^2). \end{aligned}$$

Proof. The joint pdf will be the product of conditional and marginal pdfs: $\phi(x_1, x_2) = \phi_1(x_1)h_2(x_2|x_1)$. Now $X_1 = \mu_1 + \sigma_1 Z_1$ where $Z_1 \sim \mathcal{N}(0, 1)$, so

$$\phi_1(x_1) = f(z_1)/\sigma_1,$$

with $z_1 = (x_1 - \mu_1)/\sigma_1$. Turn to $X_2 = \mu_2 + \sigma_2 Z_2$. For given $X_1 = x_1$, that is, for given $Z_1 = (x_1 - \mu_1)/\sigma_1 = z_1$, we see that

$$Z_2|z_1 \sim \mathcal{N}(\rho z_1, 1 - \rho^2),$$

and that X_2 is a linear function of Z_2. So $X_2|x_1 \sim$ normal, with

$$\begin{aligned} E(X_2|x_1) &= \mu_2 + \sigma_2 E(Z_2|z_1) = \mu_2 + \rho\sigma_2 z_1, \\ V(X_2|x_1) &= \qquad\quad \sigma_2^2 V(Z_2|z_1) = \sigma_2^2(1 - \rho^2). \end{aligned}$$

That is,

$$X_2|x_1 \sim \mathcal{N}[\mu_2 + \rho\sigma_2 z_1, \sigma_2^2(1 - \rho^2)],$$

with $z_1 = (x_1 - \mu_1)/\sigma_1$. So the conditional pdf is

$$h_2(x_2|x_1) = f(u_2)/[\sigma_2\sqrt{(1 - \rho^2)}],$$

with $u_2 = [x_2 - (\mu_2 + \rho\sigma_2 z_1)]/[\sigma_2\sqrt{(1 - \rho^2)}]$. So the joint pdf is

$$\phi(x_1, x_2) = \phi_1(x_1)h_2(x_2|x_1) = f(z_1)f(u_2)/[\sigma_1\sigma_2\sqrt{(1 - \rho^2)}].$$

Multiply out and rearrange. ■

The pdf in Eqs. (7.3)–(7.4) defines a five-parameter family, the (general) *bivariate normal*, or BVN, *distribution*, with parameters μ_1, μ_2, σ_1^2,

σ_2^2, and $\sigma_{12} = \rho\sigma_1\sigma_2$. We write this as $(X_1, X_2) \sim \text{BVN}(\mu_1, \mu_2, \sigma_1^2, \sigma_2^2, \sigma_{12})$. It is easy to verify that the derivation reverses, so that

$$(X_1, X_2) \sim \text{BVN}(\mu_1, \mu_2, \sigma_1^2, \sigma_2^2, \sigma_{12}) \quad \text{iff} \quad (Z_1, Z_2) \sim \text{SBVN}(\rho),$$

where $Z_1 = (X_1 - \mu_1)/\sigma_1$, $Z_2 = (X_2 - \mu_2)/\sigma_2$, $\rho = \sigma_{12}/(\sigma_1\sigma_2)$.

Referring back to Section 7.2, we have an equivalent statement:

$$(X_1, X_2) \sim \text{BVN}(\mu_1, \mu_2, \sigma_1^2, \sigma_2^2, \sigma_{12})$$

$$\text{iff } U_1 \text{ and } U_2 \text{ are independent } \mathcal{N}(0, 1) \text{ variables,}$$

where

$$U_1 = (X_1 - \mu_1)/\sigma_1,$$

$$U_2 = [X_2 - (\mu_2 + \rho\sigma_2 Z_1)]/[\sigma_2\sqrt{(1 - \rho^2)}],$$

$$\rho = \sigma_{12}/(\sigma_1\sigma_2).$$

7.4. Properties of Bivariate Normal Distribution

As a result of the derivation and its reversal, we can directly deduce properties of the general bivariate normal distribution. The marginal distribution of X_1 is normal: $X_1 \sim \mathcal{N}(\mu_1, \sigma_1^2)$. The conditional distribution of X_2 given x_1 is normal: $X_2|x_1 \sim \mathcal{N}[\mu_2 + \rho\sigma_2 z_1, \sigma_2^2(1 - \rho^2)]$, with $z_1 = (x_1 - \mu_1)/\sigma_1$.

Let $\sigma_{12} = \rho\sigma_1\sigma_2$, and write

$$\begin{aligned} \mu_2 + \rho\sigma_2 z_1 &= \mu_2 + \rho\sigma_2(x_1 - \mu_1)/\sigma_1 \\ &= [\mu_2 - (\rho\sigma_2/\sigma_1)\mu_1] + (\rho\sigma_2/\sigma_1)\, x_1 \\ &= \alpha + \beta x_1, \end{aligned}$$

where $\alpha = \mu_2 - \beta\mu_1$, $\beta = \rho\sigma_2/\sigma_1 = \sigma_{12}/\sigma_1^2$. Also,

$$\sigma_2^2(1 - \rho^2) = \sigma_2^2[1 - \sigma_{12}^2/(\sigma_1^2\sigma_2^2)] = \sigma_2^2 - \beta^2\sigma_1^2 = \sigma^2,$$

say. So we can write

$$X_2|x_1 \sim \mathcal{N}(\alpha + \beta x_1, \sigma^2).$$

As for the moments, we have found that

$$E(X_1) = \mu_1 + \sigma_1 E(Z_1) = \mu_1,$$
$$V(X_1) = \sigma_1^2 V(Z_1) = \sigma_1^2,$$
$$C(X_1, X_2) = \sigma_1\sigma_2 C(Z_1, Z_2) = \sigma_1\sigma_2\rho = \sigma_{12}.$$

And we also have seen that $\sigma_{12} = 0 \Rightarrow \rho = 0 \Rightarrow Z_1$ and Z_2 independent $\Rightarrow X_1$ and X_2 independent.

We restate these properties for reference.

If $(X_1, X_2) \sim \text{BVN}(\mu_1, \mu_2, \sigma_1^2, \sigma_2^2, \sigma_{12})$, *then*:

P1. The expectations, variances, and covariance are:

$$E(X_1) = \mu_1, \qquad E(X_2) = \mu_2,$$
$$V(X_1) = \sigma_1^2, \qquad V(X_2) = \sigma_2^2, \qquad C(X_1, X_2) = \sigma_{12},$$

thus justifying the symbols used for the parameters.

P2. The marginal distribution of X_1 is normal:

$$X_1 \sim \mathcal{N}(\mu_1, \sigma_1^2).$$

P3. The conditional distribution of X_2 given X_1 is normal:

$$X_2|X_1 \sim \mathcal{N}(\alpha + \beta X_1, \sigma^2),$$

where

$$\beta = \sigma_{12}/\sigma_1^2, \qquad \alpha = \mu_2 - \beta\mu_1, \qquad \sigma^2 = \sigma_2^2 - \beta^2\sigma_1^2.$$

Observe that the CEF is linear and the conditional variance is constant.

P4. Uncorrelatedness implies stochastic independence: If $\sigma_{12} = 0$, then X_1 and X_2 are independent.

Of course, the roles of X_1 and X_2 can be reversed throughout. We can also see that

P5. A (nontrivial) pair of linear functions of X_1 and X_2 is distributed bivariate normal: If $Y_1 = a_1 + b_1X_1 + c_1X_2$, and $Y_2 = a_2 + b_2X_1 + c_2X_2$, where the a's, b's, and c's are constants, with $b_1c_2 - b_2c_1 \neq 0$, then

$(Y_1, Y_2) \sim$ BVN. The condition $b_1c_2 - b_2c_1 \neq 0$ rules out constancy of either variable or perfect correlation between the variables. Some writers drop the condition and refer to "degenerate bivariate normal distributions."

Proof. It suffices to show that Y_1 and Y_2 can be expressed as linear functions of W_1 and W_2 where $(W_1, W_2) \sim \text{SBVN}(\rho)$. ■

We can calculate the probabilities of various events in a BVN distribution by translating into a standard normal event.

Example. Suppose $(X, Y) \sim \text{BVN}(2, 4, 6, 5.5, 3)$. In order to find $\Pr(Y \leq 2|x = 2)$, calculate $\beta = 3/6 = 0.5$, $\alpha = 4 - 0.5(2) = 3$, $\sigma^2 = 5.5 - (0.5)^2\, 6 = 4$. We know that $Z = [Y - (3 + 0.5x)]/\sqrt{4} \sim \mathcal{N}(0, 1)$, so at $x = 2$, we have $Z = (Y - 4)/2$, and $Y \leq 2 \Leftrightarrow Z \leq -1$. We find $\Pr(Y \leq 2|x = 2) = F(-1) = 1 - F(1) = 1 - 0.841 = 0.159$.

7.5. Remarks

• Because it has linear CEF's, constant conditional variances, normal marginals, and normal conditionals, the BVN is convenient for illustration of theoretical concepts. There is another reason for our interest: the BVN arises as the limiting joint distribution of sample means in random sampling from any bivariate distribution (see Chapter 10).

• There is a lot more to a BVN distribution than marginal normality of its components. Figure 7.5 plots a non-BVN distribution that has normal marginals. The joint pdf is:

$$\phi(x, y) = 2zf(x)f(y), \quad \text{where } z = 1 \text{ if } xy > 0,\ z = 0 \text{ if } xy \leq 0,$$

and $f(\cdot)$ denotes the $\mathcal{N}(0, 1)$ pdf. The joint pdf is nonzero only in the NE and SW quadrants. The marginal pdf of X is

$$\phi_1(x) = \int_{-\infty}^{\infty} \phi(x, y)\, dy = \int_{-\infty}^{\infty} 2zf(x)f(y)\, dy = 2\, f(x) \int_{-\infty}^{\infty} zf(y)\, dy.$$

For $x > 0$, $zf(y) = f(y)$ for $y > 0$, and $zf(y) = 0$ for $y \leq 0$. So for $x > 0$,

$$\int_{-\infty}^{\infty} zf(y)\, dy = \int_{0}^{\infty} f(y)\, dy = 1/2.$$

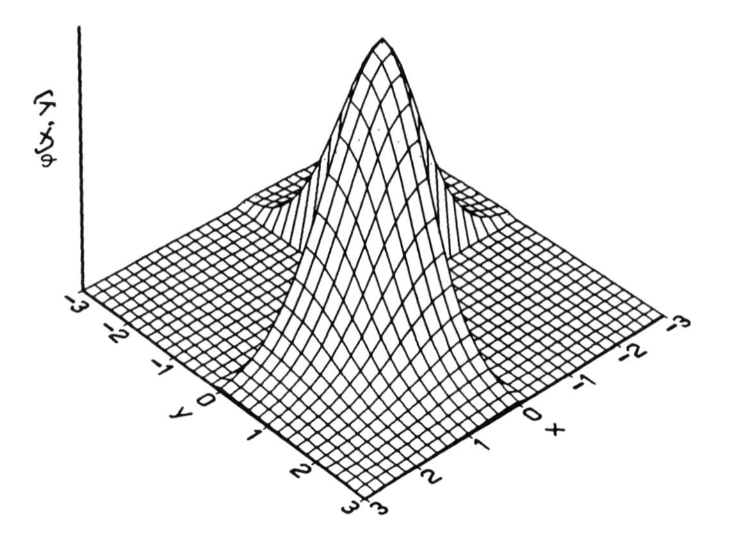

Figure 7.5 Non-BVN distribution with normal marginals.

Similarly for $x \leq 0$. So $\phi_1(x) = 2f(x)(1/2) = f(x)$. That is, $X \sim \mathcal{N}(0, 1)$. By symmetry, $Y \sim \mathcal{N}(0, 1)$, so both marginals are normal. But the joint distribution is not BVN, and indeed the CEF's are not linear, and the conditional pdf's are not normal.

• In a BVN distribution, uncorrelatedness implies independence. But two univariate normal variables may be uncorrelated without being independent. After all, their joint distribution may not be bivariate normal.

Exercises

7.1 The random variable X is distributed $\mathcal{N}(3, 16)$. Calculate each of the following.

(a) $\Pr(X \leq 7)$. (b) $\Pr(X > 5)$. (c) $\Pr(X = 3)$.
(d) $\Pr(-1 < X < 11)$. (e) $\Pr(X \geq 0)$. (f) $\Pr(|X| \leq 3)$.

7.2 The random variable X is distributed $\mathcal{N}(3, 16)$. Let $Y = 3 - X/4$. Calculate $\Pr(3.25 < Y < 4.25)$, and $\Pr(Y > X)$.

7.3 The pair of random variables X and Y is bivariate–normally distributed with parameters $\mu_X = 3$, $\mu_Y = 4$, $\sigma_X^2 = 9$, $\sigma_Y^2 = 20$, and $\sigma_{XY} = 6$. Calculate each of the following.

(a) $E(Y|x = 3)$. (b) $E(Y|x = 6)$. (c) $V(Y|x = 3)$.
(d) $V(Y|x = 6)$. (e) $\Pr(Y \leq 8|x = 3)$. (f) $\Pr(Y \leq 8|x = 6)$.
(g) $\Pr(Y \leq 8)$.

7.4 For the setup of Exercise 7.3, let $U = Y + X$. Calculate $E(U)$, $V(U)$, and $\Pr(U \geq 3)$.

7.5 Suppose that $Z \sim \mathcal{N}(42, 2500)$, $W \sim \mathcal{N}(0, 500)$, that Z and W are independent, and that $X = Z + W$. Calculate the conditional expectation function $E(Z|X)$. How do you know that the CEF is linear?

7.6 Suppose that $Y_1 = X + Z_1$, $Y_2 = X + Z_2$, where X, Z_1, Z_2 are independent random variables with $E(X) = 100$, $V(X) = 100$, $E(Z_1) = 0$, $V(Z_1) = 20$, $E(Z_2) = 0$, $V(Z_2) = 40$.

(a) Find the best linear predictor of Y_2 given Y_1.
(b) Find the expected squared prediction error that results when that BLP is used.
(c) Suppose that Y_1 and Y_2 are bivariate–normally distributed. Could you improve on your predictor? If so, how? If not, why not?

7.7 Suppose that X = wage income and U = nonwage income are bivariate–normally distributed with $E(X) = 30$, $E(U) = 5$, $V(X) = 185$, $V(U) = 35$, $C(X, U) = 15$. Also, total income is $Y = X + U$. All variables are measured in thousands of dollars. A person reports her total income to be $y = 20$. Calculate the probability that her wage income is less than 20.

8 *Sampling Distributions: Univariate Case*

8.1. Random Sample

We have been discussing probability distributions, that is, populations. We now turn to samples, which are sets of observations drawn from a probability distribution. We continue to treat the population probability distribution as known, deferring until Chapter 11 the practical problem that really concerns us, namely how to use a single sample to estimate features of an unknown population.

We start with a univariate probability distribution for the random variable X. Let $X_1, \ldots, X_n$ denote independent drawings from that population. That is, they are the random variables that are the outcomes when the same experiment is repeated n times independently. Then the random vector $\mathbf{X} = (X_1, \ldots, X_n)'$ is called a *random sample* of size n on the variable X, or from the population of X, or from the probability distribution of X. The values that $\mathbf{X}$ takes on will be denoted as $\mathbf{x} = (x_1, \ldots, x_n)'$. (Note: Boldface type is used to denote vectors, which are generally to be thought of as column vectors, and the prime, ′, is used to denote transposition.) The concept of (stochastic) independence, introduced in Section 6.2 for the bivariate case, is now being applied to the multivariate case: a set of random variables is independent iff their joint pdf (or pmf) factors into the product of their marginal pdf's (or pmf's).

Strictly speaking, the term "random sample" refers to the random vector $\mathbf{X}$, but in common usage a single draw, $\mathbf{x}$, on that vector is also called a random sample. Observe that in the term "random sample," the adjective "random" has a much stronger meaning than it did in the term "random variable."

If $\mathbf{X} = (X_1, \ldots, X_n)'$ is a random sample on X, then the X_i's are *independent and identically distributed.* If $f(x)$ is the pmf or pdf of X, then the joint pmf or pdf for the random sample $\mathbf{X}$ is

$$g_n(\mathbf{x}) = g_n(x_1, \ldots, x_n) = f_1(x_1) \cdots f_n(x_n) = f(x_1) \cdots f(x_n) = \prod_i f(x_i),$$

using $f_i(\cdot) = f(\cdot)$ for all i, and independence across i. (Note: The symbol Π_i is shorthand for $\Pi_{i=1}^n$.)

Here are some examples of the joint pmf or pdf for random samples.

(1) *Bernoulli*(p). The pmf of the random variable X is

$$f(x) = p^x(1-p)^{(1-x)} \qquad \text{for } x = 0, 1,$$

with $f(x) = 0$ elsewhere. Then for any $n \times 1$ vector $\mathbf{x}$ whose elements are all either 0's or 1's,

$$\begin{aligned} g_n(\mathbf{x}) &= \prod_i \left[p^{x_i}(1-p)^{(1-x_i)}\right] = \left[\prod_i p^{x_i}\right]\left[\prod_i (1-p)^{(1-x_i)}\right] \\ &= p^{(x_1+\cdots+x_n)}(1-p)^{(1-x_1+1-x_2+\cdots+1-x_n)} \\ &= p^{\Sigma_i x_i}(1-p)^{(n-\Sigma_i x_i)}, \end{aligned}$$

where Σ_i is shorthand for $\Sigma_{i=1}^n$. For any other $\mathbf{x}$, $g_n(\mathbf{x}) = 0$.

(2) *Normal*(μ, σ^2). The pdf of the random variable X is

$$f(x) = (2\pi\sigma^2)^{-1/2} \exp\{-[(x-\mu)/\sigma]^2/2\}.$$

So for any $n \times 1$ vector $\mathbf{x}$,

$$g_n(\mathbf{x}) = (2\pi\sigma^2)^{-n/2} \exp\left\{-\sum_i [(x_i-\mu)/\sigma]^2/2\right\}.$$

(3) *Exponential*(λ). The pdf of the random variable X is

$$f(x) = \lambda \exp(-\lambda x) \quad \text{for } x > 0,$$

with $f(x) = 0$ elsewhere. So for any $n \times 1$ vector $\mathbf{x}$ whose elements are all positive,

$$g_n(\mathbf{x}) = \lambda^n \exp\left(-\lambda \sum_i x_i\right).$$

For any other $\mathbf{x}$, $g_n(\mathbf{x}) = 0$.

8.2. Sample Statistics

Let $T = h(X_1, \ldots, X_n) = h(\mathbf{X})$ be a scalar function of the random sample. Then T is called a *sample statistic*. The values that $T = h(\mathbf{X})$ takes on will be denoted as $t = h(\mathbf{x})$. Sample statistics include the *sample mean*,

$$\overline{X} = (X_1 + \cdots + X_n)/n = (1/n) \sum_i X_i,$$

the *sample variance*,

$$S^2 = (1/n) \sum_i (X_i - \overline{X})^2,$$

the *sample raw moments* (for nonnegative integers r),

$$M'_r = (1/n) \sum_i X_i^r,$$

and the *sample moments about the sample mean* (for nonnegative integers r),

$$M_r = (1/n) \sum_i (X_i - \overline{X})^r.$$

The formulas used here are convenient when the observed data come in a list, the elements of which need not be distinct. If the data come in the form of a frequency distribution, one may use equivalent formulas as in Chapter 1. (Caution: Here i runs from 1 to n, the number of observations; in Chapter 1, i ran from 1 to I, the number of distinct values.)

Other sample statistics include the *sample maximum*, and the *sample proportion* having X less than or equal to some specified value c.

Any sample statistic $T = h(\mathbf{X})$ is a random variable, because its value is determined by the outcome of an experiment. In random sampling, the probability distribution of T, known as its *sampling distribution*, is completely determined by $h(\cdot)$, $f(x)$, and n.

Evidently it is possible to derive the sampling distribution of $T = h(\mathbf{X})$ from knowledge of $f(x)$ and n. As an illustration, consider the sample mean in random sampling of size 2 from a continuous distribution whose pdf and cdf are $f(x)$ and $F(x)$. The cdf of $T = (X_1 + X_2)/2$ is

$$\begin{aligned} G(t) &= \Pr(T \leq t) = \Pr(X_1 + X_2 \leq 2t) = \Pr(X_2 \leq 2t - X_1) \\ &= \int_{-\infty}^{\infty} \int_{-\infty}^{2t-x_1} f(x_1)f(x_2)\, dx_2\, dx_1 = \int_{-\infty}^{\infty} f(x_1)\left[\int_{-\infty}^{2t-x_1} f(x_2)\, dx_2\right] dx_1 \end{aligned}$$

$$= \int_{-\infty}^{\infty} f(x_1)F(2t - x_1)\, dx_1 = \int_{-\infty}^{\infty} F(2t - x)f(x)\, dx,$$

say. So the pdf of T is $g(t) = \partial G(t)/\partial t = 2 \int_{-\infty}^{\infty} f(2t - x)f(x)\, dx$.

Example. Suppose that $X \sim$ exponential(λ), so that its pdf is $f(x) = \lambda \exp(-\lambda x)$ for $x > 0$ and $f(x) = 0$ elsewhere. Then for $t > 0$,

$$\begin{aligned} g(t) &= 2 \int_0^{2t} \lambda \exp[-\lambda(2t - x)]\, \lambda \exp(-\lambda x)\, dx \\ &= 2\lambda^2 \exp(-2\lambda t) \int_0^{2t} dx \\ &= 4\lambda^2 t \exp(-2\lambda t), \end{aligned}$$

with $g(t) = 0$ elsewhere. The integration here is confined to the interval $0 < x < 2t$ because either $f(x)$ or $f(2t - x)$ is zero elsewhere.

The same logic applies to other sample statistics, and extends to random sampling with $n > 2$. In many cases, we will report rather than derive the sampling distributions.

8.3. The Sample Mean

We report the sampling distributions of the sample mean in random sampling, sample size n, for three populations:

(1) If $X \sim$ Bernoulli(p), then $Y \sim$ binomial(n, p), where $Y = n\bar{X}$.
(2) If $X \sim \mathcal{N}(\mu, \sigma^2)$, then $\bar{X} \sim \mathcal{N}(\mu, \sigma^2/n)$.
(3) If $X \sim$ exponential(λ), then $W \sim$ chi-square(k), where $k = 2n$ and $W = k\lambda\bar{X}$.

The chi-square distribution (with parameter k) will be discussed in Section 8.5. Its cdf is tabulated in Table A.2.

How does one use such information to calculate the probabilities of events defined in terms of $\bar{X}$, that is, to get points on the cdf $F_n(c) = \Pr(\bar{X} \leq c)$? The approach is familiar: translate the event into one whose probability is directly tabulated. We illustrate the procedure with our three examples.

(1) *Bernoulli*(p). For given n and p:

$$\bar{X} \leq c \quad \Leftrightarrow \quad Y = n\bar{X} \leq nc = c^*,$$

say. Go to a table, found in some statistics texts, which gives the binomial (n, p) cdf $G(\cdot; n, p)$, say, and read off $F_n(c) = G(c^*; n, p)$. Alternatively, use the binomial(n, p) pmf formula and sum up the appropriate terms.

(2) *Normal*(μ, σ^2). For given μ and σ^2:

$$\bar{X} \leq c \quad \Leftrightarrow \quad Z = (\bar{X} - \mu)/(\sigma/\sqrt{n}) \leq (c - \mu)/(\sigma/\sqrt{n}) = c^*,$$

say. Go to Table A.1, which gives the standard normal cdf $\Phi(\cdot)$, say, and read off $F_n(c) = \Phi(c^*)$.

(3) *Exponential*(λ). For given n and λ, calculate $k = 2n$. Now

$$\bar{X} \leq c \quad \Leftrightarrow \quad W = k\lambda\bar{X} \leq k\lambda c = c^*,$$

say. Go to Table A.2, which gives the chi-square(k) cdf $G_k(\cdot)$, say, and read off $F_n(c) = G_k(c^*)$.

We see that the sampling distribution of the sample mean differs as the population differs, and as the sample size differs. What can be said in general—that is, without reference to the specific form of the parent population—about the expectation and the variance of the sample mean? Now

$$\bar{X} = (1/n)X_1 + \cdots + (1/n)X_n$$

is a linear function of the n random variables X_i. Extending what we know about expectations and variances of linear functions of two variables (T5, Section 5.1) to the $n > 2$ case, we calculate

$$\begin{aligned} E(\bar{X}) &= (1/n)E(X_1) + \cdots + (1/n)E(X_n) = (1/n)\mu + \cdots + (1/n)\mu = \mu, \\ V(\bar{X}) &= (1/n^2)V(X_1) + \cdots + (1/n^2)V(X_n) + (1/n^2)C(X_1, X_2) + \cdots \\ &= (1/n^2)\sigma^2 + \ldots + (1/n^2)\sigma^2 = (n/n^2)\sigma^2 = \sigma^2/n. \end{aligned}$$

This establishes our key result:

SAMPLE MEAN THEOREM. In random sampling, sample size n, from any population with $E(X) = \mu$ and $V(X) = \sigma^2$, the sample mean $\bar{X}$ has $E(\bar{X}) = \mu$ and $V(\bar{X}) = \sigma^2/n$.

8.4. Sample Moments

The Sample Mean Theorem must cover other sample statistics as well, in random sampling.

Sample Raw Moments

Recall from Section 3.2 the definition of the population rth raw moment: $E(X^r) = \mu_r'$, say. The corresponding sample raw moment is

$$M_r' = (1/n) \sum_i X_i^r.$$

Let $Y = X^r$, so $Y_i = X_i^r$. Then $M_r' = (1/n)\Sigma_i Y_i = \bar{Y}$ is a sample mean, and $\mathbf{Y} = (Y_1, \ldots, Y_n)'$ is a random sample on the variable Y. So the theorem must apply to $M_r' = \bar{Y}$. Now

$$E(Y) = E(X^r) = \mu_r',$$

$$V(Y) = E(Y^2) - E^2(Y) = E(X^{2r}) - [E(X^r)]^2 = \mu_{2r}' - (\mu_r')^2.$$

So

$$E(M_r') = \mu_r', \qquad V(M_r') = [\mu_{2r}' - (\mu_r')^2]/n.$$

Sample Moments about the Population Mean

Recall also the definition of the population rth central moment: $E(X - \mu)^r = \mu_r$, say. The corresponding sample statistic is

$$M_r^* = (1/n) \sum_i (X_i - \mu)^r.$$

Let $Y = (X - \mu)^r$, so $Y_i = (X_i - \mu)^r$. Then $M_r^* = (1/n)\Sigma_i Y_i = \bar{Y}$ is a sample mean in random sampling on the variable Y, so the theorem must apply. Thus

$$E(M_r^*) = \mu_r, \qquad V(M_r^*) = (\mu_{2r} - \mu_r^2)/n.$$

For example, for $r = 2$, we have $M_2^* = (1/n)\Sigma_i(X_i - \mu)^2$, with

$$E(M_2^*) = \mu_2, \qquad V(M_2^*) = (\mu_4 - \mu_2^2)/n.$$

We refer to M_2^* as the *ideal sample variance*: in practice M_2^* cannot be computed, because μ is unknown.

Sample Moments about the Sample Mean

Consider the rth *sample moment about the sample mean*:

$$M_r = (1/n) \sum_i (X_i - \bar{X})^r.$$

Let $Y = (X - \bar{X})^r$, so $Y_i = (X_i - \bar{X})^r$, and $M_r = \bar{Y}$. However, the Y_i's are not independent random variables. To see this, let

$$U_1 = X_1 - \bar{X}, \qquad U_2 = X_2 - \bar{X},$$

and calculate

$$\begin{aligned} C(U_1, U_2) &= C(X_1, X_2) + C(\bar{X}, \bar{X}) - C(X_1, \bar{X}) - C(X_2, \bar{X}) \\ &= \quad 0 \quad + V(X)/n - V(X)/n - V(X)/n = -V(X)/n. \end{aligned}$$

The U_i's are correlated and hence cannot be independent. So the $Y_i = U_i^r$ are presumably not independent either. If so, the Sample Mean Theorem does not apply to sample moments about the sample mean. (To reinforce the presumption, consider the case $n = 2$: there $U_1 = -U_2$, so $Y_1 = \pm Y_2$.)

The expectation and variance can still be obtained by brute force. We confine attention to the sample variance

$$S^2 = M_2 = (1/n) \sum_i (X_i - \bar{X})^2.$$

As a matter of algebra,

$$\begin{aligned} \sum_i (X_i - \bar{X})^2 &= \sum_i [(X_i - \mu) - (\bar{X} - \mu)]^2 \\ &= \sum_i (X_i - \mu)^2 + n(\bar{X} - \mu)^2 - 2(\bar{X} - \mu) \sum_i (X_i - \mu) \\ &= \sum_i (X_i - \mu)^2 - n(\bar{X} - \mu)^2. \end{aligned}$$

So $M_2 = M_2^* - (\bar{X} - \mu)^2$, whence

$$\begin{aligned} E(M_2) &= E(M_2^*) - E(\bar{X} - \mu)^2 = \mu_2 - V(\bar{X}) = \mu_2 - \mu_2/n \\ &= \sigma^2(1 - 1/n). \end{aligned}$$

In similar fashion it can be shown that

$$\begin{aligned} V(M_2) &= (n - 1)^2\{\mu_4 - [(n - 3)/(n - 1)]\mu_2^2\}/n^3 \\ &= (\mu_4 - \mu_2^2)/n - 2(\mu_4 - 2\mu_2^2)/n^2 + (\mu_4 - 3\mu_2^2)/n^3. \end{aligned}$$

A couple of remarks on the relation between the sample variance and the ideal sample variance are useful here. First, because $M_2 = M_2^* - (\bar{X} - \mu)^2$, it follows that $M_2 \leq M_2^*$ in every sample. Second, if n is large, then

$$E(M_2) \cong \mu_2 = \sigma^2 = E(M_2^*), \qquad V(M_2) \cong (\mu_4 - \mu_2^2)/n = V(M_2^*),$$

suggesting that when the sample size is large, the distributions of the statistics M_2 and M_2^* will be quite similar. We will formalize this suggestion in Section 9.6.

8.5. Chi-square and Student's *t* Distributions

As a preliminary to further discussion of sampling from a univariate normal population, we introduce two other univariate distributions.

(1) *Chi-square Distribution.* If $Z_1, \ldots, Z_k$ are independent $\mathcal{N}(0, 1)$ variables, and $W = \Sigma_{i=1}^k Z_i^2$, then the pdf of W is

$$g_k(w) = (1/2)(w/2)^{k/2-1} \exp(-w/2)/\Gamma(k/2) \quad \text{for } w > 0,$$

with $g_k(w) = 0$ elsewhere. Here $\Gamma(n)$ is the gamma function:

$$\Gamma(1/2) = \sqrt{\pi}, \qquad \Gamma(1) = 1, \qquad \Gamma(n) = (n-1)\Gamma(n-1).$$

We write this situation as $W \sim \chi^2(k)$. The pdf defines the chi-square distribution, a one-parameter family. Figures 8.1 and 8.2 plot the pdf for selected values of k, while the cdf is tabulated in Table A.2.

The derivation reverses: if $W \sim \chi^2(k)$, then W can be expressed as the sum of squares of k independent $\mathcal{N}(0, 1)$ variables. Recall from Section 7.1 that

$$Z \sim \mathcal{N}(0, 1) \Rightarrow E(Z) = 0, E(Z^2) = 1, E(Z^4) = 3.$$

So $V(Z^2) = 3 - 1^2 = 2$. Because the Z_i's are independent $\mathcal{N}(0, 1)$ variables, we have

$$E(W) = \sum_{i=1}^k E(Z_i^2) = k, \qquad V(W) = \sum_{i=1}^k V(Z_i^2) = 2k.$$

The parameter k, traditionally called "the degrees of freedom," is simply the expectation of the variable W.

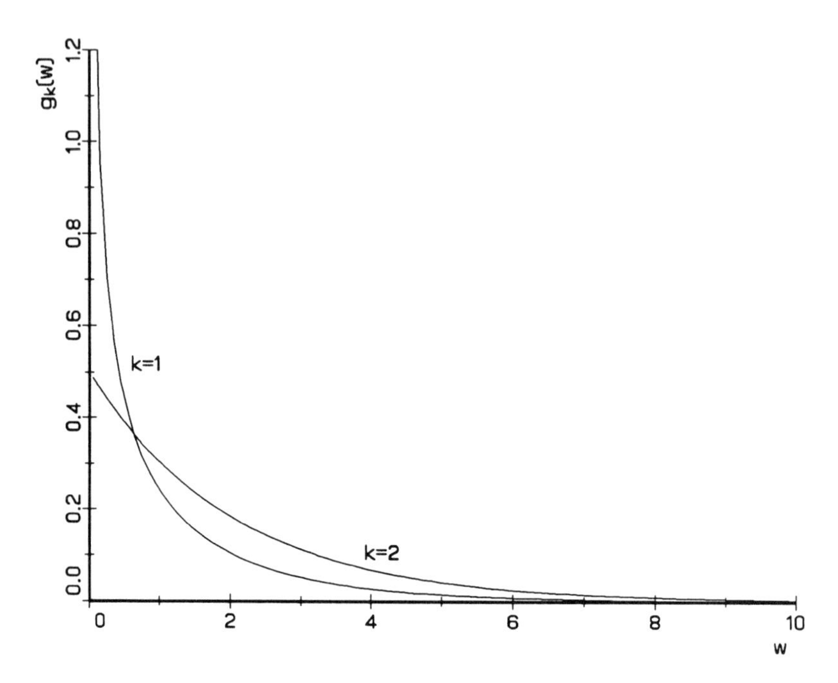

Figure 8.1 Chi-square pdf's: $k = 1, 2$.

For future reference, we also report that

$$E(1/W) = 1/(k-2) \quad \text{for } k > 2,$$

$$E(1/W^2) = 1/[(k-2)(k-4)] \quad \text{for } k > 4.$$

(2) *Student's t-distribution*. If $Z \sim \mathcal{N}(0, 1)$, $W \sim \chi^2(k)$, with Z and W being independent, and $U = Z/\sqrt{(W/k)}$, then the pdf of U is

$$g_k(u) = \frac{\Gamma[(k+1)/2]}{\sqrt{k}\,\Gamma(k/2)\Gamma(1/2)}\,(1 + u^2/k)^{-[(k+1)/2]}.$$

We write this situation as $U \sim t(k)$. This pdf defines the Student's t-distribution, a one-parameter family. The parameter k is again called "the degrees of freedom." The pdf is symmetric, centered at zero, and similar in shape to a standard normal pdf. The cdf is tabulated in many texts.

The derivation reverses: if $U \sim t(k)$, then U can be expressed as $Z/\sqrt{(W/k)} = \sqrt{k}Z/\sqrt{W}$, where $Z \sim \mathcal{N}(0, 1)$ is independent of $W \sim \chi^2(k)$.

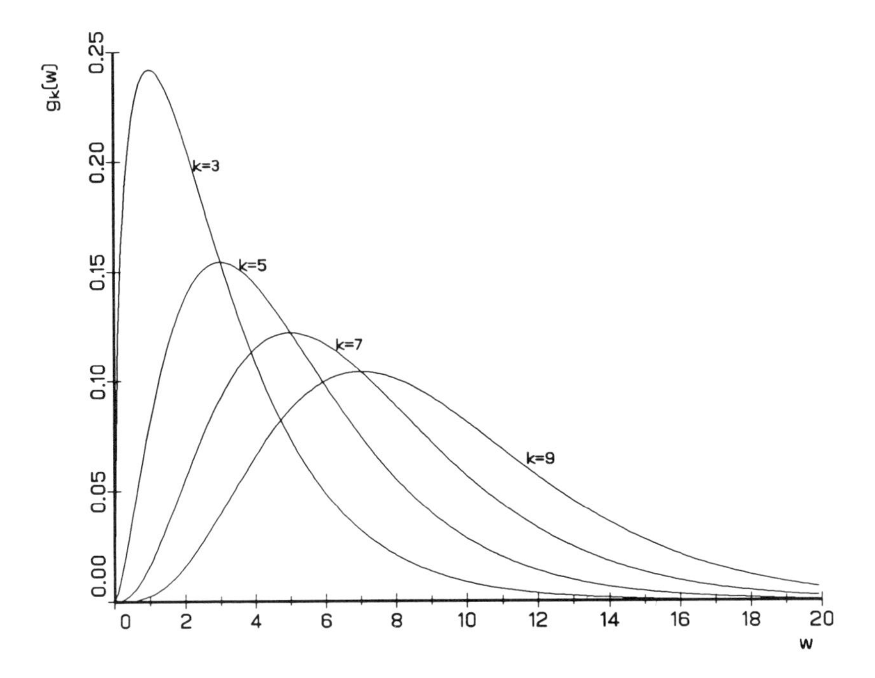

Figure 8.2 Chi-square pdf's: $k = 3, 5, 7, 9$.

So some of the moments can be calculated conveniently from those of the standard normal and chi-square distributions. In particular:

$$E(U) = \sqrt{k}E(Z)E(1/\sqrt{W}) = \sqrt{k}0E(1/\sqrt{W}) = 0 \quad \text{for } k > 1,$$

$$V(U) = E(U^2) = E(kZ^2/W) = kE(Z^2)E(1/W)$$

$$= k/(k - 2) \quad \text{for } k > 2.$$

Observe that $E(U) = 0$, and that for large k, $V(U) \cong 1$. Indeed, for $k \geq 30$, the $t(k)$ distribution is practically indistinguishable from a $\mathcal{N}(0, 1)$ distribution. More formally, as $k \to \infty$, the cdf of the Student's t-distribution, namely $G_k(\cdot)$, converges to the standard normal cdf $F(\cdot)$. More on this in Section 9.6.

8.6. Sampling from a Normal Population

Now we can fully characterize the joint sampling distribution of the sample mean and sample variance, in random sampling from a normal population. We have $X \sim \mathcal{N}(\mu, \sigma^2)$, and $X_1, \ldots, X_n$ as the random sample.

Standard Normal Population

First suppose that $X \sim \mathcal{N}(0, 1)$. For convenience, use Y rather than X as the name of the variable. So the random sample is $Y_1, \ldots, Y_n$, the sample mean is $\bar{Y} = \Sigma_i Y_i/n$, and the sample variance is $V^* = \Sigma_i(Y_i - \bar{Y})^2/n$.

The joint distribution of $\bar{Y}$ and V^* has these features:

F1*. $\sqrt{n}\bar{Y} \sim \mathcal{N}(0, 1)$,

F2*. $nV^* \sim \chi^2(n - 1)$,

F3*. $\bar{Y}$ and V^* are independent,

F4*. $\sqrt{(n - 1)}\bar{Y}/\sqrt{V^*} \sim t(n - 1)$.

Proof. Evidently all four items will be established iff we can write

$$\sqrt{n}\bar{Y} = Z_1, \qquad nV^* = Z_2^2 + \ldots + Z_n^2, \tag{8.1}$$

where $Z_1, Z_2, \ldots, Z_n$ are independent $\mathcal{N}(0, 1)$ variables. Then F1*–F3* follow immediately, while F4* follows by

$$\sqrt{(n - 1)}\bar{Y}/\sqrt{V^*} = \sqrt{n}\bar{Y}/\sqrt{[nV^*/(n - 1)]} = Z_1/\sqrt{[nV^*/(n - 1)]}.$$

A more general version of Eq. (8.1) will be established later: see Exercise 21.1. For now, we cover only the case $n = 2$. Let

$$Z_1 = (Y_1 + Y_2)/\sqrt{2}, \qquad Z_2 = (Y_1 - Y_2)/\sqrt{2}.$$

Since Y_1, Y_2 are BVN, it follows that Z_1, Z_2 are BVN. Calculating means, variances, and covariances then shows that Z_1 and Z_2 are independent $\mathcal{N}(0, 1)$ variables. Now $Z_1 = \sqrt{2}\bar{Y}$, and

$$Y_1 - \bar{Y} = Y_1 - (Y_1 + Y_2)/2 = (Y_1 - Y_2)/2 = Z_2/\sqrt{2},$$

$$Y_2 - \bar{Y} = -Z_2/\sqrt{2},$$

so $2V^* = Z_2^2/2 + Z_2^2/2 = Z_2^2$. ■

General Normal Population

Now suppose $X \sim \mathcal{N}(\mu, \sigma^2)$. The sample mean is $\bar{X} = \Sigma_i X_i/n$, and the sample variance is $S^2 = \Sigma_i(X_i - \bar{X})^2/n$. Let $Y = (X - \mu)/\sigma$, so $X = \mu + \sigma Y$, with $Y \sim \mathcal{N}(0, 1)$. Then

$$\bar{X} = \mu + \sigma\bar{Y} \Rightarrow \bar{X} - \mu = \sigma\bar{Y} \Rightarrow \sqrt{n}(\bar{X} - \mu)/\sigma = \sqrt{n}\bar{Y},$$

$$X_i - \bar{X} = \sigma(Y_i - \bar{Y}) \Rightarrow \sum_i (X_i - \bar{X})^2 = \sigma^2 \sum_i (Y_i - \bar{Y})^2,$$

$$nS^2 = \sigma^2 nV^* \Rightarrow W = nS^2/\sigma^2 = nV^*,$$

$$S = \sigma\sqrt{V^*} \Rightarrow U = \sqrt{(n-1)}(\bar{X} - \mu)/S = \sqrt{(n-1)}\bar{Y}/\sqrt{V^*}.$$

Because F1*–F4* hold for the special $\mathcal{N}(0, 1)$ population, we conclude that:

F1. $\bar{X} \sim \mathcal{N}(\mu, \sigma^2/n)$,

F2. $W = nS^2/\sigma^2 \sim \chi^2(n - 1)$,

F3. $\bar{X}$ and S^2 are independent,

F4. $U = \sqrt{(n-1)}(\bar{X} - \mu)/S \sim t(n - 1)$.

These features completely specify the joint distribution of $\bar{X}$ and S^2 in random sampling from a normal population.

Here are a few further remarks.

• Observe the contrast between these two results:

$$Z = \sqrt{n}(\bar{X} - \mu)/\sigma \sim \mathcal{N}(0, 1),$$

$$U = \sqrt{(n-1)}(\bar{X} - \mu)/S \sim t(n - 1).$$

It is sometimes said that the first gives the distribution of $\bar{X}$ when σ^2 is known, while the second gives the distribution of $\bar{X}$ when σ^2 is unknown. But this, of course, cannot be correct. Actually the first gives the distribution of a certain linear function of $\bar{X}$ (whether or not σ^2 is known), while the second gives the distribution of a certain function of $\bar{X}$ and S^2 (whether or not σ^2 is known). The practical distinction is rather that the first is usable for inference about μ when σ^2 is known, while the second, as will be seen in Section 11.5, is usable for that purpose even when σ^2 is unknown.

• Recalling that for large k, the $t(k)$ distribution is virtually indistinguishable from the $\mathcal{N}(0, 1)$ distribution, we may, when n is large, treat

U as if it were Z. Indeed, we may use the "limiting distribution" for U, namely the standard normal, as an approximation for the distribution of U, even for moderate n. More on this in Chapter 9.

• Observe the contrast between the sample variance $M_2 = S^2$ and the ideal sample variance M_2^*. Because $M_2^* = (1/n)\Sigma_i(X_i - \mu)^2$ is the mean of the squares of n independent $\mathcal{N}(0, \sigma^2)$ variables, we know that $nM_2^*/\sigma^2 \sim \chi^2(n)$, while we have seen that $nM_2/\sigma^2 \sim \chi^2(n - 1)$. It is sometimes said that "one degree of freedom is lost" when $\bar{X}$ is used in place of μ, a remark that sounds like punishment for a crime. A less dramatic statement is that the expectation is reduced by one.

Exercises

8.1 Suppose that X_1, X_2 are independent drawings from a population in which the pdf of X is $f(x) = 1$ for $0 \leq x \leq 1$, with $f(x) = 0$ elsewhere. Let $T = (X_1 + X_2)/2$. Find the pdf of T.

8.2 Suppose that Y_1, Y_2, Y_3 are independent drawings from a population in which the pdf of Y is

$$f(y) = (2 + y)/16 \quad \text{for } 0 \leq y \leq 4,$$

with $f(y) = 0$ elsewhere. Let $Z = (Y_1 + Y_2 + Y_3)/3$. Calculate $E(Z)$ and $V(Z)$.

8.3 Consider these alternative populations for a random variable X:

(a) Bernoulli with parameter $p = 0.5$.
(b) Normal with parameters $\mu = 0.5$, $\sigma^2 = 0.25$.
(c) Exponential with parameter $\lambda = 2$.

Let A be the event $\{0.4 < X \leq 0.6\}$. For each population, find $E(X)$, $V(X)$, and $\Pr(A)$.

8.4 Now consider random sampling, sample size 10, from the populations in Exercise 8.3. Let $\bar{X}$ denote the sample mean, and let B be the event $\{0.4 < \bar{X} \leq 0.6\}$. For each population, find $E(\bar{X})$, $V(\bar{X})$, and $\Pr(B)$. Comparing these results with those in Exercise 8.3, comment on the effect of increasing sample size on the distribution of sample means.

8.5 Let $\bar{X}$ and S^2 denote the sample mean and sample variance in random sampling, sample size 17, from a $\mathcal{N}(10, 102)$ population. Find the probability of each of these events:

$$A = \{\bar{X} \le 14.9\} \quad B = \{5.1 \le \bar{X} \le 14.9\} \quad C = \{S^2 \le 92.04\}$$
$$D = B \cap C \quad E = \{4(\bar{X} - 10)/S \le 1.746\} \quad F = \{\bar{X} \le 10 + 0.53S\}$$

(Note: For E and F, a Student's t-table will be required.)

8.6 The random variable X has the exponential distribution with parameter $\lambda = 2$.

(a) Use the definite integral formula

$$\int_0^\infty t^n e^{-at}\, dt = n!/a^{n+1} \quad \text{for } a > 0 \text{ and } n \text{ positive integer,}$$

to show that

$$E(X) = 1/2, \qquad E(X^2) = 1/2, \qquad E(X^3) = 3/4, \qquad E(X^4) = 3/2.$$

(b) Consider random sampling, sample size 20, from this population, and let $T = (X_1^2 + \cdots + X_{20}^2)/20$. Calculate $E(T)$ and $V(T)$.

8.7 Let $\bar{X}$ denote the sample mean in random sampling, sample size n, from a population in which $X \sim$ exponential(λ). So $E(\bar{X}) = 1/\lambda$. Consider the sample statistic $T = 1/\bar{X}$.

(a) Is $E(T)$ greater than, equal to, or less than λ? Justify your answer by reference to Jensen's Inequality (Section 3.5).
(b) Suppose $\lambda = 2$ and $n = 10$. Calculate $E(T)$ and $V(T)$.

8.8 Recall that if $X \sim$ exponential(λ), then in random sampling, sample size n, one has $W = 2n\lambda\bar{X} \sim \chi^2(2n)$. So it must be the case that $W = \Sigma_{i=1}^n V_i$ where the V_i's are independent $\chi^2(2)$ variables. To reconcile these two results, show by reference to their pdf's that the chi-square(2) distribution is the same as the exponential(1/2) distribution.

9 *Asymptotic Distribution Theory*

9.1. Introduction

As we have seen in Chapter 8, the probability distribution of the sample mean in random sampling depends on the parent population and on the sample size. For three specific parent populations, we have reported the distribution of the sample mean. For any parent population, we have shown that $E(\bar{X}) = \mu$ and $V(\bar{X}) = \sigma^2/n$. Now we develop additional information about the distribution of $\bar{X}$ that is valid for all parent populations. The information concerns *asymptotic properties* of the distribution of the sample mean.

As n gets large, $E(\bar{X})$ stays at μ while $V(\bar{X}) = \sigma^2/n$ goes to zero. So it is plausible that the distribution of the sample mean becomes degenerate at the point μ as n goes to infinity. On the other hand, consider the *standardized sample mean*

$$Z = [\bar{X} - E(\bar{X})]/[V(\bar{X})]^{1/2} = (\bar{X} - \mu)/(\sigma/\sqrt{n}) = \sqrt{n}(\bar{X} - \mu)/\sigma.$$

By linear function rules, we see that $E(Z) = 0$ and $V(Z) = 1$ for every n. So it is plausible that the distribution of the standardized sample mean approaches a nondegenerate limit as n goes to infinity. If so, we might want to use that limiting distribution to approximate the distribution of Z even when the sample size is modest. If we do that, we will be approximating the distribution of $\bar{X}$ itself, because (for given n, μ, σ) an event that is defined in terms of $\bar{X}$ can be translated into an event that is defined in terms of Z.

These remarks will be formalized as follows. In random sampling from any population with $E(X) = \mu$ and $V(X) = \sigma^2$:

- Law of Large Numbers. The *probability limit* of $\bar{X}$ is μ.
- Central Limit Theorem. The *limiting distribution* of Z is $\mathcal{N}(0, 1)$.
- The *asymptotic distribution* of $\bar{X}$ is $\mathcal{N}(\mu, \sigma^2/n)$.

To clarify the situation, consider a set of charts that refer to random sampling from the exponential(1) population, a situation where we know the exact sampling distribution of $\bar{X}$ (and hence of Z). Figures 9.1 and 9.2 show the pdf's and cdf's of $\bar{X}$ for $n = 5$, $n = 30$, and $n = 90$. Observe how the distribution becomes increasingly concentrated at the point $\mu = 1$ as n increases. In contrast, Figures 9.3 and 9.4 show the pdf's and cdf's of Z for $n = 5$, $n = 30$, and $n = 90$. Observe how the distribution becomes stabilized as n increases, taking on the appearance of the $\mathcal{N}(0, 1)$ distribution. As we shall see, those asymptotic properties prevail regardless of the population.

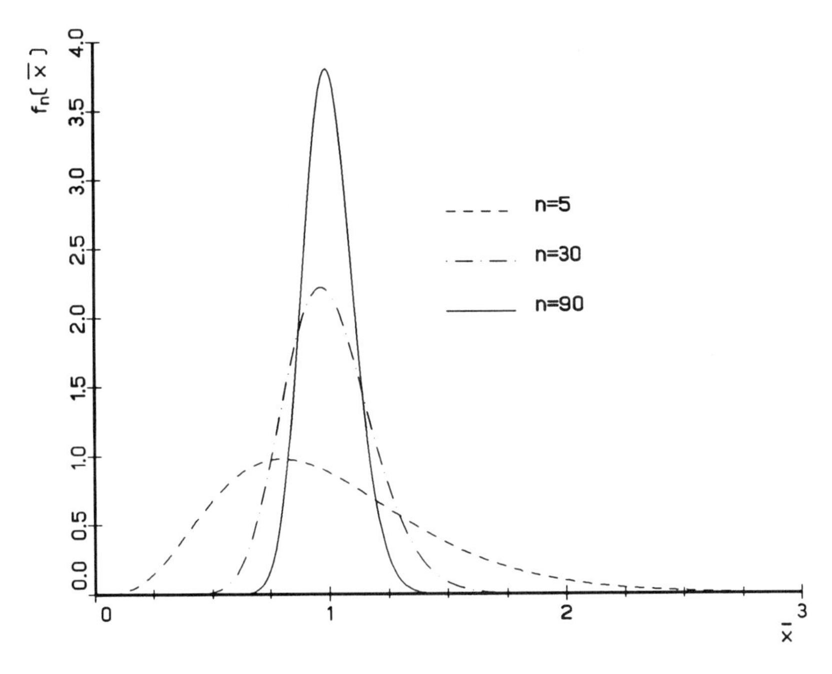

Figure 9.1 Sample mean pdf's: exponential(1) population.

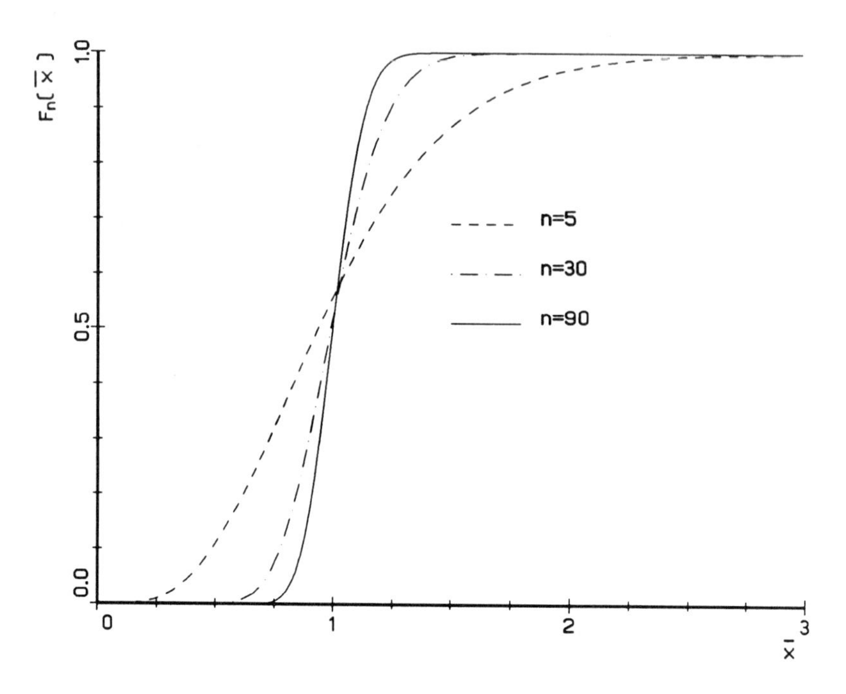

Figure 9.2 Sample mean cdf's: exponential(1) population.

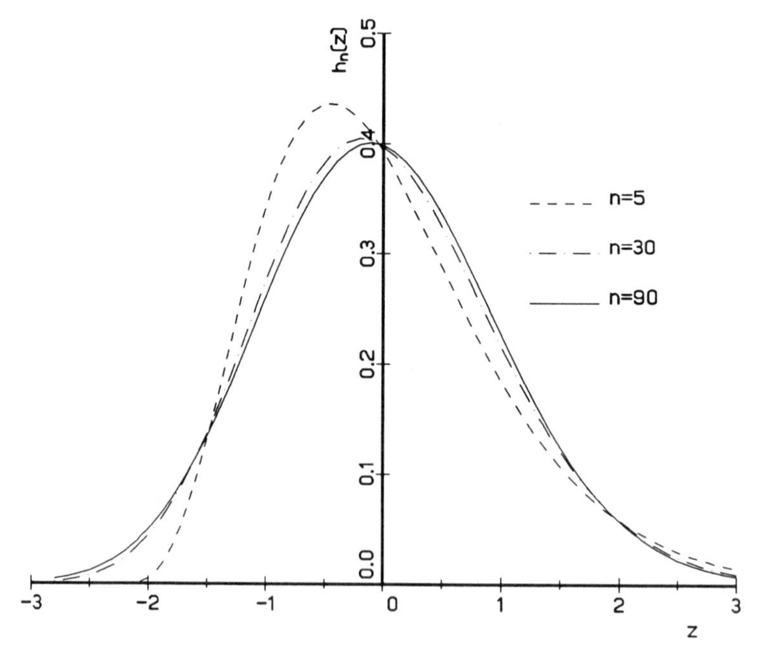

Figure 9.3 Standardized sample mean pdf's: exponential(1) population.

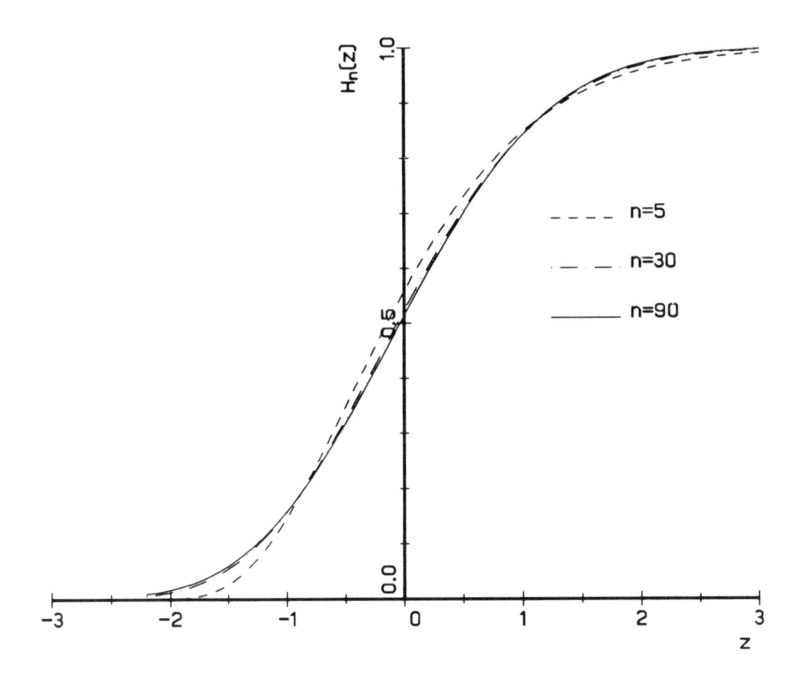

Figure 9.4 Standardized sample mean cdf's: exponential(1) population.

9.2. Sequences of Sample Statistics

To proceed systematically, think of a sequence of sample statistics, indexed by sample size. For example: $\bar{X}_1$ = sample mean in random sampling, sample size 1; $\bar{X}_2$ = sample mean in random sampling, sample size 2; . . . ; $\bar{X}_n$ = sample mean in random sampling, sample size n. Each of these random variables has its own pdf (or pmf), cdf, expectation, variance, and so forth.

More generally, let T_n be a sequence of random variables, with cdf's $G_n(t) = \Pr(T_n \leq t)$, expectations $E(T_n)$, and variances $V(T_n)$. In what follows, T_n may refer to the nth variable in the sequence, or to the sequence as a whole. We will use "lim" throughout as shorthand for "limit as $n \to \infty$."

We define three types of convergence.

Convergence in Probability. If there is a constant c such that $\lim G_n(t) = 0$ for all $t < c$ and $\lim G_n(t) = 1$ for all $t \geq c$, then we say that (the sequence) T_n *converges in probability* to c, or equivalently that the *probability limit* of T_n is c. We write this as $T_n \xrightarrow{\text{P}} c$, or as plim $T_n = c$. Let $A_n = \{|T_n - c| \geq \epsilon\}$ where $\epsilon > 0$, so

$$\Pr(A_n) = 1 - G_n(c + \epsilon) + \Pr(T_n = c + \epsilon) + G_n(c - \epsilon).$$

Iff T_n converges in probability to c, then $\lim \Pr(A_n) = 1 - 1 + 0 + 0 = 0$. So an equivalent way to define convergence in probability of T_n to c is that

$$\lim \Pr(|T_n - c| \geq \epsilon) = 0 \quad \text{for all } \epsilon > 0.$$

Convergence in Mean Square. If there is some constant c such that $\lim E(T_n - c)^2 = 0$, then we say that (the sequence) T_n *converges in mean square* to c. Two consequences are immediate:

C1. If T_n is a sequence of random variables with $\lim E(T_n) = c$ and $\lim V(T_n) = 0$, then T_n converges in mean square to c.

Proof. $E(T_n - c)^2 = V(T_n) + [E(T_n) - c]^2$. Take limits. ■

C2. If T_n converges in mean square to c, then T_n converges in probability to c.

Proof. Let $A_n = \{|T_n - c| \geq \epsilon\}$ where $\epsilon > 0$. Applying Chebyshev Inequality #1 (Section 3.5) gives $0 \leq \Pr(A_n) \leq E(T_n - c)^2/\epsilon^2$. Taking limits gives $0 \leq \lim \Pr(A_n) \leq 0$, whence $\lim \Pr(A_n) = 0$. ■

Convergence in Distribution. If there is some fixed cdf $G(t)$ such that $\lim G_n(t) = G(t)$ for all t at which $G(\cdot)$ is continuous, then we say that (the sequence) T_n *converges in distribution* to $G(\cdot)$, or equivalently that the *limiting distribution* of T_n is $G(\cdot)$. We write this as $T_n \xrightarrow{\text{D}} G(\cdot)$. Evidently convergence in probability is the special case of convergence in distribution in which the limiting distribution is degenerate.

9.3. Asymptotics of the Sample Mean

We apply these concepts to the sequence of sample means in random sampling from any population.

LAW OF LARGE NUMBERS, or LLN. In random sampling from any population with $E(X) = \mu$ and $V(X) = \sigma^2$, the *sample mean converges in probability to the population mean.*

Proof. We have $E(\bar{X}_n) = \mu$ and $V(\bar{X}_n) = \sigma^2/n$, so $\lim E(\bar{X}_n) = \mu$ and $\lim V(\bar{X}_n) = 0$. So $\bar{X}_n$ converges in mean square to μ by C1, hence $\bar{X}_n$ converges in probability to μ by C2. ■

CENTRAL LIMIT THEOREM, or CLT. In random sampling from any population with $E(X) = \mu$ and $V(X) = \sigma^2$, the *standardized sample mean* $Z = \sqrt{n}(\bar{X} - \mu)/\sigma$ *converges in distribution* to $\mathcal{N}(0, 1)$. Equivalently, $\sqrt{n}(\bar{X} - \mu)$ converges in distribution to $\mathcal{N}(0, \sigma^2)$.

Proof. See DeGroot (1975, pp. 227–233).

Associated with the CLT is an approximation procedure. The limiting cdf of $Z_n = \sqrt{n}(\bar{X}_n - \mu)/\sigma$ will be used to approximate the exact cdf of Z_n for sample size n. If the cdf of Z_n is $H_n(c^*) = \Pr(Z_n \leq c^*)$, then we will approximate $H_n(c^*)$ by $\Phi(c^*)$, where $\Phi(\cdot)$ is the $\mathcal{N}(0, 1)$ cdf. This procedure uses the limit of a sequence as an approximation to a term in the sequence; the error in this approximation is arbitrarily small if n is large enough. This is quite analogous to using $1/(1 - b)$ to approximate the finite sum $1 + b + b^2 + \cdots + b^n$ in Keynesian multiplier analysis (where $0 < b < 1$). Of course the approximation may be poor when n is small.

Approximating the cdf of the standardized sample mean Z_n by the $\mathcal{N}(0, 1)$ cdf amounts to approximating the cdf of the sample mean $\bar{X}_n$ by the $\mathcal{N}(\mu, \sigma^2/n)$ cdf. For,

$$
\begin{aligned}
F_n(c) &= \Pr(\bar{X}_n \leq c) = \Pr[\sqrt{n}(\bar{X}_n - \mu)/\sigma \leq \sqrt{n}(c - \mu)/\sigma] \\
&= \Pr(Z_n \leq c^*) = H_n(c^*),
\end{aligned}
$$

with $c^* = \sqrt{n}(c - \mu)/\sigma$. When we use the approximation $F_n(c) \cong \Phi(c^*)$ we are in effect treating $\bar{X}_n$ as if it were distributed $\mathcal{N}(\mu, \sigma^2/n)$. So we say that the *asymptotic distribution* of $\bar{X}_n$ is $\mathcal{N}(\mu, \sigma^2/n)$, and write

$$
\bar{X}_n \overset{A}{\sim} \mathcal{N}(\mu, \sigma^2/n).
$$

More generally, whenever we have a sample statistic T_n and parameters θ and ϕ^2 (that do not involve n), such that the standardized statistic $\sqrt{n}(T_n - \theta)/\phi$ converges in distribution to the $\mathcal{N}(0, 1)$ distribution, we will say that T_n is asymptotically distributed $\mathcal{N}(\theta, \phi^2/n)$, and may refer to θ and ϕ^2/n as the *asymptotic expectation* and *asymptotic variance* of T_n.

Observe again the distinction between the *limiting distribution* of the sample mean, which is degenerate at μ, and the *asymptotic distribution* of the sample mean, which is $\mathcal{N}(\mu, \sigma^2/n)$. Clearly the latter provides more useful information.

It is tempting to be cynical about the relevance of asymptotic theory to empirical work, where the sample size may be modest. But in fact the approximations are typically quite accurate.

Example. Consider random sampling, sample size 30, on the random variable X, where $X \sim \chi^2(1)$. Find $\Pr(A)$ where $A = \{\bar{X} \leq 1.16\}$. Since $E(X) = 1$ and $V(X) = 2$, we have $\bar{X} \overset{A}{\sim} \mathcal{N}(1, 2/30)$, whence $\Pr(A) \cong \Phi(c^*)$, where $c^* = (1.16 - 1)/\sqrt{(2/30)} = 0.62$, and $\Phi(\cdot)$ is the $\mathcal{N}(0, 1)$ cdf. From the standard normal table, $\Phi(c^*) = 0.73$. For the exact calculation, rely on the fact that $W = n\bar{X} = \Sigma_i X_i \sim \chi^2(30)$. From the $\chi^2(30)$ table, $\Pr(A) = \Pr(W \leq 34.8) = 0.75$. The approximation is very good even though the sample size is modest.

9.4. Asymptotics of Sample Moments

The asymptotic results for the sample mean must apply directly to the entire class of statistics that can be interpreted as sample means in random sampling. As in Section 8.4, that class includes the sample raw moments $M'_r = \bar{Y}$, where $Y = X^r$. In particular, for the sample second raw moment, $M'_2 = \Sigma_i X_i^2/n$, we have

$$M'_2 \xrightarrow{P} \mu'_2,$$

$$\sqrt{n}(M'_2 - \mu'_2)/\sqrt{(\mu'_4 - \mu_2'^2)} \xrightarrow{D} \mathcal{N}(0, 1),$$

$$M'_2 \overset{A}{\sim} \mathcal{N}[\mu'_2, (\mu'_4 - \mu_2'^2)/n],$$

where $\mu'_r = E(X^r)$ denotes the population rth raw moment.

The class also includes the sample moments about the population mean $M_r^* = \bar{Y}$, where $Y = (X - \mu)^r$. In particular, for the ideal sample variance $M_2^* = \Sigma_i(X_i - \mu)^2/n$, we have

$$M_2^* \xrightarrow{P} \mu_2,$$
$$\sqrt{n}(M_2^* - \mu_2)/\sqrt{(\mu_4 - \mu_2^2)} \xrightarrow{D} \mathcal{N}(0, 1),$$
$$M_2^* \stackrel{A}{\sim} \mathcal{N}[\mu_2, (\mu_4 - \mu_2^2)/n],$$

where $\mu_r = E(X - \mu)^r$ denotes the population rth central moment.

9.5. Asymptotics of Functions of Sample Moments

This would be the end of the story except that we are often concerned with sample statistics that are not interpretable as sample means in random sampling. Typically, the statistics are functions of such sample means. So we require rules for getting probability limits and asymptotic distributions for functions of sample means. For a linear function, there is no problem: if $T_n = a + b\bar{X}_n$ where a and b are constants that do not involve n, then $T_n = \bar{Y}_n$ is itself a sample mean in random sampling on the variable $Y = a + bX$, whence

$$T_n \xrightarrow{P} \theta, \qquad \sqrt{n}(T_n - \theta)/\phi \xrightarrow{D} \mathcal{N}(0, 1), \qquad T_n \stackrel{A}{\sim} \mathcal{N}(\theta, \phi^2/n),$$

with $\theta = a + b\mu$ and $\phi^2 = b^2\sigma^2$.

But the sample statistics are not always that simple. For example, we may be concerned with $T = 1/\bar{X}$, which is a nonlinear function of $\bar{X}$. Or we may be concerned with the sample variance

$$S^2 = \sum_i (X_i - \bar{X})^2/n = M_2^* - (\bar{X} - \mu)^2,$$

which is a function of M_2^* and $\bar{X}$ (nonlinear in the latter). Another example is the *sample t-ratio*

$$U = \sqrt{n}(\bar{X} - \mu)/S,$$

which is again a function of M_2^* and $\bar{X}$ (nonlinear in both arguments).

To derive the asymptotics of functions of sample moments, the key tools are the *Slutsky Theorems*. Here T_n, V_n, and W_n are sequences of random variables, while the functions $h(\ )$ and the constants c do not involve n. The theorems are:

S1. If $T_n \xrightarrow{P} c$ and $h(T_n)$ is continuous at c, then $h(T_n) \xrightarrow{P} h(c)$.

S2. If $V_n \xrightarrow{P} c_1$ and $W_n \xrightarrow{P} c_2$, and $h(V_n, W_n)$ is continuous at (c_1, c_2), then $h(V_n, W_n) \xrightarrow{P} h(c_1, c_2)$.

S3. If $V_n \xrightarrow{P} c$ and W_n has a limiting distribution, then the limiting distribution of $(V_n + W_n)$ is the same as that of $c + W_n$.

S4. If $V_n \xrightarrow{P} c$ and W_n has a limiting distribution, then the limiting distribution of $(V_n W_n)$ is the same as that of cW_n.

Theorems S1 and S2 say that the probability limit of a continuous function is the function of the probability limits. Further, S3 and S4 refer to the situation in which one variable has a probability limit and another variable has a limiting distribution. In their sum or product, the first variable can be treated as a constant as far as limiting distributions are concerned. For example, if $V_n \xrightarrow{P} c$ and $W_n \xrightarrow{D} \mathcal{N}(0, \sigma^2)$, then $(V_n + W_n) \xrightarrow{D} \mathcal{N}(c, \sigma^2)$ and $(V_n W_n) \xrightarrow{D} \mathcal{N}(0, c^2\sigma^2)$.

The other key tool is

S5. *Delta Method.* If $\sqrt{n}(T_n - \theta) \xrightarrow{D} \mathcal{N}(0, \phi^2)$ and $U_n = h(T_n)$ is continuously differentiable at θ, then

$$\sqrt{n}[U_n - h(\theta)] \xrightarrow{D} \mathcal{N}\{0, [h'(\theta)]^2\phi^2\}.$$

Equivalently, if $T_n \overset{A}{\sim} \mathcal{N}(\theta, \phi^2/n)$ and $U_n = h(T_n)$ is continuously differentiable at θ, then

$$U_n \overset{A}{\sim} \mathcal{N}\{h(\theta), [h'(\theta)]^2\, \phi^2/n\}.$$

Here again the understanding is that the function $h(\cdot)$ does not involve n. What S5 says is that the asymptotic distribution of $U_n = h(T_n)$ is the same as that of its linear approximation at θ, namely $U_n^* = h(\theta) + h'(\theta)(T_n - \theta)$.

For proofs of S1–S5, see Rao (1973, pp. 122, 124, 385–386). Here is an intuitive argument for S1: By continuity, $h(T_n)$ will be confined to a neighborhood of $h(c)$ provided that T_n is confined to a neighborhood of c. The probability of the latter event can be made arbitrarily close to 1 by taking n sufficiently large, so the same is true of the former event. And here is some intuition for S5: By the mean value theorem of calculus, $[U_n - h(\theta)] = h'(T_n^\circ)(T_n - \theta)$, where T_n° lies in the interval between T_n and θ. So $\sqrt{n}[U_n - h(\theta)] = h'(T_n^\circ)\sqrt{n}(T_n - \theta)$. Because T_n converges in probability to θ, so does T_n°, and hence by continuity $h'(T_n^\circ)$ converges in probability to $h'(\theta)$. Then it is not surprising that asymptotically, U_n behaves like U_n^*.

9.6. Asymptotics of Some Sample Statistics

We apply the theory to two sample statistics.

Sample Variance. Recall that

$$M_2 = M_2^* - (\bar{X} - \mu)^2 = h(M_2^*, \bar{X}), \tag{9.1}$$

say, where $M_2 = S^2 = \Sigma_i(X_i - \bar{X})^2/n$ is the sample variance, and $M_2^* = \Sigma_i(X_i - \mu)^2/n$ is the ideal sample variance. By the LLN,

$$M_2^* \xrightarrow{P} \mu_2, \qquad \bar{X} \xrightarrow{P} \mu.$$

So by S2,

$$M_2 = h(M_2^*, \bar{X}) \xrightarrow{P} h(\mu_2, \mu) = \mu_2 - (\mu - \mu)^2 = \mu_2.$$

That is, the probability limit of S^2 is the same as that of M_2^*, namely σ^2. Next, rewrite Eq. (9.1) as

$$\sqrt{n}(M_2 - \mu_2) = \sqrt{n}(M_2^* - \mu_2) - U^2, \tag{9.2}$$

where $U = \sqrt[4]{n}(\bar{X} - \mu)$. By the CLT,

$$\sqrt{n}(M_2^* - \mu_2) \xrightarrow{D} \mathcal{N}(0, \mu_4 - \mu_2^2).$$

By linear function rules, $E(U) = \sqrt[4]{n}E(\bar{X} - \mu) = 0$, and $V(U) = \sqrt{n}V(\bar{X}) = \sigma^2/\sqrt{n}$. Because $\lim E(U) = 0$ and $\lim V(U) = 0$, U converges in mean square to 0 by C1, and hence $U \xrightarrow{P} 0$ by C2. So $U^2 \xrightarrow{P} 0$ by S1, whence by S3 we conclude that

$$\sqrt{n}(M_2 - \mu_2) \xrightarrow{D} \mathcal{N}(0, \mu_4 - \mu_2^2).$$

That is, the limiting distribution of $\sqrt{n}(S^2 - \sigma^2)$ is the same as that of $\sqrt{n}(M_2^* - \sigma^2)$, namely $\mathcal{N}(0, \mu_4 - \mu_2^2)$. Equivalently, the asymptotic distribution of S^2 is the same as that of M_2^*, namely $\mathcal{N}[\sigma^2, (\mu_4 - \mu_2^2)/n]$.

Sample t-ratio. Let $U = \sqrt{n}(\bar{X} - \mu)/S$ and $Z = \sqrt{n}(\bar{X} - \mu)/\sigma$, so $U = (\sigma/S)Z$. Now $Z \xrightarrow{D} \mathcal{N}(0, 1)$ by the CLT, and $\sigma/S \xrightarrow{P} 1$ (using $S^2 \xrightarrow{P} \sigma^2$, and S1). So by S4, $U \xrightarrow{D} \mathcal{N}(0, 1)$. That is, the limiting distribution of U is the same as that of Z, namely $\mathcal{N}(0, 1)$. The same conclusion will follow if one uses $\sqrt{(n - 1)}$ instead of $\sqrt{n}$ in defining the sample t-ratio, as is sometimes done.

Exercises

9.1 Discuss Exercise 8.4 in the light of the asymptotic theory of this chapter.

9.2 In random sampling, sample size $n = 20$, on the variable X where $X \sim$ exponential(2), let $T = \Sigma_{i=1}^{n} X_i^2/n$. In Exercise 8.6, you found $E(T)$ and $V(T)$. Use the CLT to approximate the probability that T is less than or equal to 1.

9.3 As in Exercise 8.7, let $\bar{X}$ denote the sample mean in random sampling, sample size n, from a population in which the random variable $X \sim$ exponential(λ). For convenience, let $\theta = E(X) = 1/\lambda$. So $E(\bar{X}) = \theta$, $V(\bar{X}) = \theta^2/n$, plim $\bar{X} = \theta$, and the limiting distribution of $\sqrt{n}(\bar{X} - \theta)$ is $\mathcal{N}(0, \theta^2)$. Consider the sample statistic $U = 1/\bar{X}$.

(a) Use a Slutsky theorem to show that plim $U = \lambda$.
(b) Use the Delta method to find the limiting distribution of $\sqrt{n}(U - \lambda)$.
(c) Use your result to approximate $\Pr(U \leq 5/2)$ in random sampling, sample size 16, from an exponential population with $\lambda = 2$.
(d) Find the exact $\Pr(U \leq 5/2)$.

9.4 The probability distribution of the random variable X is given by $\Pr(X = 1) = 1/3$, $\Pr(X = 2) = 2/3$. In random sampling sample size n, let $T = \Sigma_{i=1}^{n} X_i^3/n$. For $n = 98$, approximate $\Pr(5 \leq T \leq 6)$.

9.5 In a population, the random variable X = length of unemployment (in months) has the exponential distribution with parameter $\lambda = 2$. Consider random sampling, sample size $n = 21$. Let T = proportion of the sampled persons who have been unemployed between 0.4158 and

1 months. Approximate the probability that T lies between 0.4 and 0.5. Hint: Define the random variable

$$U = \begin{cases} 1 & \text{if } 0.4158 \le X \le 1, \\ 0 & \text{otherwise.} \end{cases}$$

9.6 In a certain population the random variable Y has variance equal to 490. Two independent random samples, each of size 20, are drawn. The first sample mean is used as the predictor of the second sample mean.

(a) Calculate the expectation, expected square, and variance, of the prediction error.
(b) Approximate the probability that the prediction error is less than 14 in absolute value.

10 Sampling Distributions: Bivariate Case

10.1. Introduction

Having acquired considerable information about the distributions of sample statistics in random sampling from a univariate population, we proceed to bivariate populations. Consider a bivariate population in which the pmf or pdf of (X, Y) is $f(x, y)$. The first and second moments include

$$E(X) = \mu_X, \qquad E(Y) = \mu_Y,$$

$$V(X) = \sigma_X^2, \qquad V(Y) = \sigma_Y^2, \qquad C(X, Y) = \sigma_{XY}.$$

And for nonnegative integers (r, s), the raw and central moments are

$$E(X^r Y^s) = \mu'_{rs}, \qquad E(X^{*r} Y^{*s}) = \mu_{rs},$$

where $X^* = X - \mu_X$, $Y^* = Y - \mu_Y$. For example, $\mu_{20} = \sigma_X^2$ (formerly called μ_2), $\mu_{02} = \sigma_Y^2$, and $\mu_{11} = \sigma_{XY}$.

A random sample of size n from this population consists of n independent drawings: the random vectors (X_i, Y_i) for $i = 1, \ldots, n$ are independently and identically distributed. Of course X_i and Y_i need not be independent; indeed $C(X_i, Y_i) = \sigma_{XY}$. Independence runs *across* the n observations, not *within* each observation. The joint pmf or pdf of the random sample is

$$g_n(x_1, y_1, x_2, y_2, \ldots, x_n, y_n) = \prod_{i=1}^{n} f(x_i, y_i).$$

Sample statistics are functions of the random sample. They include not only the single-variable statistics $\bar{X}$, $\bar{Y}$, S_X^2, S_Y^2, but also the joint statistics that involve both components of the random vector (X, Y). A leading example is the *sample covariance*

$$S_{XY} = (1/n) \sum_i (X_i - \bar{X})(Y_i - \bar{Y}).$$

We may also be concerned with the joint distribution of several statistics. For example, the two sample means

$$\bar{X} = (1/n) \sum_i X_i, \qquad \bar{Y} = (1/n) \sum_i Y_i,$$

are a pair of random variables that have a joint sampling distribution. We already have their expectations and variances. We can calculate their covariance by T5, the linear function rule (Section 5.1), extended to n variables:

$$C(\bar{X}, \bar{Y}) = (1/n^2) \sum_i \sum_h C(X_i, Y_h) = (1/n^2) \sum_i C(X_i, Y_i)$$
$$= (1/n^2) n\sigma_{XY} = \sigma_{XY}/n,$$

using the independence and identically-distributed features of random sampling. This result on the covariance of two sample means is quite analogous to the result on the variance of a single sample mean, $V(\bar{X}) = \sigma_X^2/n$.

All previous conclusions for the univariate case apply to the single-variable statistics, but we have additional conclusions as well. We confine attention to general results that are applicable regardless of the form of the population.

10.2. Sample Covariance

The theory for the sampling distribution of the sample covariance,

$$S_{XY} = (1/n) \sum_i (X_i - \bar{X})(Y_i - \bar{Y}) = M_{11},$$

say, runs quite parallel to that for the sample variance.

Ideal Sample Covariance

Consider first the *ideal sample covariance*, namely the sample second joint moment about the population means:

$$M_{11}^* = (1/n) \sum_i (X_i - \mu_X)(Y_i - \mu_Y) = (1/n) \sum_i V_i = \bar{V},$$

say, where $V = X^*Y^*$, so $V_i = X_i^*Y_i^*$. Now $V_1, \ldots, V_n$ is a random sample on the variable $V = X^*Y^*$, so $M_{11}^* = \bar{V}$ is a sample mean in random sampling on the variable V. Hence the earlier theory for sample means applies.

The population mean and variance of V are

$$E(V) = E(X^*Y^*) = C(X, Y) = \sigma_{XY} = \mu_{11},$$
$$V(V) = E(V^2) - E^2(V) = E(X^{*2}Y^{*2}) - E^2(X^*Y^*) = \mu_{22} - \mu_{11}^2.$$

So we have the exact results

$$E(M_{11}^*) = E(\bar{V}) = E(V) \quad = \mu_{11} = \sigma_{XY},$$
$$V(M_{11}^*) = V(\bar{V}) = V(V)/n = (\mu_{22} - \mu_{11}^2)/n,$$

and also the asymptotic results

$$M_{11}^* \xrightarrow{P} \mu_{11},$$
$$\sqrt{n}(M_{11}^* - \mu_{11}) \xrightarrow{D} \mathcal{N}(0, \mu_{22} - \mu_{11}^2),$$
$$M_{11}^* \overset{A}{\sim} \mathcal{N}[\mu_{11}, (\mu_{22} - \mu_{11}^2)/n].$$

Sample Covariance

Now turn to the sample covariance itself, namely the sample second joint moment about the sample means. As a matter of algebra,

$$(10.1) \quad S_{XY} = M_{11} = M_{11}^* - (\bar{X} - \mu_X)(\bar{Y} - \mu_Y).$$

So

$$E(S_{XY}) = \sigma_{XY} - C(\bar{X}, \bar{Y}) = \sigma_{XY} - \sigma_{XY}/n = (1 - 1/n)\sigma_{XY},$$

which is quite parallel to $E(S^2) = (1 - 1/n)\sigma^2$. And direct calculation gives

$$V(S_{XY}) = (n - 1)^2(\mu_{22} - \mu_{11}^2)/n^3 + 2(n - 1)(\mu_{20}\mu_{02})/n^3,$$

which is quite parallel to the exact result for $V(S^2)$ in Section 8.4. As for asymptotics, we obtain

$$S_{XY} \xrightarrow{P} \sigma_{XY}, \qquad \sqrt{n}(S_{XY} - \sigma_{XY}) \xrightarrow{D} \mathcal{N}(0, \mu_{22} - \mu_{11}^2).$$

Proof. First, in Eq. (10.1), $M_{11}^* \xrightarrow{P} \sigma_{XY}$, $(\bar{X} - \mu_X) \xrightarrow{P} 0$, and $(\bar{Y} - \mu_Y) \xrightarrow{P} 0$, so S_{XY} converges in probability to σ_{XY}, using Slutsky Theorem S2 (Section 9.5) twice. Next, rewrite Eq. (10.1) as

$$(10.2) \qquad \sqrt{n}(M_{11} - \mu_{11}) = \sqrt{n}(M_{11}^* - \mu_{11}) - UW,$$

where

$$U = \sqrt[4]{n}(\bar{X} - \mu_X), \qquad W = \sqrt[4]{n}(\bar{Y} - \mu_Y).$$

Since $U \xrightarrow{P} 0$ and $W \xrightarrow{P} 0$, we have $(UW) \xrightarrow{P} 0$ by S2. So by S3 the limiting distribution of $\sqrt{n}(S_{XY} - \sigma_{XY})$ is the same as the limiting distribution of $\sqrt{n}(M_{11}^* - \mu_{11})$. ■

We conclude that the asymptotic distribution of the sample covariance is the same as that of the ideal sample covariance:

$$S_{XY} \overset{A}{\sim} \mathcal{N}[\sigma_{XY}, (\mu_{22} - \mu_{11}^2)/n],$$

which is quite parallel to the asymptotic result for the sample variance in Section 9.6.

10.3. Pair of Sample Means

Now turn to the joint distribution of the pair of sample means, $\bar{X}$ and $\bar{Y}$. We proceed to asymptotics. For a random vector, convergence in probability means that each component of the vector converges in probability, and convergence in distribution means that the sequence of joint cdf's has as its limit some fixed joint cdf. For convenience we drop the subscript "n" that identifies a sequence. The key theorems are:

BIVARIATE LAW OF LARGE NUMBERS. In random sampling from any bivariate population, the sample mean vector $(\bar{X}, \bar{Y})$ converges in probability to the population mean vector (μ_X, μ_Y).

BIVARIATE CENTRAL LIMIT THEOREM. In random sampling from any bivariate population, the standardized sample mean vector, $[\sqrt{n}(\bar{X} - \mu_X)/\sigma_X, \sqrt{n}(\bar{Y} - \mu_Y)/\sigma_Y]$, converges in distribution to the SBVN(ρ) distribution, where $\rho = \sigma_{XY}/(\sigma_X\sigma_Y)$. Equivalently, in random sampling from any bivariate population, $(\bar{X}, \bar{Y}) \overset{A}{\sim} \mathcal{N}(\mu_X, \mu_Y, \sigma_X^2/n, \sigma_Y^2/n, \sigma_{XY}/n)$.

These theorems apply directly, of course, to any pair of sample moments that can be interpreted as a pair of sample means in random sampling. But again we need tools for deriving asymptotics for functions of sample means. The Slutsky Theorems S1–S4 extend in the obvious way, while S5 generalizes to the

BIVARIATE DELTA METHOD. If $(T_1, T_2) \stackrel{A}{\sim} \text{BVN}(\theta_1, \theta_2, \phi_1^2/n, \phi_2^2/n, \phi_{12}/n)$ and $U = h(T_1, T_2)$ is continuously differentiable at the point (θ_1, θ_2), then $U \stackrel{A}{\sim} \mathcal{N}[h(\theta_1, \theta_2), \phi^2/n]$, where

$$\phi^2 = h_1^2\phi_1^2 + h_2^2\phi_2^2 + 2h_1h_2\phi_{12},$$

$$h_1 = h_1(\theta_1, \theta_2) = \partial h(T_1, T_2)/\partial T_1 \quad \text{evaluated at } T_1 = \theta_1, T_2 = \theta_2,$$

$$h_2 = h_2(\theta_1, \theta_2) = \partial h(T_1, T_2)/\partial T_2 \quad \text{evaluated at } T_1 = \theta_1, T_2 = \theta_2.$$

In other words, the asymptotic distribution of $U = h(T_1, T_2)$ is the same as that of its linear approximation at the point (θ_1, θ_2), namely

$$U^* = h(\theta_1, \theta_2) + h_1(\theta_1, \theta_2)(T_1 - \theta_1) + h_2(\theta_1, \theta_2)(T_2 - \theta_2).$$

The understanding is that the function $h(\cdot, \cdot)$ does not involve n.

10.4. Ratio of Sample Means

To illustrate the application of this theory, consider the ratio of sample means, $T = \bar{X}/\bar{Y}$, with the proviso that $\mu_Y \neq 0$. The asymptotic joint distribution of $\bar{X}$ and $\bar{Y}$ is given by the Bivariate CLT, and the analysis for T starts with

$$T = \bar{X}/\bar{Y} = h(\bar{X}, \bar{Y}), \qquad h(\mu_X, \mu_Y) = \mu_X/\mu_Y = \theta,$$

say, where $h(\cdot, \cdot)$ is the ratio function. We recognize that

$$E(T) = E(\bar{X}/\bar{Y}) \neq E(\bar{X})/E(\bar{Y}) = \mu_X/\mu_Y = \theta.$$

But $\bar{X} \xrightarrow{P} \mu_X$ and $\bar{Y} \xrightarrow{P} \mu_Y$ by the LLN (Section 9.3), so that $T \xrightarrow{P} \theta$ by S2.

Proceeding, we calculate

$$h_1(\bar{X}, \bar{Y}) = \partial h/\partial \bar{X} = 1/\bar{Y}, \qquad h_1(\mu_X, \mu_Y) = 1/\mu_Y,$$

$$h_2(\bar{X}, \bar{Y}) = \partial h/\partial \bar{Y} = -\bar{X}/\bar{Y}^2, \qquad h_2(\mu_X, \mu_Y) = -\mu_X/\mu_Y^2 = -\theta/\mu_Y.$$

So the Bivariate Delta method gives

$$T \stackrel{A}{\sim} \mathcal{N}(\theta, \phi^2/n),$$

where ϕ^2/n is the (asymptotic) variance of the linear approximation to T at the point (μ_X, μ_Y), namely

$$T^* = \theta + (1/\mu_Y)(\bar{X} - \mu_X) - (\theta/\mu_Y)(\bar{Y} - \mu_Y)$$
$$= \theta + (1/\mu_Y)[(\bar{X} - \mu_X) - \theta(\bar{Y} - \mu_Y)].$$

That is,

$$\phi^2 = (1/\mu_Y)^2(\sigma_X^2 + \theta^2\sigma_Y^2 - 2\theta\sigma_{XY}). \tag{10.3}$$

10.5. Sample Slope

In Section 5.4, we introduced the population linear projection (BLP) of Y on X in a bivariate distribution, namely the line $E^*(Y|X) = \alpha + \beta X$, with

$$\beta = \sigma_{XY}/\sigma_X^2, \qquad \alpha = \mu_Y - \beta\mu_X.$$

The corresponding feature in a sample is the *sample linear projection* (or sample LP) of Y on X, namely the line $\hat{Y} = A + BX$, with

$$B = S_{XY}/S_X^2 = M_{11}/M_{20}, \qquad A = \bar{Y} - B\bar{X}.$$

To further illustrate application of our theory, we seek the asymptotic distribution of the statistic B, the *sample slope*, in random sampling from any bivariate population.

Ideal Sample Slope

We first treat a simpler statistic, the ratio of the corresponding sample moments taken about the population means, $B^* = M_{11}^*/M_{20}^*$, which we refer to as the *ideal sample slope*. Now

$$M_{11}^* = (1/n) \sum_i X_i^*Y_i^* = \bar{V}, \qquad M_{20}^* = (1/n) \sum_i X_i^{*2} = \bar{W},$$

say, where $V = X^*Y^*$, $W = X^{*2}$, $X^* = X - \mu_X$, $Y^* = Y - \mu_Y$. So

$$B^* = M_{11}^*/M_{20}^* = \bar{V}/\bar{W}$$

is a ratio of sample means in random sampling on (V, W), and the machinery of Section 10.4 applies directly.

Because $\mu_V = E(X^*Y^*) = \sigma_{XY}$ and $\mu_W = E(X^{*2}) = \sigma_X^2$, we have $\mu_V/\mu_W = \beta$, so

$$B^* \xrightarrow{P} \sigma_{XY}/\sigma_X^2 = \beta.$$

And similarly, we have

$$B^* \overset{A}{\sim} \mathcal{N}(\beta, \phi^2/n),$$

with

$$\phi^2 = (1/\mu_W)^2[\sigma_V^2 + \beta^2\sigma_W^2 - 2\beta\sigma_{VW}]. \tag{10.4}$$

It remains to express the variances and covariance of V and W in terms of the parameters of the bivariate distribution of X and Y. Calculate

$$\begin{aligned}\sigma_V^2 &= V(V) = E(V^2) - E^2(V)\\ &= E(X^{*2}Y^{*2}) - E^2(X^*Y^*) = \mu_{22} - \mu_{11}^2,\\ \sigma_W^2 &= V(W) = E(W^2) - E^2(W)\\ &= E(X^{*4}) - E^2(X^{*2}) = \mu_{40} - \mu_{20}^2,\\ \sigma_{VW} &= C(V, W) = E(VW) - E(V)E(W)\\ &= E(X^{*3}Y^*) - E(X^*Y^*)E(X^{*2}) = \mu_{31} - \mu_{11}\mu_{20}.\end{aligned}$$

So in Eq. (10.4) the term in square brackets can be written as

$$\begin{aligned}&\mu_{22} - \mu_{11}^2 + \beta^2(\mu_{40} - \mu_{20}^2) - 2\beta(\mu_{31} - \mu_{11}\mu_{20})\\ &\quad = \mu_{22} + \beta^2\mu_{40} - 2\beta\mu_{31},\end{aligned}$$

using $\mu_{20}\beta = \mu_{11}$. Thus

$$\phi^2 = (\mu_{22} + \beta^2\mu_{40} - 2\beta\mu_{31})/\mu_{20}^2. \tag{10.5}$$

Sample Slope

Now return to $B = M_{11}/M_{20}$, the sample slope itself. We know that $M_{11} \xrightarrow{P} \mu_{11}$ and $M_{20} \xrightarrow{P} \mu_{20}$, so it follows immediately by S2 that

$$B \xrightarrow{P} \mu_{11}/\mu_{20} = \beta.$$

Next write

$$B - \beta = (B^* - \beta) + (B - B^*).$$

As a matter of algebra,

$$B - B^* = (1/M_{20}^*)[(M_{11} - M_{11}^*) - (M_{11}/M_{20})(M_{20} - M_{20}^*)],$$

so

$$\sqrt{n}(B - B^*) = (1/M_{20}^*)[\sqrt{n}(M_{11} - M_{11}^*) - (M_{11}/M_{20})\sqrt{n}(M_{20} - M_{20}^*)].$$

Since $M_{20}^* \xrightarrow{P} \mu_{20}$, $(M_{11}/M_{20}) \xrightarrow{P} \mu_{11}/\mu_{20}$, $\sqrt{n}(M_{11} - M_{11}^*) \xrightarrow{P} 0$, and $\sqrt{n}(M_{20} - M_{20}^*) \xrightarrow{P} 0$, it follows that $\sqrt{n}(B - B^*) \xrightarrow{P} 0$. So by S3 the limiting distribution of $\sqrt{n}(B - \beta)$ is the same as that of $\sqrt{n}(B^* - \beta)$. Equivalently, the asymptotic distribution of B is the same as that of B^*. We conclude that

$$B \overset{A}{\sim} \mathcal{N}(\beta, \phi^2/n),$$

with ϕ^2 as given in Eq. (10.5).

10.6. Variance of Sample Slope

Because the sample slope is very commonly used to measure the sample relation between two variables, we should learn more about its sampling distribution. In Eq. (10.5), the denominator of ϕ^2 is $\mu_{20}^2 = V^2(X)$, while the numerator can be written as

$$\begin{aligned}\mu_{22} + \beta^2\mu_{40} - 2\beta\mu_{31} &= E(X^{*2}Y^{*2}) + \beta^2E(X^{*4}) - 2\beta E(X^{*3}Y^*) \\ &= E[X^{*2}(Y^* - \beta X^*)^2] = E(X^{*2}U^2),\end{aligned}$$

say, where

$$U = Y^* - \beta X^* = (Y - \mu_Y) - \beta(X - \mu_X) = Y - (\alpha + \beta X)$$

is the deviation from the population BLP. So

$$\phi^2 = E(X^{*2}U^2)/V^2(X). \tag{10.6}$$

A special case arises if $E(U^2|X)$ is constant over X. Let σ^2 denote that constant value of $E(U^2|X)$, and calculate

$$E(X^{*2}U^2) = E_X[E(X^{*2}U^2|X)] = E_X[X^{*2}E(U^2|X)] = E(X^{*2}\sigma^2)$$
$$= \sigma^2 E(X^{*2}) = \sigma^2 V(X).$$

So in this special case, $\phi^2 = \sigma^2/\sigma_X^2$, which may look familiar to those who have studied "linear regression analysis."

Perhaps the only situation that ensures the constancy of $E(U^2|X)$ is a population in which the conditional expectation function $E(Y|X)$ is linear in X (so that $U = Y - E(Y|X)$, with $E(U^2|X) = V(Y|X)$), *and* the conditional variance function $V(Y|X)$ is constant over X. (We know one population, the bivariate normal, that has those features.) In this situation we can say that, for given sample size n, the asymptotic variance of the sample slope will be large if the conditional variance of Y is large and/or the marginal variance of X is small.

Exercises

10.1 Given a data set with n paired observations (x_i, y_i), let $e_i = y_i - (a + bx_i)$, where a and b are constants to be chosen to minimize $\Sigma_i e_i^2$.

(a) Show that the solution values are

$$b = \left[\sum_i (x_i - \bar{x})(y_i - \bar{y})\right] \Big/ \left[\sum_i (x_i - \bar{x})^2\right], \qquad a = \bar{y} - b\bar{x}.$$

(b) Referring to Section 10.5, show that the sample linear projection $\hat{y} = A + Bx$ has this *least-squares* property.

10.2 These statistics were obtained in a sample of 30 observations from a bivariate population $f(x, y)$: $\Sigma_i x_i = 30$, $\Sigma_i y_i = 120$, $\Sigma_i x_i^2 = 150$, $\Sigma_i y_i^2 = 1830$, $\Sigma_i x_i y_i = 480$. Here Σ_i denotes summation over i from 1 to 30.

(a) Calculate the sample means, sample variances, and sample covariance.

(b) Find $\hat{y} = a + bx$, the sample linear projection of y on x.

10.3 For the savings rate–income data of Chapter 1:

(a) Calculate the sample linear projection of y = savings rate on x = income. Hint: The sample covariance can be calculated as

$$S_{XY} = \sum_{i=1}^{10} [x_i m_{Y|x_i} p(x_i)] - m_X m_Y.$$

(b) Comment on the relation between this line and the conditional mean function that was plotted in Figure 1.1.

10.4 A bivariate population $f(x, y)$ has these moments:

$$\begin{array}{lll} E(X) = 1 & E(Y) = 4 & E(X^*Y^*) = 360 \\ E(X^{*2}) = 120 & E(Y^{*2}) = 1350 & E(X^*Y^{*3}) = 0 \\ E(X^{*3}) = 10 & E(Y^{*3}) = -40 & E(X^{*3}Y^*) = 320 \\ E(X^{*4}) = 14880 & E(Y^{*4}) = 12000 & E(X^{*2}Y^{*2}) = 1596000 \end{array}$$

Here $X^* = X - E(X)$, $Y^* = Y - E(Y)$.

(a) Find the BLP, $E^*(Y|X) = \alpha + \beta X$.

(b) Consider random sampling, sample size 30, from this population. Let $\bar{X}$ be the sample mean of X, S_X^2 be the sample variance of X, and B be the slope in the sample linear projection of Y on X. Approximate the probability of each of these events:

$$A_1 = \{\bar{X} \leq 3/2\}, \qquad A_2 = \{S_X^2 \leq 128\}, \qquad A_3 = \{B \leq 5\}.$$

11 Parameter Estimation

11.1. Introduction

We have accumulated considerable information about the probability distributions of sample statistics in random sampling from univariate and bivariate populations. In some cases we have the complete exact sampling distribution, in some cases only its expectation and variance. In many cases, we have the asymptotic sampling distribution. The information was obtained by deducing features of the distributions of functions of random variables (the sample statistics) from knowledge of the distribution of the original variables (in the population).

The practical problem is, of course, quite the opposite: it calls for inferring (or guessing, or estimating) features of the population from knowledge of a single sample. This problem is not trivial, because the same sample might arise in sampling from many different populations. We are turning from deduction to inference.

We will have a single random sample $(y_1, \ldots, y_n)'$ drawn from an unknown population $Y \sim f(y)$. We are interested in some feature of the population, a parameter θ, say. Our task is to find an *estimate* of θ, a single number that will serve as our guess of the value of the parameter. Naturally the estimate will be a function of the sample data. How shall we process the sample data to come up with an estimate? That is, what function $h(y_1, \ldots, y_n)$ shall we choose as our estimate?

Now the sample $\mathbf{y} = (y_1, \ldots, y_n)'$ is a single observation on the random vector $\mathbf{Y} = (Y_1, \ldots, Y_n)'$. For a function $h(\mathbf{y})$, the estimate that we calculate, $t = h(y_1, \ldots, y_n)$, will be a single observation on the random variable $T = h(\mathbf{Y})$. The random variable T is referred to as the *estimator*, as distinguished from the value t that it happens to take on, which is the estimate.

11.2. The Analogy Principle

Perhaps the most natural rule for selecting an estimator is the *analogy principle*. A population parameter is a feature of the population. To estimate it, use the corresponding feature of the sample.

Here are some applications of this principle, that is, examples of *analog estimators*.

• To estimate a population moment, use the corresponding sample moment. For $\theta = \mu$, use $T = \bar{Y}$. For $\theta = \sigma^2$, use $T = S^2$.

• To estimate a function of population moments, use the function of the sample moments. For the population BLP slope $\beta = \sigma_{XY}/\sigma_X^2$, use $B = S_{XY}/S_X^2$. For the population BLP intercept $\alpha = \mu_Y - \beta\mu_X$, use $A = \bar{Y} - B\bar{X}$. For σ^2/n, use S^2/n. This sort of application of the analogy principle is also referred to as the *method of moments*.

• To estimate $\Pr(Y \leq c)$, use the sample proportion of observations that have $Y \leq c$.

• To estimate the population median, use the sample median.

• To estimate the population maximum, use the sample maximum.

• To estimate the population BLP $E^*(Y|X) = \alpha + \beta X$ (the line that minimizes expected squared deviations in the population), use the sample least-squares line $\hat{Y} = A + BX$ (the line that minimizes the mean of squared deviations in the sample). As shown in Exercise 10.1, this gives the same answers for A and B as above.

• To estimate a population CEF $\mu_{Y|X}$, use the sample conditional mean function $m_{Y|X}$, discussed in Chapter 1.

The analogy principle is constructive as well as natural. Once we decide which feature of the population is of interest to us, we will almost inevitably recognize an estimator for it. Adopting the analogy principle as the starting point in a search for estimators leads immediately to some questions. Are analog estimators sensible from a statistical point of view? How reliable are they? What shall we do when an analog estimator is unreliable, or inadequate? What shall we do when there are several analog estimators of the same parameter? (For example, if the population $f(x)$ is symmetric, then the population mean and median coincide, but the sample mean and median are distinct.) For a comprehensive development of the theory, see Manski (1988).

To address such questions here, we turn to the classical criteria for evaluating estimators. We will begin with criteria that refer to exact sampling distributions.

11.3. Criteria for an Estimator

Let $T = h(Y_1, \ldots, Y_n)$ be a sample statistic with a pdf (or pmf) $g(t)$, and moments $E(T)$, $V(T)$, etc. We may be sampling from a univariate population (with each Y_i a scalar) or from a bivariate population (with each Y_i a two-element vector).

Choosing an estimator T amounts to choosing a sampling distribution from which to get a single draw. So the issue becomes: what probability distribution would we like to draw from? When we choose an estimator, we are buying a lottery ticket; in which lottery would we prefer to participate? Presume that the prize in the lottery is higher the closer T is to θ. We would like $T = \theta$, but that ideal is unattainable. The only sample statistics with degenerate distributions are the trivial ones, those for which $h(\mathbf{Y}) \equiv$ constant. It is easy enough to pick such a function, for example, $h(\mathbf{Y}) \equiv 3$, but that is hardly an attractive estimator unless $\theta = 3$. What we ask is that T be close to θ, whatever the value of θ might be.

A natural measure of distance between the random variable T and the parameter θ is the *mean squared error* (MSE), $E(T - \theta)^2$. We used the MSE measure in discussing prediction of a random variable in Section 3.4. Now the target is a fixed parameter rather than a random variable, but again it seems desirable to have a small value for $E(T - \theta)^2$. According to T3 (Section 3.3), the MSE of T about θ can be written as

$$E(T - \theta)^2 = V(T) + [E(T) - \theta]^2.$$

Define the *bias* of T as an estimator of θ as $E(T) - \theta = E(T - \theta)$. So the MSE of T as an estimator of θ equals the variance of T plus the squared bias of T as an estimator of θ. In general, both variance and bias depend on the unknown parameter θ, and it is not feasible to find a T that minimizes $E(T - \theta)^2$ for all θ. Still, small MSE is desirable.

It also seems desirable to have an estimator for which the expected deviation from the parameter is zero:

DEFINITION. T is an *unbiased estimator* of θ iff $E(T - \theta) = 0$, for all θ.

For unbiased estimators, MSE = variance, which leads to a popular criterion, namely minimum variance unbiasedness:

DEFINITION. T is a *minimum variance unbiased estimator*, or MVUE, of θ iff

(i) $E(T - \theta) = 0$ for all θ, and
(ii) $V(T) \leq V(T^*)$ for all T^* such that $E(T^* - \theta) = 0$.

No other member of the class of unbiased estimators of θ has a variance that is smaller than the variance of T.

The MVUE criterion may be operational even when minimizing MSE is not. At least, it may be operational if we restrict the class of estimators.

Estimation of Population Mean

Suppose that we are random sampling on the variable Y, where $E(Y) = \mu$ and $V(Y) = \sigma^2$ are unknown. To estimate the population mean μ, the analogy principle suggests that we use the sample mean $\bar{Y}$. Now

$$\bar{Y} = (1/n) \sum_i Y_i, \qquad E(\bar{Y}) = E(Y) = \mu.$$

So $\bar{Y}$ is a linear unbiased estimator of μ. But there are many other linear unbiased estimators of μ. Let $T = \Sigma_i c_i Y_i$, where the c's are constants. Then

$$E(T) = E\left(\sum_i c_i Y_i\right) = \sum_i c_i E(Y_i) = \mu \sum_i c_i,$$

$$V(T) = \sum_i c_i^2 V(Y_i) = \sigma^2 \sum_i c_i^2.$$

So any linear function of the Y_i's with intercept equal to 0 and sum of slopes equal to 1 will be unbiased for μ. To find the best of these, choose the c_i's to minimize $\Sigma_i c_i^2$ subject to $\Sigma_i c_i = 1$. Let

$$Q = \sum_i c_i^2 - \lambda\left(\sum_i c_i - 1\right),$$

where λ is a Lagrangean multiplier. Then

$$\partial Q/\partial c_i = 2c_i - \lambda \qquad (i = 1, \ldots, n)$$

$$\partial Q/\partial \lambda = -\left(\sum_i c_i - 1\right).$$

Setting the derivatives at zero gives $c_i = \lambda/2$ for all i, and $\Sigma_i c_i = 1$. So $\Sigma_i c_i = n\lambda/2$, whence $\lambda = 2/n$, whence $c_i = 1/n$ for all i. It can be confirmed that the first-order conditions locate a minimum, so the optimal choice is $T = \Sigma_i(1/n)Y_i = \bar{Y}$.

We have established the following:

THEOREM. In random sampling, sample size n, from any population, the sample mean is the *minimum variance linear unbiased estimator*, or MVLUE, of the population mean.

This result is strong in that it applies to any population, but weak in that only linear functions of the observations are considered.

Let us evaluate some other analog estimators with respect to the criteria introduced here.

Population Raw Moment. For $\theta = \mu_s' = E(Y^s)$, the analogy principle suggests $M_s' = (1/n)\Sigma_i Y_i^s$ as the estimator. Since this is the sample mean in random sampling on the variable $W = Y^s$, we conclude that M_s' is the MVLUE of μ_s', where "linear" now means linear in the Y^s's. Similarly, in the bivariate case, M_{rs}' is the MVLUE of μ_{rs}', where "linear" now means linear in the X^rY^s's.

Population Variance. For $\theta = \mu_2 = E(Y - \mu_Y)^2 = \sigma^2$ (with μ_Y unknown), the analogy principle suggests $M_2 = S^2$, the sample variance, as the estimator. As we have seen (Section 8.4), S^2 cannot be interpreted as a sample mean in random sampling. And indeed, this analog estimator is biased: $E(S^2) = \sigma^2(1 - 1/n)$. But the bias is easily removed. Define the *adjusted sample variance*

$$S^{*2} = \sum_i (Y_i - \bar{Y})^2/(n - 1) = nS^2/(n - 1).$$

Then $E(S^{*2}) = \sigma^2$, so S^{*2} is unbiased. But we have no general result here on minimum variance unbiasedness.

Population Covariance. For estimating the population covariance, the adjusted sample covariance $S_{XY}^* = nS_{XY}/(n - 1)$ is unbiased, but not necessarily minimum variance unbiased.

Population Maximum. For $\theta = \max(Y)$, the analogy principle suggests $T = \max(Y_1, \ldots, Y_n)$ as the estimator. But $T \leq \theta$ and $\Pr(T = \theta) < 1$, so $E(T) < \theta$. This analog estimator is biased, and there is no obvious way to remove the bias.

Population Linear Projection Slope. For $\beta = \sigma_{XY}/\sigma_X^2$ (with μ_X and μ_Y unknown), the analogy principle suggests $B = S_{XY}/S_X^2$ as the estimator. This is a nonlinear function of sample moments. We can adjust the numerator and denominator to remove their biases, but this will not remove bias in the ratio, because the expectation of a ratio is not generally equal to the ratio of expectations. Indeed, this analog estimator will in general be biased, and there is no obvious way to remove the bias. (An important exception occurs when the population CEF is linear: see Section 13.2.)

11.4. Asymptotic Criteria

As we have seen, evaluating analog estimators on the basis of their exact expectation and variance may run into an impasse, for the exact distribution may well depend on specifics of the population. Progress is available if we rely on asymptotic, that is, approximate, sampling distributions. We put an n subscript on T, as in Chapter 9, to emphasize the dependence upon sample size, and introduce two classical criteria.

DEFINITION. T_n is a *consistent* estimator of θ iff $T_n \xrightarrow{P} \theta$. Equivalently, T_n is a consistent estimator of θ iff $\lim \Pr(|T_n - \theta| \geq \epsilon) = 0$ for all $\epsilon > 0$.

Consistency is attractive because it says that as the sample size increases indefinitely, the distribution of the estimator becomes entirely concentrated at the parameter value.

The sample mean is a consistent estimator of the population mean in random sampling. The Law of Large Numbers says precisely that, taking $\theta = \mu$, $T = \bar{Y}$. By the same law, any sample raw moment is a consistent estimator of the corresponding population raw moment in random sampling. Further, by the Slutsky Theorems S1 and S2, any continuous function of the sample moments is, in random sampling, a consistent estimator of the corresponding function of the population moments. For example, $S^2 = M_2 = M_2^* - (\bar{X} - \mu)^2 = h(M_2^*, \bar{X})$ is a consistent estimator of $\sigma^2 = \mu_2 = h(\mu_2, \mu)$, and $B = S_{XY}/S_X^2$ is a consistent estimator of $\beta = \sigma_{XY}/\sigma_X^2$.

There is a general presumption that analog estimators are consistent in random sampling. The intuition runs as follows. The analog estimator is a function of the frequency distribution in the sample. The parameter is the same function of the probability distribution. The sample frequencies will, as n gets large, converge to the population frequencies, that is, to the probability distribution.

Typically there are many consistent estimators of the same parameter. For example, in sampling from a symmetric population, both the sample mean and the sample median are consistent estimators of $\mu = E(Y)$. To discriminate among consistent estimators of the same parameter, we turn from degenerate limiting distributions to asymptotic distributions. Typically the consistent estimators will have asymptotic normal distributions, centered at the parameter value.

DEFINITION. T_n is a *best asymptotically normal*, or BAN, *estimator* of θ iff

(i) $T_n \overset{A}{\sim} \mathcal{N}(\theta, \phi^2/n)$, and
(ii) $\phi^2 \leq \phi^{*2}$ for all T_n^* such that $T_n^* \overset{A}{\sim} \mathcal{N}(\theta, \phi^{*2}/n)$.

No other member of the class of consistent, asymptotically normal estimators of θ has an asymptotic variance that is smaller than the asymptotic variance of T_n.

In effect, the BAN criterion is the asymptotic version of the MVUE criterion. Throughout econometrics, except for those situations in which exact results are available, best asymptotic normality, sometimes labeled *asymptotic efficiency*, is the customary criterion of choice.

11.5. Confidence Intervals

In reporting an estimate of a parameter, it is a good idea to accompany it with some information about the reliability of the estimator, or rather about its unreliability—the extent to which it varies from sample to sample. The natural measure is the standard deviation of the estimator. Often that information is presented in the form of a confidence interval.

Suppose that $Y \sim \mathcal{N}(\mu, \sigma^2)$. Then in random sampling, sample size n, the sample mean $\bar{Y}$ is distributed $\mathcal{N}(\mu, \sigma^2/n)$, so $Z = \sqrt{n}(\bar{Y} - \mu)/\sigma \sim \mathcal{N}(0, 1)$. Let

$$A = \{|Z| \leq 1.96\},$$

so $\Pr(A) = F(1.96) - F(-1.96) = 0.975 - 0.025 = 0.95$, where $F(\cdot)$ is the $\mathcal{N}(0, 1)$ cdf. Now

$$A = \{|\bar{Y} - \mu| \leq 1.96\sigma/\sqrt{n}\}$$
$$= \{\mu - 1.96\sigma/\sqrt{n} \leq \bar{Y} \leq \mu + 1.96\sigma/\sqrt{n}\}$$
$$= \{\bar{Y} - 1.96\sigma/\sqrt{n} \leq \mu \leq \bar{Y} + 1.96\sigma/\sqrt{n}\}.$$

So the statement, "The parameter μ lies in the interval $\bar{Y} \pm 1.96\sigma/\sqrt{n}$," is true with probability 95%. We say that $\bar{Y} \pm 1.96\sigma/\sqrt{n}$ is a 95% *confidence interval* for the parameter μ.

To get a confidence interval for μ at a different confidence level, just use a different cutoff point: for example, $\bar{Y} \pm 1.645\sigma/\sqrt{n}$ is a 90% confidence interval for μ. For a given confidence level, a narrow confidence interval is desirable: it indicates that the sample has said a lot about the parameter value. What produces a narrow interval, obviously, is a small standard deviation, $\sigma/\sqrt{n}$, that is, a small σ^2 and/or a large n.

Now suppose that the random variable Y is distributed (not necessarily normally) with $E(Y) = \mu$ and $V(Y) = \sigma^2$. Then in random sampling, sample size n, the sample mean $\bar{Y}$ is asymptotically distributed $\mathcal{N}(\mu, \sigma^2/n)$. So, by the logic above, $\Pr(A) \cong 0.95$, and we say that $\bar{Y} \pm 1.96\sigma/\sqrt{n}$ is an *approximate* 95% *confidence interval* for the parameter μ.

In practice these results will not be operational, because σ^2 is unknown. Consider the sample t-ratio $U = \sqrt{n}(\bar{Y} - \mu)/S$, where S^2 is the sample variance. Recall from Section 9.6 that $U \xrightarrow{D} \mathcal{N}(0, 1)$. So by the logic above, $\bar{Y} \pm 1.96S/\sqrt{n}$ is an approximate 95% confidence interval for the parameter μ. The statistic $S/\sqrt{n}$ is called the *standard error* of $\bar{Y}$, as distinguished from its *standard deviation*, $\sigma/\sqrt{n}$.

The logic extends to construct approximate confidence intervals for parameters other than the sample mean. Suppose that T_n is a sample statistic used to estimate a parameter θ, and that $T_n \overset{A}{\sim} \mathcal{N}(\theta, \phi^2/n)$. If ϕ^2 is known, then $T_n \pm 1.96\phi/\sqrt{n}$ provides an approximate 95% confidence interval for the parameter θ. More practically, when we have $\hat{\phi}^2$, a consistent estimator of ϕ^2, then $T_n \pm 1.96\hat{\phi}/\sqrt{n}$ provides an approximate 95% confidence interval for the parameter θ.

Example. The sample variance S^2 is used to estimate the population variance σ^2. Recall from Section 9.6 that $S^2 \overset{A}{\sim} \mathcal{N}(\sigma^2, \phi^2/n)$, with $\phi^2 = \mu_{40} - \mu_{20}^2$. So let $\hat{\phi}^2 = M_{40} - M_{20}^2$, and report $S^2 \pm 1.96\hat{\phi}/\sqrt{n}$ as the approximate 95% confidence interval for σ^2.

The reasoning is the same as that used above to get $\bar{Y} \pm 1.96S/\sqrt{n}$ as an approximate 95% confidence interval for μ. The statistic $\hat{\phi}/\sqrt{n}$ is called the (asymptotic) standard error of T_n.

A few remarks on confidence intervals:

• Perhaps the only proper exception to using the 1.96 rule for 95% confidence arises when Y is normally distributed. For then we know (Section 8.6) that the exact distribution of $\sqrt{(n-1)}(\bar{Y}-\mu)/S$ is $t(n-1)$, so the Student's t-table can be consulted for the critical value to replace 1.96. There may be other exceptional cases in which the exact distribution of the sample t-ratio is known. But, common practice notwithstanding, there is no good reason to rely routinely on a t-table rather than a normal table unless Y itself is normally distributed.

• It is good practice to report the standard error of a parameter estimate along with the estimate itself. Conventionally, the standard error is put in parentheses underneath the estimate. Readers can then construct (approximate) confidence intervals as they see fit.

• It is common practice, but not good practice, to report the "t-statistic" (ratio of an estimate to its standard error) instead of (or even in addition to) the standard error. More on this in Section 21.3.

Exercises

11.1 In random sampling, sample size n, from a univariate population, let $T = c\bar{Y}$, where $\bar{Y}$ is the sample mean.

(a) Choose c to minimize the MSE of T as an estimator of $\mu = E(Y)$.
(b) Comment on the practical usefulness of your result.

11.2 In size-n random sampling from a bivariate population $f(x, y)$, suppose that the objective is to estimate the parameter $\theta = \mu_Y - \mu_X$. For example, the population may consist of married couples, with Y = husband's earnings and X = wife's earnings. The sample statistics $\bar{X}$, $\bar{Y}$, S_X^2, S_Y^2, and S_{XY} are available.

(a) Propose a statistic T that is an unbiased estimator of θ. Show that it is unbiased.
(b) Find its variance $V(T)$ in terms of the population variances and covariance of X and Y.
(c) For the practical case, in which those population variances and covariance are unknown, propose an unbiased estimator of $V(T)$. Show that it is unbiased.

(d) What statistic would you report in practice as a standard error for T?

11.3 In one population, $E(Y_1) = \mu$ and $V(Y_1) = \sigma_1^2$; in a second population, $E(Y_2) = \mu$ and $V(Y_2) = \sigma_2^2$. The population variances are known, but their common expectation μ is unknown. Random samples of sizes n_1 and n_2 respectively are drawn from the two populations. The two samples are independent. It is proposed to combine the sample means $\bar{Y}_1$ and $\bar{Y}_2$ linearly into a single estimator of the common mean μ.

(a) Consider all possible linear combinations of $\bar{Y}_1$ and $\bar{Y}_2$. Determine the one that is minimum variance unbiased as an estimator of μ.

(b) Verify that the variance of that estimator is less than the variance of each of the two sample means.

11.4 You are interested in estimating $\theta = \mu_1 - \mu_2$, where $Y_1 \sim \mathcal{N}(\mu_1, 50)$ and $Y_2 \sim \mathcal{N}(\mu_2, 100)$. You can afford a total of 100 observations. Determine how many you should draw on Y_1 and how many on Y_2.

11.5 Suppose that $Y_1 = X + U_1$ and $Y_2 = X + U_2$, where X = permanent income, Y_1 = current income in year 1, and Y_2 = current income in year 2. It is known that U_1 and U_2 have zero expectations and are uncorrelated with X. It is also known that $V(X) = 400$, $V(U_1) = 200$, $V(U_2) = 100$, and $C(U_1, U_2) = 60$. A random sample of size 10 is drawn from the joint probability distribution of Y_1 and Y_2. The objective is to estimate $E(X)$, which is unknown. The sample means are $\bar{Y}_1$ and $\bar{Y}_2$. Consider all linear combinations of the sample means that are unbiased estimators of $E(X)$, and find the one that has minimum variance.

11.6 A random sample from a population has $n = 30$, $\Sigma_i x_i = 120$, $\Sigma_i x_i^2 = 8310$.

(a) Calculate unbiased estimates of the population mean, the population variance, and the variance of the sample mean.

(b) Provide an approximate 95% confidence interval for the population mean.

11.7 A random sample from a Bernoulli population has 35 observations with $Y = 1$, and 65 observations with $Y = 0$.

(a) Calculate unbiased estimates of the population mean, the population variance, and the variance of the sample mean.
(b) Provide an approximate 95% confidence interval for the population mean.

11.8 A random sample from an exponential population with unknown parameter λ has $n = 50$, $\Sigma_i x_i = 30$.

(a) Calculate an unbiased estimate of the population mean. Is that estimate consistent? Explain briefly.
(b) Calculate a consistent estimate of the parameter λ. Is that estimate unbiased? Explain briefly.
(c) Provide an approximate 95% confidence interval for λ.

11.9 These statistics were calculated in a random sample of size 100 from the joint distribution of X and Y:

$$\bar{X} = 2, \quad S_X^2 = 5, \quad \bar{Y} = 1, \quad S_Y^2 = 4, \quad S_{XY} = 3.$$

Construct an approximate 95% confidence interval for the parameter $\theta = E(X)/E(Y)$.

11.10 We are interested in estimating the proportion of the population whose incomes are below the poverty line, a prespecified level of income. Let Y = income and c = poverty line, so the parameter of interest is $\theta = \Pr(Y \leq c) = G(c)$, where $G(\cdot)$ is the unknown cdf of income. For random sampling, sample size n, from the population, the analogy principle suggests that we estimate θ by T = proportion of the sample observations having $Y \leq c$.

(a) Find $E(T)$ and $V(T)$. Is T unbiased? Is T consistent? Explain.
(b) Show that $T \overset{A}{\sim} \mathcal{N}[\theta, \theta(1 - \theta)/n]$.

11.11 For the setup in Exercise 11.10, suppose now that it is known that Y is normally distributed, with known variance but unknown mean. So $\theta = G(c) = F[(c - \mu)/\sigma]$, where $F(\cdot)$ is the standard normal cdf and σ is known. Because θ is a function of the population moment μ, the analogy principle suggests an alternative estimator of θ, namely $U = F[(c - \bar{Y})/\sigma]$, where $\bar{Y}$ is the sample mean in random sampling, sample size n.

(a) Show that U is consistent. Is it unbiased? Explain.
(b) Find the asymptotic distribution of U.

(c) On the basis of their asymptotic distributions, which estimator of θ would you prefer to use, T or U? Hint: Two useful facts about the standard normal pdf and cdf, $f(z)$ and $F(z)$, are

$$\partial F(z)/\partial z = f(z),$$

$$[f(z)]^2/\{F(z)[1 - F(z)]\} < 0.64 \quad \text{for all values of } z.$$

12 *Advanced Estimation Theory*

12.1. The Score Variable

Our discussion of parameter estimation has been general with respect to the population: we have not assumed knowledge of the family to which the population belongs. Now we turn to estimation in more completely specified situations, where the family, that is, the form, of the pdf or pmf is known up to a parameter of interest. The value of that parameter is then a missing link needed to complete the specification of the population.

Suppose that the random variable Y has pmf or pdf $f(y; \theta)$, where the function is known except for the parameter value θ. Define the *log-likelihood variable*

$$L = \log f(Y; \theta) = L(Y; \theta),$$

and the *score variable*

$$Z = \partial \log f(Y; \theta)/\partial\theta = \partial L/\partial\theta = z(Y; \theta).$$

We write Y rather than y as the argument to emphasize that both L and Z, being functions of the random variable Y, are themselves random variables. The score variable plays several roles in the theory. First we establish:

ZERO EXPECTED SCORE (or ZES) RULE. The expected value of the score variable is zero.

Proof. For convenience treat the continuous case. Since Z is a function of Y, its expectation is

$$E(Z) = \int z(y;\, \theta) f(y;\, \theta)\, dy = \int zf\, dy.$$

(Note: In this chapter, $\int$ is shorthand for $\int_{-\infty}^{\infty}$, and the arguments of functions may be omitted for convenience.) Because $f(y;\, \theta)$ is a pdf, it must be true that

$$\int f(y;\, \theta)\, dy = 1 \quad \text{for all } \theta.$$

Differentiating both sides with respect to θ gives

$$\int (\partial f / \partial \theta)\, dy = 0.$$

(Note: Here and subsequently it is assumed that the range of integration does not depend on θ.) But

$$\partial f / \partial \theta = (\partial \log f / \partial \theta) f = z(y;\, \theta) f = zf,$$

so $\int zf\, dy = 0$. ■

12.2. Cramér-Rao Inequality

One role of the score variable is to set a standard for unbiased estimation of θ in random sampling. We show:

CRAMÉR-RAO INEQUALITY, or CRI. In random sampling, sample size n, from an $f(y;\, \theta)$ population, if $T = h(Y_1, \ldots, Y_n)$ and $E(T) = \theta$ for all θ, then $V(T) \geq 1/[nV(Z)]$.

Proof. First, consider the case $n = 1$. Here $T = h(Y)$ with $E(T) = \theta$ for all θ. That is,

$$\int h(y) f(y;\, \theta)\, dy = \theta \quad \text{for all } \theta.$$

Differentiating both sides with respect to θ gives

$$\int h(y) zf\, dy = 1,$$

which says that $E(TZ) = 1$. Because $E(Z) = 0$, it follows that $C(T, Z) = 1$. Recall the Cauchy-Schwarz Inequality (Section 6.6), which says that squared correlations cannot exceed unity: it must be the case that $V(T)V(Z) \geq C^2(T, Z)$. So here $V(T)V(Z) \geq 1$, or $V(T) \geq 1/V(Z)$, which is the CRI for $n = 1$.

Proceed to the case $n > 1$. Let $g = g_n(y_1, \ldots, y_n)$ denote the joint pdf of the random sample. Here $T = h(Y_1, \ldots, Y_n)$ with $E(T) = \theta$ for all θ. That is,

$$\int \cdots \int h(y_1, \ldots, y_n)\, g_n(y_1, \ldots, y_n; \theta)\, dy_n \cdots dy_1 = \theta.$$

Differentiating both sides with respect to θ gives

$$\int \cdots \int h\ (\partial g/\partial\theta)\, dy_n \cdots dy_1 = 1.$$

But

$$\partial g/\partial\theta = (\partial \log g/\partial\theta)g, \qquad g = \prod_{i=1}^{n} f(y_i; \theta),$$

so $\log g = \Sigma_i \log f(y_i; \theta) = \Sigma_i \log f_i$, say. Thus

$$\partial \log g/\partial\theta = \sum_i (\partial \log f_i/\partial\theta) = \sum_i z(y_i; \theta) = \sum_i z_i = n\bar{Z},$$

say, where $\bar{Z}$ is the sample mean of the score variable. So

$$\int \cdots \int hn\bar{z}g\, dy_n \cdots dy_1 = 1,$$

which says that $nE(T\bar{Z}) = 1$. Because $E(\bar{Z}) = E(Z) = 0$, it follows that $nC(T, \bar{Z}) = 1$, so $C(T, \bar{Z}) = 1/n$. By the Cauchy-Schwarz Inequality, $V(T)V(\bar{Z}) \geq 1/n^2$, so $V(T) \geq 1/[n^2V(\bar{Z})]$. But $V(\bar{Z}) = V(Z)/n$, so $V(T) \geq 1/[nV(Z)]$. ■

The CRI does not provide us with an estimator, but rather sets a standard against which unbiased estimators can be assessed. If we happen to know, or have located, an unbiased estimator T with $V(T) = 1/[nV(Z)]$, then we can stop searching for a better (that is, lower-variance) unbiased estimator, because the CRI tells us that there is no T^* such that $E(T^*) = \theta$ and $V(T^*) < V(T)$.

Example. Suppose that $Y \sim$ Bernoulli(θ), so its pmf is

$$f(y; \theta) = \theta^y(1 - \theta)^{(1-y)} \quad \text{for } y = 0, 1.$$

As a random variable

$$f(Y; \theta) = \theta^{Y}(1 - \theta)^{(1-Y)}.$$

So

$$L = \log f = Y \log \theta + (1 - Y) \log(1 - \theta),$$

$$Z = \partial L/\partial\theta = Y/\theta - (1 - Y)/(1 - \theta) = (Y - \theta)/[\theta(1 - \theta)].$$

We know (see Table 3.1) that $E(Y) = \theta$ and $V(Y) = \theta(1 - \theta)$. So

$$E(Z) = E(Y - \theta)/[\theta(1 - \theta)] = 0,$$

$$V(Z) = V(Y)/[\theta(1 - \theta)]^2 = \theta(1 - \theta)/[\theta(1 - \theta)]^2 = 1/[\theta(1 - \theta)].$$

The expectation illustrates the ZES rule. The variance formula implies by the CRI that if T is an unbiased estimator of θ, then

$$V(T) \geq 1/[nV(Z)] = \theta(1 - \theta)/n.$$

The sample mean $\bar{Y}$ has $E(\bar{Y}) = \theta$ and $V(\bar{Y}) = V(Y)/n = \theta(1 - \theta)/n = 1/[nV(Z)]$. It follows that $\bar{Y}$ is the MVUE of θ in random sampling from a Bernoulli population. This conclusion is considerably stronger than the previous general result that $\bar{Y}$ is MVLUE (see Section 11.3), for now the class of estimators considered is no longer confined to linear functions of the observations.

For the normal distribution, as well as for the Bernoulli, the sample mean is the MVUE of the population mean.

There is another way to state the CRI. Recall that $Z = \partial \log f/\partial\theta$, with $E(Z) = \int z(y; \theta)f(y; \theta)\, dy = 0$. Define the *information variable*

$$W = -\partial Z/\partial\theta = -\partial^2 \log f/\partial\theta^2.$$

This too is a random variable. We show:

INFORMATION RULE. The expectation of the information variable is equal to the variance of the score variable.

Proof. Differentiate $E(Z) = \int zf\, dy = 0$ with respect to θ:

$$\int (\partial zf/\partial\theta)\, dy = 0.$$

Now

$$\partial(zf)/\partial\theta = z(\partial f/\partial\theta) + (\partial z/\partial\theta)f = z(\partial \log f/\partial\theta)f - wf$$
$$= z^2 f - wf = (z^2 - w)f.$$

So $\int (z^2 - w)f\, dy = 0$. That is, $E(Z^2 - W) = 0$, so $E(Z^2) = E(W)$. But $E(Z) = 0$, so $V(Z) = E(Z^2) = E(W)$. ■

Example. For the Bernoulli distribution we have seen that the score variable, $Z = (Y - \theta)/[\theta(1 - \theta)]$, has $V(Z) = 1/[\theta(1 - \theta)]$. The information variable is $W = -\partial Z/\partial\theta = (1 + Z - 2\theta Z)/[\theta(1 - \theta)]$. With $E(Z) = 0$, we see that $E(W) = 1/[\theta(1 - \theta)]$.

With $E(W) = V(Z)$, we can restate the CRI conclusion as $V(T) \geq 1/[nE(W)]$. This restatement is useful because for some distributions, $E(W)$ is easier to calculate than $V(Z)$. It also accounts for the label "information variable": the larger the expected information variable is, the more precise the unbiased estimation of a parameter may be.

12.3. ZES-Rule Estimation

A second role of the score variable is to provide an estimator of θ. Recall an analogy that suggested the sample mean as an estimator of the population mean (Section 11.2). An instructive way to restate that analogy is as follows. Because $E(Y - \mu) = 0$, we can characterize μ as the value for c that makes $E(Y - c) = 0$. Now the sample analog of the population average $E(Y - c)$ is the sample average $(1/n)\Sigma_i(Y_i - c)$, so let us estimate μ by the value for c that makes $(1/n)\Sigma_i(Y_i - c) = 0$. The result is $c = (1/n)\Sigma_i Y_i = \bar{Y}$.

With that in mind, we will use the ZES rule to obtain an estimator of θ. Suppose that we are drawing a random sample from a population in which the pdf or pmf $f(Y; \theta)$ is known except for the value of the parameter θ. The score variable is $Z = z(Y; \theta)$. Because $E(Z) = 0$, we can characterize θ as the value for c that makes $E[z(Y; c)] = 0$. Now, the sample analog of $E[z(Y; c)]$ is $(1/n)\Sigma_i z(Y_i; c)$, so let us estimate θ by the value for c that makes $(1/n)\Sigma_i z(Y_i; c) = 0$, or equivalently that makes $\Sigma_i z(Y_i; c) = 0$. Changing notation somewhat, let T be the solution value, and let $\hat{Z}_i = z(Y_i; T)$. Then by construction, the *ZES-rule estimator* T satisfies $\Sigma_i \hat{Z}_i = 0$.

Example. Suppose that $Y \sim$ exponential(λ), so $f(Y; \lambda) = \lambda \exp(-\lambda Y)$. Then $L = \log \lambda - \lambda Y$, and $Z = (1/\lambda) - Y$. Let $\hat{Z}_i = (1/T) - Y_i$. Then $\Sigma_i \hat{Z}_i = (n/T) - \Sigma_i Y_i = n[(1/T) - \bar{Y}]$. Setting this at zero gives $T = 1/\bar{Y}$ as the ZES-rule estimator of λ.

We sketch a derivation of the asymptotic distribution of ZES-rule estimators. Using a linear approximation at the point θ, and recalling the definition of the information variable $W = -\partial Z/\partial\theta$, write

$$\hat{Z}_i = z(Y_i; T) \cong z(Y_i; \theta) + [\partial z(Y_i; \theta)/\partial\theta](T - \theta) = Z_i - W_i(T - \theta),$$

where $W_i = -\partial Z_i/\partial\theta$. So

$$\sum_i \hat{Z}_i \cong \sum_i Z_i - (T - \theta) \sum_i W_i.$$

By construction, $\Sigma_i \hat{Z}_i = 0$, whence

$$T - \theta \cong \sum_i Z_i / \sum_i W_i = \bar{Z}/\bar{W},$$

where $\bar{Z} = (1/n)\Sigma_i Z_i$ and $\bar{W} = (1/n)\Sigma_i W_i$. We may neglect the approximation error. Then, because $\bar{Z}$ and $\bar{W}$ are sample means in random sampling on the variables Z and W, the problem amounts to finding the asymptotic distribution of a ratio of sample means, as in Section 10.4. By the LLN, $\bar{Z} \xrightarrow{P} E(Z)$ and $\bar{W} \xrightarrow{P} E(W) = V(Z)$, so that $(T - \theta) \xrightarrow{P} 0$. We conclude that T, which may or may not be unbiased, is a consistent estimator of θ. Further, we have

$$\sqrt{n}(T - \theta) \cong \sqrt{n}\bar{Z}/\bar{W} = (1/\bar{W})\sqrt{n}\bar{Z}.$$

By the CLT and LLN, we know that $\sqrt{n}\bar{Z} \xrightarrow{D} \mathcal{N}[0, V(Z)]$ and $\bar{W} \xrightarrow{P} E(W)$, so by S4,

$$\sqrt{n}(T - \theta) \xrightarrow{D} \mathcal{N}(0, \phi^2),$$

with $\phi^2 = V(Z)/[E^2(W)] = 1/V(Z)$. Equivalently,

$$T \overset{A}{\sim} \mathcal{N}\{\theta, 1/[nV(Z)]\}.$$

Observe that the asymptotic variance of T is at the lower bound for unbiased estimation of θ, which is an attractive property. Indeed there is an asymptotic version of the Cramér-Rao Inequality that says that the asymptotic variance of a consistent estimator cannot be less than the CRI lower bound. So the ZES-rule estimator is a BAN estimator. In this sense, of all the analogies to draw on for an estimator of θ, the ZES rule is the best. Of course, to use it, we must have knowledge of the form of the population pdf or pmf.

Example. Continue the preceding exponential example. Recall (Section 8.3) that $W = k\lambda\bar{Y} \sim \chi^2(k)$ with $k = 2n$. (Caution: Do not confuse this W with the information variable.) Using the results for $E(1/W)$ and $E(1/W^2)$ reported in Section 8.5, we find

$$E(T) = n\lambda/(n - 1), \qquad V(T) = n^2\lambda^2/[(n - 1)^2(n - 2)].$$

While $E(T) > \lambda$ and $V(T) > \lambda^2/n$, we see for large n that $E(T) \cong \lambda$ and also $V(T) \cong \lambda^2/n$, which is the CRI lower bound.

12.4. Maximum Likelihood Estimation

There is another approach, which is better known, that produces ZES-rule estimation.

Consider a population in which the random variable Y has pdf or pmf $f(y; \theta)$, with the function f known but the parameter θ unknown. Under random sampling, sample size n, the joint pdf for the sample is

$$g_n(y_1, \ldots, y_n; \theta) = \prod_i f(y_i; \theta).$$

We are accustomed to reading this as a function of $y_1, \ldots, y_n$ for given θ, but mathematically it can also be read as a function of θ for given $y_1, \ldots, y_n$. When that is done we refer to it as the *likelihood function* for θ:

$$\mathcal{L} = \mathcal{L}(\theta; y_1, \ldots, y_n) = g_n(y_1, \ldots, y_n; \theta) = \prod_i f(y_i; \theta).$$

The *maximum likelihood*, or ML, *estimator* of θ is the value for θ that maximizes the sample likelihood function $\mathcal{L}$. Now to maximize $\mathcal{L}$ we may as well maximize its logarithm,

$$\log \mathcal{L} = \sum_i \log f(y_i; \theta) = \sum_i L_i,$$

where $L_i = \log f(y_i; \theta)$. Differentiating with respect to θ gives

$$\partial \log \mathcal{L}/\partial\theta = \sum_i \partial L_i/\partial\theta = \sum_i Z_i = \sum_i z(Y_i; \theta).$$

Setting this at zero gives $\Sigma_i z(y_i; T) = 0$, which remains to be solved for the ML estimator T. Observe that this first-order condition (FOC) is precisely the equation for ZES-rule estimation.

Example. Suppose $Y \sim$ Bernoulli(θ), so $Z = (Y - \theta)/[\theta(1 - \theta)]$. Then

$$\sum_i Z_i = \sum_i (Y_i - \theta)/[\theta(1 - \theta)].$$

The FOC chooses T to make

$$\sum_i \hat{Z}_i = \sum_i (Y_i - T)/[T(1 - T)] = 0,$$

that is, to make $\Sigma_i(Y_i - T) = 0$. So $\Sigma_i Y_i = nT$, whence $T = \bar{Y}$. It can be confirmed that this locates a maximum. So the ML estimator of the Bernoulli parameter θ is the sample mean.

We have seen how the ML principle, or, for that matter, the analogy principle applied to the ZES rule, constructively provides an estimator for an unknown parameter when the population family is known.

Indeed, there is another analogy that leads directly to maximizing the logarithm of the likelihood function. In the population, θ can be characterized as the value for c that maximizes the expectation of the log-likelihood variable $L = \log f(Y; c)$. The argument runs as follows. Let

$$D(c) = \log f(y; c) - \log f(y; \theta) = \log [f(y; c)/f(y; \theta)].$$

Because logarithm is a convex function, we see by Jensen's Inequality (Section 3.5) that

$$E[D(c)] \leq \log E[f(y; c)/f(y; \theta)]$$

with equality if $c = \theta$. But

$$E[f(y; c)/f(y; \theta)] = \int [f(y; c)/f(y; \theta)]f(y; \theta)\, dy = \int f(y; c)\, dy = 1,$$

using the fact that $f(y; c)$ is a pdf. So $E[D(c)] \leq \log(1) = 0$, with equality if $c = \theta$. This says that θ is the value for c that maximizes the population mean log-likelihood variable. As we have seen, the ML estimator has the corresponding property in the sample: the ML estimator T is the value for c that maximizes the sample sum (hence the sample mean) of the log-likelihood variable.

An advantage of this alternative analogy is that it resolves a choice that arises when the FOC has multiple solutions, as may happen in nonlinear cases. The choice is resolved by taking the solution that globally maximizes the sample mean log-likelihood variable.

We restate the properties of ML estimators. Consider random sampling from a population whose score variable is $Z = \partial \log f(Y; \theta)/\partial\theta$ and whose information variable is $W = -\partial Z/\partial\theta$.

If T is the ML estimator of θ, then:

$$T \xrightarrow{P} \theta,$$

$$\sqrt{n}(T - \theta) \xrightarrow{D} \mathcal{N}(0, \phi^2), \text{ where } \phi^2 = 1/V(Z) = 1/E(W),$$

$$T \overset{A}{\sim} \mathcal{N}(\theta, \phi^2/n),$$

T is a BAN estimator of θ.

A convenient property of ML estimation is *invariance*: If $\alpha = h(\theta)$ is a monotonic function of θ, and T is the ML estimator of θ, then $A = h(T)$ is the ML estimator of α. Examples: When $\bar{Y}$ is the ML estimator of μ, then (provided that $\mu \neq 0$) $1/\bar{Y}$ is the ML estimator of $1/\mu$; when S^2 is the ML estimator of σ^2, then S is the ML estimator of σ.

Exercises

12.1 The random variable X has the power distribution on the interval $[0, 1]$. That is, the pdf of X is

$$f(x; \theta) = \theta x^{\theta-1} \quad \text{for } 0 \leq x \leq 1,$$

with $f(x; \theta) = 0$ elsewhere. The parameter θ is unknown. Consider random sampling, sample size n.

(a) Show that the maximum likelihood estimator of θ is $T = 1/\bar{Y}$, where $Y = -\log X$. (As usual, "log" denotes natural logarithm.)
(b) Find the asymptotic distribution of T, in terms of θ and n only.

12.2 The random variable Y has the exponential distribution with parameter λ.

(a) Recalling (from Table 3.1) what is known about $E(Y)$ and $V(Y)$, calculate $E(Z)$, $V(Z)$, and $E(W)$. Do your results satisfy the rule $E(W) = V(Z)$?
(b) Complete the sentence: In random sampling, sample size n, from this population, for T to be an unbiased estimator of λ, its variance must be greater than or equal to ———.

12.3 The random variable Y has the Poisson distribution with parameter λ.

(a) Explain why the sample mean and the sample variance are distinct analog estimators of λ.
(b) Determine which of them is the ZES-rule estimator.

12.4 Consider random sampling, sample size n, from the exponential(λ) distribution. Let $T = 1/\bar{Y}$ and $T^* = (n - 1)T/n$.

(a) Find $E(T)$, $V(T)$, $E(T^*)$, and $V(T^*)$.
(b) Comment on your results in the light of the CRI.
(c) Find the MSE of T and T^* as estimators of λ.

13 *Estimating a Population Relation*

13.1. Introduction

We are now prepared to address systematically the questions raised in Chapter 1 about estimating the relation between Y and X in a bivariate population.

By way of review, first consider estimating a population mean in random sampling from a univariate population. We have learned that the sample mean $\bar{Y}$ is an unbiased and consistent estimator of the population mean μ, and that $\bar{Y} \overset{A}{\sim} \mathcal{N}(\mu, \sigma^2/n)$. At least two analogies lead to $\bar{Y}$ as the estimator of μ. First, the population mean is the best constant predictor of Y in the population: μ is the value for c that minimizes $E(U^2)$, where $U = Y - c$. The sample mean has the analogous property in the sample: $\bar{Y}$ is the value for c that minimizes $\Sigma_i u_i^2/n$, where $u_i = y_i - c$. Second, μ is the value for c that makes $E(U) = 0$ in the population. The sample mean has the analogous property in the sample: $\bar{Y}$ is the value for c that makes $\Sigma_i u_i/n = 0$.

Next consider a bivariate population, in which the population linear projection is $E^*(Y|X) = \alpha + \beta X$, with

$$\beta = \sigma_{XY}/\sigma_X^2, \qquad \alpha = \mu_Y - \beta\mu_X.$$

In random sampling, consider the sample linear projection, $\hat{Y} = A + BX$, with

$$B = S_{XY}/S_X^2, \qquad A = \bar{Y} - B\bar{X}.$$

We have learned (Sections 10.5 and 10.6) that the sample slope B consistently estimates β, and that

$$B \overset{A}{\sim} \mathcal{N}(\beta, \phi^2/n),$$

where $\phi^2 = E(X^{*2}U^2)/V^2(X)$ and $U = Y - E^*(Y|X)$. A similar result holds for the intercept A. At least two analogies lead to the line $\hat{Y} = A + BX$ as the estimator of $E^*(Y|X) = \alpha + \beta X$. First, the population linear projection is the best linear predictor of Y given X in the population: it minimizes $E(U^2)$, where now $U = Y - (a + bX)$. The sample linear projection has the analogous property in the sample: it minimizes $\Sigma_i u_i^2/n$, where now $u_i = y_i - (a + bx_i)$. This may be referred to as the *least-squares analogy*. Second, the population LP is the line such that $E(U) = 0$ and $E(XU) = 0$ in the population. The sample LP has the analogous properties in the sample: it makes $\Sigma_i u_i/n = 0$ and $\Sigma_i x_i u_i/n = 0$. This may be referred to as the *instrumental-variable*, or orthogonality, *analogy*.

With this background, we can proceed to estimation of the conditional expectation function $E(Y|X)$. We will suppose that the functional form of the CEF is known; that is, $E(Y|X) = h(X; \boldsymbol{\theta})$, where the function $h(X; \boldsymbol{\theta})$ is known up to the values of one or more parameters, the element(s) of the vector $\boldsymbol{\theta}$.

13.2. Estimating a Linear CEF

Suppose that the population CEF is known to be linear. Then the CEF coincides with the BLP: that is, $E(Y|X) = \alpha + \beta X$, with $\beta = \sigma_{XY}/\sigma_X^2$ and $\alpha = \mu_Y - \beta\mu_X$. The two analogies apply, so the natural estimator is again the sample LP, namely $\hat{Y} = A + BX$, for which the asymptotic distribution is as above.

In fact, when $E(Y|X)$ is linear, A and B are *unbiased* estimators of α and β. We show

THEOREM. In random sampling from a population in which $E(Y|X) = \alpha + \beta X$, the sample intercept A and slope B are unbiased estimators of α and β.

This is a surprising result because A and B are nonlinear functions of the sample moments.

Proof. Begin with some algebra:

$$S_X^2 = (1/n) \sum_i (X_i - \bar{X})^2 = \sum_i X_i^2/n - \bar{X}^2 = (1/n) \sum_i (X_i - \bar{X})X_i,$$

$$S_{XY} = (1/n) \sum_i (X_i - \bar{X})(Y_i - \bar{Y}) = \sum_i X_iY_i/n - \bar{X}\bar{Y}$$

$$= (1/n) \sum_i (X_i - \bar{X})Y_i,$$

$$B = S_{XY}/S_x^2 = \sum_i (X_i - \bar{X})Y_i \Big/ \sum_i (X_i - \bar{X})^2$$

$$= \sum_i \left[(X_i - \bar{X}) \Big/ \sum_{h=1}^{n} (X_h - \bar{X})^2\right] Y_i$$

$$= \sum_i W_iY_i,$$

say, where the random variables

$$W_i = (X_i - \bar{X}) \Big/ \sum_{h=1}^{n} (X_h - \bar{X})^2 \qquad (i = 1, \ldots, n)$$

are functions only of the X_i's. As a matter of algebra,

$$\sum_i W_i = 0, \qquad \sum_i W_iX_i = 1, \qquad \sum_i W_i^2 = 1 \Big/ \left[\sum_i (X_i - \bar{X})^2\right].$$

Now condition on a given set of observations on the X's, that is, condition on $\mathbf{X} = \mathbf{x} = (x_1, \ldots, x_n)'$. Conditional on $\mathbf{X} = \mathbf{x}$, the values of the W_i are constants, which we write as

$$w_i = (x_i - \bar{x}) \Big/ \sum_{h=1}^{n} (x_h - \bar{x})^2 \qquad (i = 1, \ldots, n).$$

The expectation of the slope B conditional on $\mathbf{x}$ is

$$E(B|\mathbf{x}) = E\left(\sum_i W_iY_i|\mathbf{x}\right) = \sum_i E(W_iY_i|\mathbf{x}) = \sum_i w_iE(Y_i|\mathbf{x})$$

$$= \sum_i w_i(\alpha + \beta x_i) = \alpha \sum_i w_i + \beta \sum_i w_ix_i = \beta,$$

because $\Sigma_i w_i = 0$ and $\Sigma_i w_ix_i = 1$ by the algebra above. We have $E(B|\mathbf{x}) = \beta$ for all $\mathbf{x}$, so B is mean-independent of $\mathbf{X}$, and $E(B) = \beta$.

Similarly, for the intercept $A = \bar{Y} - B\bar{X}$, we have

$$E(A|\mathbf{x}) = E(\bar{Y}|\mathbf{x}) - E(B|\mathbf{x})\bar{x} = (\alpha + \beta\bar{x}) - \beta\bar{x} = \alpha.$$

We have $E(A|\mathbf{x}) = \alpha$ for all $\mathbf{x}$, so $E(A) = \alpha$. ■

There is a subtlety in the derivation, namely the step that equates $E(Y_i|\mathbf{x})$ to $E(Y_i|x_i) = \alpha + \beta x_i$. This step is justified by the random sampling assumption. To clarify what is involved, it suffices to show why $E(Y_1|x_1, x_2) = E(Y_1|x_1)$. Consider the conditional pdf of Y_1 given both x_1 and x_2:

$$g(y_1|x_1, x_2) = f(y_1, x_1, x_2)/f_2(x_1, x_2).$$

Independence across the observations implies that the trivariate density in the numerator factors into

$$f(y_1, x_1, x_2) = g_1(y_1, x_1)f_1(x_2),$$

and, in conjunction with "identically distributed," implies that the bivariate density in the denominator factors into

$$f_2(x_1, x_2) = f_1(x_1)f_1(x_2).$$

So

$$g(y_1|x_1, x_2) = g_1(y_1, x_1)/f_1(x_1) = g_2(y_1|x_1).$$

When two distributions are the same, their expectations are the same. That is,

$$E(Y_1|x_1, x_2) = E(Y_1|x_1).$$

This calculation extends to further conditioning, and of course to $i = 2, \ldots, n$. Thus the step from $E(Y_i|x_i) = \alpha + \beta x_i$ to $E(Y_i|\mathbf{x}) = \alpha + \beta x_i$ is justified under random sampling. Observe how linearity of the CEF is crucial to the argument.

As for the variance of B, under random sampling from a population with linear CEF, we have:

$$V(B|\mathbf{x}) = V\left(\sum_i W_i Y_i|\mathbf{x}\right) = \sum_i w_i^2 V(Y_i|\mathbf{x}) = \sum_i w_i^2 V(Y_i|x_i),$$

and $V(B)$ will equal the expectation (over all $\mathbf{x}$) of those conditional variances, using the Analysis of Variance formula (T10, Section 5.2), with $V_{\mathbf{X}}[E(B|\mathbf{X})] = 0$.

A particularly sharp result is obtained if Y is *variance-independent* of X, that is, if the conditional variance function $V(Y|X)$ is constant. This will be referred to as the *homoskedastic* case. If $V(Y|X) = \sigma^2$, say, for all X, then

$$V(B|\mathbf{x}) = \sum_i w_i^2\sigma^2 = \sigma^2 \sum_i w_i^2 = \sigma^2 \Big/ \sum_i (x_i - \bar{x})^2 = (\sigma^2/n)(1/s_X^2),$$

where $s_X^2 = \Sigma_i(x_i - \bar{x})^2/n$ is the sample variance of X. We conclude that

$$V(B) = E_{\mathbf{X}}[V(B|\mathbf{X})] = (\sigma^2/n)E(1/S_X^2).$$

For large n, this exact variance is approximately (σ^2/n) $(1/\sigma_X^2)$, which is indeed the asymptotic variance of B found in Section 10.6.

In this homoskedastic case, with σ^2 and $E(1/S_X^2)$ unknown, we get a standard error for B by taking the square root of $\hat{V}(B) = S^2/(nS_X^2)$, with $S^2 = \Sigma_i e_i^2/n$ and $e_i = Y_i - A - BX_i$. It can be shown that S^2 and $1/S_X^2$ are consistent for σ^2 and $1/\sigma_X^2$.

13.3. Estimating a Nonlinear CEF

The preceding theory for estimating linear CEF's applies also to some nonlinear CEF's. For example, if $E(Y|X) = \alpha + \beta X^2$, then the theory surely applies to $E(Y|Z) = \alpha + \beta Z$, where $Z = X^2$. Similarly if $E(Y|X) = \alpha + \beta/X$, then the theory applies to $E(Y|Z) = \alpha + \beta Z$, with $Z = 1/X$. What is critical, it now appears, is that the CEF be linear in the unknown parameters α, β.

But suppose that the population CEF is nonlinear in unknown parameters. For example, suppose that we know $E(Y|X) = \exp(\theta_1 + \theta_2 X)$, with θ_1 and θ_2 unknown. How shall we estimate θ_1 and θ_2?

Again we appeal to the least-squares and instrumental-variable analogies.

(1) The CEF is the best predictor. In particular, it is the best predictor of the form $h(X; c_1, c_2) = \exp(c_1 + c_2X)$. In the population, θ_1 and θ_2 are the values for c_1 and c_2 that minimize $E(U^2)$, where

$$U = Y - \exp(c_1 + c_2X) = Y - h(X; c_1, c_2).$$

So in the sample, let

$$u_i = y_i - \exp(c_1 + c_2x_i),$$

and choose c_1, c_2 to minimize $(1/n)\ \Sigma_i u_i^2$ or, equivalently, to minimize the criterion $\phi = \phi(c_1, c_2) = \Sigma_i u_i^2$. The derivatives are

$$\partial\phi/\partial c_1 = 2 \sum_i u_i(\partial u_i/\partial c_1) = -2 \sum_i u_i h_i,$$

$$\partial\phi/\partial c_2 = 2\sum_i u_i(\partial u_i/\partial c_2) = -2\sum_i u_i h_i x_i,$$

where $h_i = h(x_i;\, c_1, c_2)$. So the FOC's are

$$\sum_i h_i u_i = 0, \qquad \sum_i h_i x_i u_i = 0.$$

On the proviso that these locate a minimum, we have a pair of nonlinear equations to be solved for the *nonlinear least squares*, or NLLS, *estimators* of θ_1, θ_2.

(2) The deviations from the CEF have zero expectation and zero covariance with X. That is, let $U = Y - E(Y|X)$; then $E(U) = 0$ and $E(XU) = 0$. So let us choose as estimates of θ_1, θ_2, the values of c_1, c_2 that make $\Sigma_i u_i/n = 0$ and $\Sigma_i x_i u_i/n = 0$. Equivalently, we choose them to satisfy

$$\sum_i u_i = 0, \qquad \sum_i x_i u_i = 0.$$

This is a pair of nonlinear equations to be solved for the *instrumental-variable*, or IV, *estimators* of θ_1, θ_2.

In the linear CEF case, where $h(X;\, c_1, c_2) = c_1 + c_2 X$, the two analogies produce the same estimators, because $\partial u_i/\partial c_1 = -1$ and $\partial u_i/\partial c_2 = -x_i$. Further, in the linear CEF case we have explicit solutions. In the nonlinear CEF case, NLLS and IV estimators do not coincide, and further we will need to rely on numerical solutions. It is not hard to show that both our analog estimators are consistent (though not unbiased), and to obtain their asymptotic distributions: the derivation is similar to that used for the ZES-rule estimator (Section 12.3). Which analog estimator is preferable may depend on the population family. If the conditional distributions of Y given X are normal with constant variance (that is, if $U \sim \mathcal{N}(0, \sigma^2)$ independently of X), and the marginal distribution of X does not contain the parameters θ_1, θ_2, σ^2, then it is easy to verify that the NLLS estimators are also ML, and hence BAN.

Observe that NLLS estimation can itself be viewed as a type of IV estimation. The NLLS FOC's $\Sigma_i h_i u_i/n = 0$ and $\Sigma_i h_i x_i u_i/n = 0$ are the sample analogs of $E[h(X)U] = 0$ and $E[g(X)U] = 0$, where $g(X) = h(X)X$. Since deviations from a CEF have zero expected cross-product with *every* function of X (Section 5.3), such sample analogs are legitimate.

13.4. Estimating a Binary Response Model

To illustrate the opportunities that arise when more is specified about the population, we take up a leading example of a nonlinear CEF, namely a binary response model, more specifically the *probit model*. Here Y is a binary variable, one that takes on only the values 0 and 1, and

$$E(Y|X) = F(\theta_1 + \theta_2 X),$$

where $F(\cdot)$ is the standard normal cdf. Section 6.3 contains a story that leads to this model.

We consider three estimators for the probit model: nonlinear least squares, instrumental variables, and maximum likelihood.

For NLLS, one minimizes the criterion $\phi = \phi(c_1, c_2) = \Sigma_i u_i^2$, where

$$u_i = y_i - F_i, \qquad F_i = F(v_i), \qquad v_i = c_1 + c_2 x_i.$$

Let $f(\cdot)$ denote the standard normal pdf, so $f_i = f(v_i) = \partial F_i/\partial v_i$. The derivatives are

$$\partial\phi/\partial c_1 = 2 \sum_i u_i(\partial u_i/\partial c_1) = -2 \sum_i u_i f_i,$$

$$\partial\phi/\partial c_2 = 2 \sum_i u_i(\partial u_i/\partial c_2) = -2 \sum_i u_i f_i x_i.$$

So the FOC's are

$$\sum_i f_i u_i = 0, \qquad \sum_i f_i x_i u_i = 0.$$

On the proviso that these locate a minimum, we have a pair of nonlinear equations to be solved for the NLLS estimators of θ_1 and θ_2.

For IV estimation, we seek the values of c_1, c_2 that make $\Sigma_i u_i/n = 0$ and $\Sigma_i x_i u_i/n = 0$. Equivalently, we choose them to satisfy

$$\sum_i u_i = 0, \qquad \sum_i x_i u_i = 0.$$

This is a different pair of nonlinear equations to be solved, for the IV estimators of θ_1, θ_2.

Maximum-likelihood, or ZES-rule, estimation is available because the probit model automatically specifies the form of the conditional distribution of Y given X. Because Y is a binary variable with $E(Y|X) = F(\theta_1 + \theta_2 X)$, it is clear that conditional on X, the variable Y has a Bernoulli distribution with parameter $F(\theta_1 + \theta_2 X)$. Because the parameters θ_1 and θ_2 do not appear in the marginal distribution for X, to maximize the likelihood it suffices to maximize the conditional likelihood. Adapting an example in Section 12.2, we see that the conditional log-likelihood variable is

$$L_i = y_i \log F_i + (1 - y_i) \log(1 - F_i).$$

With two parameters being estimated, there are two score variables:

$$\begin{aligned} Z_{1i} = \partial L_i/\partial\theta_1 &= (y_i/F_i)f_i - [(1 - y_i)/(1 - F_i)]f_i \\ &= w_i(y_i - F_i) \quad = w_i u_i, \\ Z_{2i} = \partial L_i/\partial\theta_2 &= w_i x_i(y_i - F_i) = w_i x_i u_i, \end{aligned}$$

say, where $w_i = f_i/[F_i(1 - F_i)]$. Both have expectation zero. So we choose c_1 and c_2 to satisfy

$$\sum_i w_i u_i = 0, \qquad \sum_i w_i x_i u_i = 0,$$

which are yet another pair of nonlinear equations to be solved, for the ML estimators of θ_1, θ_2.

The three estimators are distinct and will differ in any sample. All three are consistent (though not unbiased) for θ_1 and θ_2. For the probit model, one can resolve the choice among the estimators by appealing to the BAN property of ML estimation. More on all this in Section 29.5.

13.5. Other Sampling Schemes

Thus far, we have confined attention to random sampling. But other sampling schemes may be relevant in practice. We explore the possibilities for estimation of the population relation between Y and X, when the observations are not randomly drawn from the bivariate population $f(x, y) = g_2(y|x)f_1(x)$.

Selective Sampling

Suppose that the sampling is *explicitly selective* on X alone in the sense that the probability that a particular (X, Y) draw will be retained depends only on X. Let $\psi(X)$ = probability of retention as a function of X. Then the relevant marginal pdf of X is no longer $f_1(x)$ but rather

$$f_1^*(x) = \psi(x)f_1(x)\Big/\int \psi(x)f_1(x)\,dx.$$

(Note: In this chapter the symbol $\int$ is shorthand for $\int_{-\infty}^{\infty}$.)

Example. For studying the relation of savings y to income x, we might be oversampling high-income households by using $\psi(x) = 0.5$ for $x \leq d$, $\psi(x) = 1.0$ for $x > d$, where d is a prespecified level of income.

By assumption, $g_2(y|x)$ is not affected by this selective sampling scheme, so the new joint pdf is

$$f^*(x, y) = g_2(y|x)f_1^*(x).$$

If the successive observations are independent, then we are in effect randomly sampling (X, Y) from a new, selected, population. Because the conditional pdf $g_2(y|x)$ has not changed, neither has the CEF $E(Y|X)$, so the theory of the preceding sections applies. Because the two CEF's are the same, estimators of the new CEF will serve as estimators of the original CEF.

This argument does not carry over to BLP estimation (unless the CEF is linear). The explicit-on-X selection produces implicit-on-Y selection. The marginal pdf of Y changes from $f_2(y)$ to

$$f_2^*(y) = \int f^*(x, y)\,dx = \int g_2(y|x)f_1^*(x)\,dx.$$

Presumably the marginal expectations and variances of both variables, and their covariance, are different in the selected population. If so, the BLP $E^*(Y|X)$ is presumably different. Another way to see this is to recall (T13, Section 5.5) the best linear approximation property of the BLP: it minimizes $E(W^2)$, where $W = E(Y|X) - (a + bX)$, and the expectation is taken over the marginal distribution of X. Because $f_1^*(x)$ differs from $f_1(x)$, one should presume that the values of a and b that minimize $\int w^2 f_1^*(x)\,dx$ differ from those that minimize $\int w^2 f_1(x)\,dx$. If the two BLP's are different, then the sample LP, which will consistently estimate the BLP of the new population, will not serve for the original BLP.

This negative conclusion also applies to the "reverse" CEF, namely $E(X|Y)$. The new conditional pdf for X given Y, namely

$$g_1^*(x|y) = f^*(x, y)/f_2^*(y),$$

presumably differs from $g_1(x|y)$. So the new $E(X|Y)$ presumably differs from the original one, and the results obtained in sampling from the selected population are inappropriate for estimating the original $E(X|Y)$.

We conclude that explicit selection on X alone affects both of the BLP's, and $E(X|Y)$ as well; these effects are sometimes labeled *selection bias*. But it does not affect $E(Y|X)$—so random sampling from the original joint probability distribution is not needed when a CEF is the target.

Varying Marginals

Another departure from random sampling arises if the marginal distribution of X changes from observation to observation. There is no longer a single bivariate population from which the sample is drawn. If the observations are independent, the joint pdf of the sample x's becomes

$$g(x_1, \ldots, x_n) = \prod_i f_{1i}(x_i),$$

where the $f_{1i}(\cdot)$ functions vary over i. Provided that the conditional pdf $g_2(y|x)$ is the same at all observations, then the CEF $E(Y|X)$ will remain the same at each observation. If so, then least squares remains appropriate, and is indeed unbiased if the CEF is linear.

Since there is no longer a single bivariate population, best linear prediction in the population is not well-defined, unless one uses $\Sigma_i f_{1i}(x_i)/n$, say, as a marginal pmf for X. To assess asymptotics, one needs further specification: how do the $f_{1i}(\cdot)$'s develop as n grows?

Nonstochastic Explanatory Variable

An extreme special case of the varying-marginals scheme arises if at each observation the marginal distribution of X is degenerate, that is to say, $X_1, \ldots, X_n$ are constants (not all equal to each other, of course). In the econometric literature, this is known as the *nonstochastic* (or nonrandom, or fixed) *explanatory variable* case. Another description is *stratified sampling*, with the values of X defining the strata, or subpopulations. If the CEF is linear, and the observations are independent, then least squares estimation is unbiased. To verify this, return to Section 13.2, and utilize the fact that there is only one possible value of the vector $\mathbf{x} = (x_1, \ldots, x_n)'$, so the conditioning can be suppressed.

There is no longer a single bivariate population, but a population BLP might be defined by using the empirical frequency distribution of the n values x_i as a marginal pmf for X. To consider asymptotic properties, some supplementary information is needed: how does the x_i series develop as n grows?

It is sometimes said that the nonstochastic explanatory variable case requires that the researcher "controls," "sets," or "manipulates" the values of the conditioning variable. This is ambiguous or misleading. For example, if X denotes gender, a researcher may decide to collect a sample consisting of 50 men, followed by 25 women. If so, the sample values of X are nonstochastic as required, but the researcher has not controlled, set, or manipulated the gender of any individual in the population.

Exercises

13.1 Consider a random sample of size n from the joint distribution of (X, Y), where $Y|X$ is Bernoulli with $E(Y|X) = F(\theta X)$, with $F(\cdot)$ being the $\mathcal{N}(0, 1)$ cdf. Determine whether the following statement is true or false: The ZES-rule estimator of θ is the value for c that satisfies $\Sigma_i\{x_i[y_i - F(cx_i)]\} = 0$.

13.2 Consider a random sample of size n from the joint distribution of (X, Y), where $Y|X$ is exponential with parameter $\lambda = 1/\exp(\theta X)$. Determine whether the following statement is true or false: The maximum likelihood estimator of θ is the value c that minimizes $\Sigma_i[y_i - \exp(cx_i)]^2$.

13.3 In Exercise 5.8, we introduced the best proportional predictor, or BPP, of Y given X. It is $E^{**}(Y|X) = \gamma X$, where $\gamma = E(XY)/E(X^2)$. For estimating γ, the analogy principle suggests the statistic $T = \Sigma_i X_i Y_i/\Sigma_i X_i^2$. Assume random sampling.

(a) Show that T is a consistent estimator of γ.
(b) Show that $T \overset{A}{\sim} \mathcal{N}(\gamma, \phi^2/n)$, where

$$\phi^2 = E(U^2X^2)/E^2(X^2) = \phi_2^2/\phi_1^2,$$

say, with

$$U = Y - \gamma X,$$

$$\phi_1^2 = E^2(X^2),$$

$$\phi_2^2 = E(U^2X^2) = E[(Y^2 - 2\gamma XY + \gamma^2X^2)X^2]$$

$$= E(X^2Y^2) - 2\gamma E(X^3Y) + \gamma^2E(X^4).$$

(c) Propose a consistent estimator of ϕ^2, for use in constructing an approximate confidence interval for γ.

13.4 Continuing Exercise 13.3, suppose that we have a random sample of size $n = 100$ from the joint distribution of (X, Y). In the population, X takes on only the values 1, 2, 3, 4, 5, and Y takes on only the values 0, 1. The sample observations are

	X				
Y	1	2	3	4	5
0	11	2	5	9	2
1	21	7	9	18	16

Construct an approximate 95% confidence interval for γ, the slope of the BPP of Y given X.

14 *Multiple Regression*

14.1. Population Regression Function

In most economic contexts, the relation of interest involves more than two variables. Economists might consider how the output of a firm is related to its inputs of labor, capital, and raw materials, or how the earnings of a worker are related to her age, education, race, region of residence, and years of work experience. So we will move from simple regression (one conditioning variable) to *multiple regression* (several conditioning variables).

The setting for this is a multivariate population in which the k-variate random vector $(Y, X_2, \ldots, X_k)$ has joint pdf (or pmf) $f(y, x_2, \ldots, x_k)$. The conditional probability distribution of Y given $X_2, \ldots, X_k$ is described by the conditional pdf (or pmf)

$$g(y|x_2, \ldots, x_k) = f(y, x_2, \ldots, x_k)/f_1(x_2, \ldots, x_k).$$

Here $f_1(x_2, \ldots, x_k)$ is the "joint-marginal" pdf (or pmf) of $X_2, \ldots, X_k$: it is "marginal" in that it is integrated (or summed) over one variable, but "joint" in that it still refers to several variables.

The conditional expectation function, or population regression function, of Y given $X_2, \ldots, X_k$ is

$$E(Y|X_2, \ldots, X_k) = \int_{-\infty}^{\infty} y g(y|x_2, \ldots, x_k)\, dy.$$

The CEF traces out the path of the conditional means of Y across subpopulations defined by the values of the X's. As in the bivariate case (Sections 5.3 and 5.4), the CEF has some distinctive characteristics in the population.

(1) The CEF is the best predictor of Y given the X's. That is, let $U = Y - h(\ )$, where $h(\ ) = h(X_2, \ldots, X_k)$ is any function of the X's; to minimize $E(U^2)$, choose $h(\ ) = E(Y|X_2, \ldots, X_k)$.

(2) The deviation from the CEF has zero expectation and is mean-independent of all of the X's. That is, let $\epsilon = Y - E(Y|X_2, \ldots, X_k)$; then $E(\epsilon|X_2, \ldots, X_k) = 0$. It follows by the Law of Iterated Expectations (T8, Section 5.2), that $E(\epsilon) = 0$ and $E(\epsilon|X_j) = 0$ for $j = 2, \ldots, k$. And so it follows that ϵ is uncorrelated with every function of the X's.

There is another feature of interest in any multivariate probability distribution, namely the population linear projection, or BLP, of Y on $X_2, \ldots, X_k$:

$$E^*(Y|X_2, \ldots, X_k) = \beta_1 + \beta_2 X_2 + \ldots + \beta_k X_k,$$

where the β's are chosen to minimize expected squared deviations of Y. That is, let $U = Y - h(\ )$, where $h(\ )$ is any *linear* function of the X's. If we choose $h(\ )$ to minimize $E(U^2)$, the solution is $E^*(Y|X_2, \ldots, X_k)$. More explicitly, write $U = Y - (c_1 + \Sigma_{j=2}^{k} c_j X_j)$. Then

$$\partial E(U^2)/\partial c_1 = -2E(U),$$

$$\partial E(U^2)/\partial c_j = -2E(X_j U) \qquad (j = 2, \ldots, k).$$

Equating these k derivatives to zero to locate the minimum gives the first-order conditions

$$E(U) = 0, \qquad E(X_j U) = 0 \qquad (j = 2, \ldots, k),$$

which taken together are equivalent to

$$E(U) = C(X_2, U) = \cdots = C(X_k, U) = 0.$$

Substituting for U gives the system of k linear equations that determine the values of the k β's in terms of population means, variances, and covariances of $Y, X_2, \ldots, X_k$.

As in the bivariate case (Sections 5.4 and 5.5), the BLP has some distinctive characteristics in the population.

(1) The BLP is the best linear approximation to the CEF. That is, let $W = r(\ ) - h(\ )$, where $r(\ )$ is the CEF and $h(\ )$ is any linear function of the X's; to minimize $E(W^2)$, choose $h(\ ) = E^*(Y|X_2, \ldots, X_k)$.

(2) The deviation from the BLP has zero expectation and is uncorrelated with each of the X's, as shown in the FOC's.

As in the bivariate case, we draw on the analogy principle to suggest an estimator of the BLP, or equivalently of a linear CEF. Suppose that we have a sample of n observations from the multivariate population. These data take the form

$$\begin{matrix} y_1 & x_{12} & \cdots & x_{1j} & \cdots & x_{1k} \\ \vdots & \vdots & & \vdots & & \vdots \\ y_i & x_{i2} & \cdots & x_{ij} & \cdots & x_{ik} \\ \vdots & \vdots & & \vdots & & \vdots \\ y_n & x_{n2} & \cdots & x_{nj} & \cdots & x_{nk} \end{matrix}$$

The first subscript indexes the observations ($i = 1, \ldots, n$); the second subscript indexes the conditioning variables ($j = 2, \ldots, k$). The aim is to process these data to get estimates of the BLP parameters, the β's. The least-squares analogy suggests that we take as the estimates of the β's, the values for the c's that minimize the criterion

$$\phi = \phi(c_1, c_2, \ldots, c_k) = \sum_{i=1}^{n} u_i^2,$$

where

$$u_i = y_i - (c_1 + c_2 x_{i2} + \cdots + c_k x_{ik}).$$

Solving that problem may be referred to as "running the LS linear regression" of Y on $X_2, \ldots, X_k$.

This minimization problem is a purely algebraic one that can be posed without reference to population CEF's or BLP's: find the best-fitting line in a body of data, where fit is measured in terms of sum of squared sample deviations. In the remainder of this chapter, we explore LS linear regression in isolation from its probability setting.

14.2. Algebra for Multiple Regression

It is convenient to define a variable X_1, called "the constant," that is equal to 1 at all observations, and thus to add a column with elements

$x_{i1} = 1$ to the display of the data above. Given the n observations on Y, $X_1, \ldots, X_k$, the criterion to be minimized is $\phi = \Sigma_{i=1}^{n} u_i^2$, where

$$u_i = y_i - (c_1 x_{i1} + c_2 x_{i2} + \cdots + c_k x_{ik}).$$

Differentiating ϕ with respect to the c_j (for $j = 1, \ldots, k$) gives

$$\partial\phi/\partial c_j = \sum_i (\partial u_i^2/\partial c_j) = \sum_i 2u_i(\partial u_i/\partial c_j) = 2 \sum_i u_i(-x_{ij})$$

$$= -2 \sum_i x_{ij} u_i.$$

So the first-order conditions for the minimum are

$$\sum_i x_{ij} u_i = 0 \qquad (j = 1, \ldots, k).$$

This is a system of k linear equations in $c_1, \ldots, c_k$.

At this point, it is convenient to adopt a vector notation. Define the $n \times 1$ vectors

$$\mathbf{y} = \{y_i\}, \qquad \mathbf{x}_1 = \{x_{i1}\}, \ldots, \mathbf{x}_k = \{x_{ik}\},$$

and the $n \times 1$ vector

$$\mathbf{u} = \mathbf{y} - (\mathbf{x}_1 c_1 + \cdots + \mathbf{x}_k c_k) = \{u_i\}.$$

(Note the convention: Typical elements of vectors are identified by curly brackets.) Then the criterion may be written as

$$\phi = \mathbf{u}'\mathbf{u} = \phi(c_1, \ldots, c_k),$$

and the FOC's may be written as

$$\mathbf{x}_j'\mathbf{u} = 0 \qquad (j = 1, \ldots, k).$$

A matrix formulation is even more convenient. Define the $n \times k$ matrix

$$\mathbf{X} = (\mathbf{x}_1, \ldots, \mathbf{x}_k),$$

and the $k \times 1$ vector $\mathbf{c} = (c_1, \ldots, c_k)'$. Then the criterion may be written as

$$\phi = \mathbf{u}'\mathbf{u} = \phi(\mathbf{c}),$$

where

$$\mathbf{u} = \mathbf{y} - \mathbf{X}\mathbf{c},$$

and the FOC's may be written as

$$\mathbf{X}'\mathbf{u} = \mathbf{0}.$$

Let $\mathbf{c}^*$ denote a solution value for $\mathbf{c}$, that is, $\mathbf{X}'(\mathbf{y} - \mathbf{X}\mathbf{c}^*) = \mathbf{0}$ or, equivalently,

$$\mathbf{X}'\mathbf{X}\mathbf{c}^* = \mathbf{X}'\mathbf{y}.$$

This system of k linear equations in the k elements of $\mathbf{c}^*$ is known as the set of *normal equations* for LS linear regression. Here $\mathbf{Q} = \mathbf{X}'\mathbf{X} = \{\mathbf{x}_h'\mathbf{x}_j\}$ is the $k \times k$ symmetric matrix of sums of squares and cross-products of the explanatory variables, while $\mathbf{X}'\mathbf{y} = \{\mathbf{x}_j'\mathbf{y}\}$ is the $k \times 1$ vector of sums of cross-products of the explanatory variables with the dependent variable.

Two cases arise when we consider solving the normal equations. *Case 1*, the *full-rank* case, holds when the $k \times k$ matrix $\mathbf{Q}$ is nonsingular, equivalently when $\mathbf{Q}$ is invertible, has rank k, has determinant $|\mathbf{Q}| \neq 0$. *Case 2*, the *short-rank* case, holds when $\mathbf{Q}$ is singular, equivalently when $\mathbf{Q}$ is not invertible, has rank less than k, has determinant $|\mathbf{Q}| = 0$.

In Case 1, the normal equations $\mathbf{Q}\mathbf{c}^* = \mathbf{X}'\mathbf{y}$ have a unique solution, which we denote as

$$\mathbf{b} = \mathbf{Q}^{-1}\mathbf{X}'\mathbf{y}.$$

The claim is that $\mathbf{b}$ is the unique minimizer of the criterion: that is, $\phi(\mathbf{b}) < \phi(\mathbf{c})$ for all $\mathbf{c} \neq \mathbf{b}$. In Case 2, the normal equations $\mathbf{Q}\mathbf{c}^* = \mathbf{X}'\mathbf{y}$ have an infinity of solutions, none of which is expressible in terms of the inverse of $\mathbf{Q}$. There the claim is weaker, namely that if $\mathbf{c}^*$ is a solution, then $\phi(\mathbf{c}^*) \leq \phi(\mathbf{c})$ for all $\mathbf{c}$, with equality iff $\mathbf{c}$ also satisfies the normal equations. Both claims will be verified in Section 14.5.

Confining attention to the full-rank case, we introduce some terminology and notation. The least-squares *coefficient vector* is

$$\mathbf{b} = \mathbf{Q}^{-1}\mathbf{X}'\mathbf{y} = \mathbf{A}\mathbf{y},$$

where $\mathbf{A} = \mathbf{Q}^{-1}\mathbf{X}'$ is $k \times n$. The *fitted-value vector* is

$$\hat{\mathbf{y}} = \mathbf{X}\mathbf{b} = \mathbf{X}\mathbf{A}\mathbf{y} = \mathbf{N}\mathbf{y},$$

where $\mathbf{N} = \mathbf{XA} = \mathbf{XQ}^{-1}\mathbf{X}'$ is $n \times n$. The *residual vector* is

$$\mathbf{e} = \mathbf{y} - \hat{\mathbf{y}} = \mathbf{Iy} - \mathbf{Ny} = (\mathbf{I} - \mathbf{N})\mathbf{y} = \mathbf{My},$$

where $\mathbf{M} = \mathbf{I} - \mathbf{N} = \mathbf{I} - \mathbf{XA} = \mathbf{I} - \mathbf{XQ}^{-1}\mathbf{X}'$ is $n \times n$. The minimized value of the criterion, the *sum of squared residuals* from the least-squares line, is $\phi(\mathbf{b}) = \mathbf{e}'\mathbf{e}$.

Here are some easily verified properties of the **Q**, **A**, **N**, and **M** matrices that prove useful in the sequel:

$$\begin{aligned}
(\mathbf{Q}^{-1})' &= \mathbf{Q}^{-1}, \\
\mathbf{AX} &= (\mathbf{Q}^{-1}\mathbf{X}')\mathbf{X} = \mathbf{Q}^{-1}\mathbf{Q} = \mathbf{I}, \\
\mathbf{AA}' &= (\mathbf{Q}^{-1}\mathbf{X}')(\mathbf{XQ}^{-1\prime}) = \mathbf{Q}^{-1}\mathbf{QQ}^{-1} = \mathbf{Q}^{-1}, \\
\mathbf{N}' &= \mathbf{A}'\mathbf{X}' = (\mathbf{XQ}^{-1\prime})\mathbf{X}' = \mathbf{XQ}^{-1}\mathbf{X}' = \mathbf{N}, \\
\mathbf{NN} &= (\mathbf{XA})\mathbf{XA} = \mathbf{X}(\mathbf{AX})\mathbf{A} = \mathbf{XIA} = \mathbf{XA} = \mathbf{N}, \\
\mathbf{M}' &= \mathbf{I} - \mathbf{N}' = \mathbf{I} - \mathbf{N} = \mathbf{M}, \\
\mathbf{MM} &= (\mathbf{I} - \mathbf{N})(\mathbf{I} - \mathbf{N}) = \mathbf{I} - \mathbf{N} - \mathbf{N} + \mathbf{NN} = \mathbf{I} - \mathbf{N} = \mathbf{M}, \\
\mathbf{NX} &= (\mathbf{XA})\mathbf{X} = \mathbf{X}(\mathbf{AX}) = \mathbf{XI} = \mathbf{X}, \\
\mathbf{MX} &= (\mathbf{I} - \mathbf{N})\mathbf{X} = \mathbf{X} - \mathbf{NX} = \mathbf{X} - \mathbf{X} = \mathbf{O}.
\end{aligned}$$

Observe that the $k \times k$ matrix $\mathbf{Q}^{-1}$ is symmetric, while the $n \times n$ matrices **M** and **N** are *idempotent*. (A square matrix **T** is said to be idempotent iff $\mathbf{T} = \mathbf{T}'$ and $\mathbf{TT} = \mathbf{T}$.)

To recapitulate the algebra of multiple regression: we are given the data **y** and $\mathbf{X} = (\mathbf{x}_1, \ldots, \mathbf{x}_k)$, and asked to find the linear combination of the columns of **X** that comes closest to **y** in the least-squares sense. Then, provided that $\mathbf{Q} = \mathbf{X}'\mathbf{X}$ is nonsingular, the vector $\mathbf{b} = \mathbf{Ay}$ is the unique coefficient vector that solves the minimization problem, the vector $\hat{\mathbf{y}} = \mathbf{Ny}$ gives the values of the linear combination, and the vector $\mathbf{e} = \mathbf{My}$ gives the residuals.

14.3. Ranks of X and Q

In discussing the normal equations, we distinguished two cases with respect to the $k \times k$ symmetric matrix $\mathbf{Q} = \mathbf{X}'\mathbf{X}$: the full-rank case where $\text{rank}(\mathbf{Q}) = k$, and the short-rank case where $\text{rank}(\mathbf{Q}) < k$. We now show that these are equivalently described in terms of the $n \times k$ matrix $\mathbf{X}$: the full-rank case has rank($\mathbf{X}$) = k, the short-rank case has rank($\mathbf{X}$) < k.

Let $\mathbf{d} = (d_1, \ldots, d_k)'$ be any $k \times 1$ vector. Then the $n \times 1$ vector

$$\mathbf{v} = \mathbf{Xd} = \mathbf{x}_1 d_1 + \cdots + \mathbf{x}_k d_k$$

is a linear combination of the columns of $\mathbf{X}$, and the scalar

$$\mathbf{v'v} = (\mathbf{d'X'})(\mathbf{Xd}) = \mathbf{d'X'Xd} = \mathbf{d'Qd}$$

is a sum of squares.

Suppose that the rank of $\mathbf{X}$ is less than k, the number of its columns. This means that there is a nontrivial linear combination of the columns of $\mathbf{X}$ that equals the zero vector. That is to say, there is a $k \times 1$ vector $\mathbf{d} \neq \mathbf{0}$ such that $\mathbf{v} = \mathbf{Xd} = \mathbf{0}$. For that same $\mathbf{d}$, we have $\mathbf{Qd} = \mathbf{X'v} = \mathbf{0}$, so that there is a nontrivial linear combination of the columns of $\mathbf{Q}$ that equals the zero vector. That is to say, the rank of $\mathbf{Q}$ is less than k, the number of its columns. Conversely, suppose that the rank of $\mathbf{Q}$ is less than the number of its columns. That is to say, there is a vector $\mathbf{d} \neq \mathbf{0}$ such that $\mathbf{Qd} = \mathbf{0}$. For that same $\mathbf{d}$, let $\mathbf{v} = \mathbf{Xd}$. Then $\mathbf{v'v} = \mathbf{d'Qd} = \mathbf{d'0} = 0$, which means that $\mathbf{v} = \mathbf{0}$ (because a sum of squares is zero iff all its elements are zero). That is to say, the rank of $\mathbf{X}$ is less than the number of its columns.

We have shown that rank($\mathbf{Q}$) < $k \Leftrightarrow$ rank($\mathbf{X}$) < k, and that rank($\mathbf{Q}$) = $k \Leftrightarrow$ rank($\mathbf{X}$) = k. (In fact it can be shown that rank($\mathbf{Q}$) = rank($\mathbf{X}$).) Therefore, the two cases may be restated as

Case 1. Full-rank case: rank($\mathbf{X}$) = k.
Case 2. Short-rank case: rank($\mathbf{X}$) < k.

This description in terms of $\mathbf{X}$ is more useful; it permits us to think directly about data situations in which the short-rank case occurs.

14.4. The Short-Rank Case

The rank of a matrix cannot exceed the number of its rows or the number of its columns. So the short-rank case is guaranteed to arise when $n < k$, that is, when the number of observations is less than the number of explanatory variables. But the short-rank case may arise even when $n \geq k$. That will happen when one of the x's is an exact linear function of the others. For example, suppose that with $n = 100$,

$k = 4$, it happens that $\mathbf{x}_4 = \mathbf{x}_1 + \mathbf{x}_2$. Then the nonzero vector $\mathbf{d} = (1, 1, 0, -1)'$ will satisfy $\mathbf{Xd} = \mathbf{0}$, so rank($\mathbf{X}$) < 4.

Why does rank($\mathbf{X}$) $< k$ rule out unique solution to the normal equations $\mathbf{Qc} = \mathbf{X}'\mathbf{y}$? From a mechanical point of view: suppose that $\mathbf{Qc}^* = \mathbf{X}'\mathbf{y}$ and that $\mathbf{d} \neq \mathbf{0}$ satisfies $\mathbf{Xd} = \mathbf{0}$ and hence $\mathbf{Qd} = \mathbf{0}$. Let $\mathbf{c}^{**} = \mathbf{c}^* + \mathbf{d}$. Then

$$\mathbf{Qc}^{**} = \mathbf{Q}(\mathbf{c}^* + \mathbf{d}) = \mathbf{Qc}^* + \mathbf{Qd} = \mathbf{Qc}^* = \mathbf{X}'\mathbf{y},$$

so $\mathbf{c}^{**}$, which is different from $\mathbf{c}^*$, also satisfies the normal equations. From a more fundamental point of view: the minimization problem seeks the coefficient vector $\mathbf{c}$ such that the linear combination $\mathbf{Xc} = \mathbf{x}_1 c_1 + \cdots + \mathbf{x}_k c_k$ comes closest to the observed vector $\mathbf{y}$ in the least-squares sense. But if rank($\mathbf{X}$) $< k$, then there is a nonzero vector $\mathbf{d}$, which when added to $\mathbf{c}$, gives a different set of coefficients $(\mathbf{c} + \mathbf{d})$ that generate the very same linear combination: $\mathbf{X}(\mathbf{c} + \mathbf{d}) = \mathbf{Xc}$. So the same best-fitting linear combination is expressible in different ways.

In the short-rank case, how does one locate a solution to the normal equations? Suppose that rank($\mathbf{X}$) $= k^* < k$, and without loss of generality suppose that rank($\mathbf{X}^*$) $= k^*$, where $\mathbf{X}^*$ consists of the first k^* columns of $\mathbf{X}$. Let $\mathbf{Q}^* = \mathbf{X}^{*\prime}\mathbf{X}^*$, and let $\mathbf{b}^* = \mathbf{Q}^{*-1}\mathbf{X}^{*\prime}\mathbf{y}$. Then $\mathbf{c}^* = (\mathbf{b}^{*\prime}, \mathbf{0}')'$, where the $\mathbf{0}$ is $(k - k^*) \times 1$, solves the original normal equations. In words, run the LS linear regression of $\mathbf{y}$ on $\mathbf{X}^*$ (a full-rank case) and assign zero values to the coefficients on the remaining columns of $\mathbf{X}$.

14.5. Second-Order Conditions

We now verify that the FOC's locate the *minimum* of $\phi(\mathbf{c}) = \mathbf{u}'\mathbf{u}$, where $\mathbf{u} = \mathbf{y} - \mathbf{Xc}$. Let $\mathbf{c}^*$ solve the FOC's, so $\mathbf{X}'\mathbf{u}^* = \mathbf{0}$ with $\mathbf{u}^* = \mathbf{y} - \mathbf{Xc}^*$. For any $\mathbf{c}$, let $\mathbf{d} = \mathbf{c} - \mathbf{c}^*$; then

$$\mathbf{u} = \mathbf{y} - \mathbf{Xc} = \mathbf{y} - \mathbf{X}(\mathbf{c}^* + \mathbf{d}) = \mathbf{y} - \mathbf{Xc}^* - \mathbf{Xd} = \mathbf{u}^* - \mathbf{Xd}.$$

Because $\mathbf{X}'\mathbf{u}^* = \mathbf{0}$, we have

$$\phi(\mathbf{c}) = \mathbf{u}'\mathbf{u} = \mathbf{u}^{*\prime}\mathbf{u}^* + \mathbf{d}'\mathbf{X}'\mathbf{Xd} = \phi(\mathbf{c}^*) + \mathbf{v}'\mathbf{v},$$

say, where $\mathbf{v} = \mathbf{Xd}$. Because $\mathbf{v}'\mathbf{v}$ is a sum of squares, we know that $\mathbf{v}'\mathbf{v} \geq 0$, with equality iff $\mathbf{v} = \mathbf{0}$. It follows that $\phi(\mathbf{c}) \geq \phi(\mathbf{c}^*)$, with equality iff $\mathbf{v} = \mathbf{0}$, that is iff $\mathbf{Xd} = \mathbf{0}$, that is iff $\mathbf{Qd} = \mathbf{0}$, that is iff $\mathbf{c}$ also solves the FOC's.

If $\text{rank}(\mathbf{X}) = k$, then the only vector $\mathbf{d}$ that satisfies $\mathbf{Xd} = \mathbf{0}$ is the zero vector, so $\phi(\mathbf{c}) = \phi(\mathbf{c}^*)$ iff $\mathbf{d} = \mathbf{0}$, that is, iff $\mathbf{c} = \mathbf{b}$. This verifies the claim that in the full-rank case, $\mathbf{b}$ uniquely minimizes ϕ: $\phi(\mathbf{c}) > \phi(\mathbf{b})$ for all $\mathbf{c} \neq \mathbf{b}$. If $\text{rank}(\mathbf{X}) < k$, there are many vectors that solve the FOC's: they differ from one another by vectors $\mathbf{d} \neq \mathbf{0}$ that satisfy $\mathbf{Xd} = \mathbf{0}$. This verifies the claim that in the short-rank case, $\phi(\mathbf{c}) \geq \phi(\mathbf{c}^*)$ with equality iff $\mathbf{c}$ also solves the FOC's.

We can draw some other implications from the fact that $\mathbf{d}'\mathbf{Qd} = \mathbf{v}'\mathbf{v}$ is a sum of squares. Recall the matrix-algebra concept of definiteness. A square symmetric matrix $\mathbf{T}$ is *nonnegative definite* iff for every $\mathbf{d}$, the scalar $\mathbf{d}'\mathbf{Td}$ is nonnegative, and is *positive definite* iff for every $\mathbf{d} \neq \mathbf{0}$, the scalar $\mathbf{d}'\mathbf{Td}$ is positive. If the sign of $\mathbf{d}'\mathbf{Td}$ depends on $\mathbf{d}$, then $\mathbf{T}$ is said to be *indefinite*. Now a sum of squares is nonnegative, and is zero iff all of its elements are zero. We conclude that $\mathbf{Q} = \mathbf{X}'\mathbf{X}$ is nonnegative definite, and further that $\mathbf{Q}$ is positive definite iff $\mathbf{Xd} = \mathbf{0}$ implies $\mathbf{d} = \mathbf{0}$, that is iff $\text{rank}(\mathbf{X}) = k$.

Exercises

14.1 Let $\mathbf{X}$ and $\mathbf{y}$ be

$$\mathbf{X} = \begin{pmatrix} 1 & 2 \\ 1 & 4 \\ 1 & 3 \\ 1 & 5 \\ 1 & 2 \end{pmatrix}, \qquad \mathbf{y} = \begin{pmatrix} 14 \\ 17 \\ 8 \\ 16 \\ 3 \end{pmatrix}.$$

Calculate the following, using fractions to maintain precision. Feel free to factor out a common denominator in displaying a matrix.

(a) $\mathbf{Q} = \mathbf{X}'\mathbf{X}, \quad |\mathbf{X}'\mathbf{X}|, \quad \mathbf{Q}^{-1}$.
(b) $\mathbf{A} = \mathbf{Q}^{-1}\mathbf{X}', \quad \mathbf{b} = \mathbf{Ay}$.
(c) $\mathbf{N} = \mathbf{XA}, \quad \hat{\mathbf{y}} = \mathbf{Ny}$.
(d) $\mathbf{M} = \mathbf{I} - \mathbf{N}, \quad \mathbf{e} = \mathbf{My}$.
(e) $\text{tr}(\mathbf{N}), \quad \text{tr}(\mathbf{M})$.

(Note: If $\mathbf{T}$ is a square matrix, then $\text{tr}(\mathbf{T}) = \text{trace}(\mathbf{T}) =$ sum of diagonal elements of $\mathbf{T}$.)

14.2 For a certain data set with $n = 100$ observations, the explanatory variables include $x_1 = 1$, $x_2 =$ a binary variable that is equal to 1 for males and equal to zero for females, and x_3 = a binary variable that is equal to 1 for females and equal to zero for males. Will the **X** matrix have full column rank? Explain.

14.3 Let **X** be an $n \times k$ matrix whose rank is k, and let

$$\mathbf{Q} = \mathbf{X}'\mathbf{X}, \qquad \mathbf{A} = \mathbf{Q}^{-1}\mathbf{X}', \qquad \mathbf{N} = \mathbf{X}\mathbf{A}, \qquad \mathbf{M} = \mathbf{I} - \mathbf{N}.$$

Recall that

$$\mathbf{b} = \mathbf{A}\mathbf{y}, \qquad \hat{\mathbf{y}} = \mathbf{N}\mathbf{y}, \qquad \mathbf{e} = \mathbf{M}\mathbf{y},$$

are the vectors of coefficients, fitted values, and residuals that result when an $n \times 1$ vector **y** is linearly regressed on **X**. Show the following, as concisely as possible.

(a) $\mathbf{AN} = \mathbf{A}, \quad \mathbf{AM} = \mathbf{O}, \quad \mathbf{MN} = \mathbf{O}, \quad \mathbf{NM} = \mathbf{O}.$
(b) $\mathbf{NX} = \mathbf{X}, \quad \mathbf{MX} = \mathbf{O}.$
(c) $\mathbf{N}\hat{\mathbf{y}} = \hat{\mathbf{y}}, \quad \mathbf{Ne} = \mathbf{0}.$
(d) $\mathbf{M}\hat{\mathbf{y}} = \mathbf{0}, \quad \mathbf{Me} = \mathbf{e}.$
(e) $\mathbf{X}'\hat{\mathbf{y}} = \mathbf{X}'\mathbf{y}.$
(f) $\mathbf{y}'\hat{\mathbf{y}} = \mathbf{y}'\mathbf{Xb} = \mathbf{b}'\mathbf{X}'\mathbf{y} = \mathbf{b}'\mathbf{Qb} = \hat{\mathbf{y}}'\hat{\mathbf{y}}.$
(g) $\mathbf{e}'\mathbf{e} = \mathbf{y}'\mathbf{My} = \mathbf{y}'\mathbf{y} - \hat{\mathbf{y}}'\hat{\mathbf{y}}.$

14.4 Show that every idempotent matrix is nonnegative definite.

14.5 Let m_{ii} and n_{ii} denote the ith diagonal elements of **M** and **N**. Show that $0 \leq m_{ii} \leq 1$ and $0 \leq n_{ii} \leq 1$.

15 Classical Regression

15.1. Matrix Algebra for Random Variables

In this chapter we will establish a model for multiple regression, that is, a population specification and sampling scheme that support running LS linear regression to estimate population parameters. In preparation, we develop a general matrix-algebra system for dealing with random variables.

Setting aside the regression application, let $Y_1, \ldots, Y_n$ be a set of n random variables whose joint pdf (or pmf) is $f(y_1, \ldots, y_n)$. The expectations, variances, and covariances are (for $i, h = 1, \ldots, n$):

$$E(Y_i) = \mu_i, \qquad V(Y_i) = \sigma_i^2 = \sigma_{ii}, \qquad C(Y_h, Y_i) = \sigma_{hi} = \sigma_{ih}.$$

It is natural to display these in an $n \times 1$ vector $\mathbf{Y}$, an $n \times 1$ vector $\boldsymbol{\mu}$, and an $n \times n$ matrix $\boldsymbol{\Sigma}$, where:

$$\mathbf{Y} = \begin{pmatrix} Y_1 \\ \cdot \\ \cdot \\ \cdot \\ Y_n \end{pmatrix}, \qquad \boldsymbol{\mu} = \begin{pmatrix} \mu_1 \\ \cdot \\ \cdot \\ \cdot \\ \mu_n \end{pmatrix}, \qquad \boldsymbol{\Sigma} = \begin{pmatrix} \sigma_{11} & \cdot & \cdot & \cdot & \sigma_{1n} \\ \cdot & & & & \cdot \\ \cdot & \cdot & \sigma_{hi} & \cdot & \cdot \\ \cdot & & & & \cdot \\ \sigma_{n1} & \cdot & \cdot & \cdot & \sigma_{nn} \end{pmatrix}.$$

At the risk of some confusion *we adopt the matrix-algebra convention of lowercase characters* for vectors, overriding the statistical convention of uppercase characters for random variables, and write the $n \times 1$ random vector and its elements as

$$\mathbf{y} = (y_1, \ldots, y_n)'.$$

The *expectation of a random vector* (or matrix) is defined to be the vector (or matrix) of expectations of its elements. The *variance matrix of a random vector* is defined to be the matrix of variances and covariances of its elements. We write

$$E(\mathbf{y}) = \boldsymbol{\mu}, \qquad V(\mathbf{y}) = \boldsymbol{\Sigma}.$$

When $\mathbf{y}$ is $n \times 1$, then $\boldsymbol{\mu}$ is $n \times 1$, and $\boldsymbol{\Sigma}$ is $n \times n$ symmetric.

Let $\boldsymbol{\epsilon} = \mathbf{y} - \boldsymbol{\mu} = \{y_i - \mu_i\}$ be the $n \times 1$ vector of deviations of the y's from their respective expectations. So

$$\boldsymbol{\epsilon\epsilon}' = (\mathbf{y} - \boldsymbol{\mu})(\mathbf{y} - \boldsymbol{\mu})' = \{(y_h - \mu_h)(y_i - \mu_i)\}$$

is an $n \times n$ symmetric random matrix whose elements are the squares and cross-products of those deviations. Then

$$E(\boldsymbol{\epsilon}) = E(\mathbf{y} - \boldsymbol{\mu}) = \{E(y_i) - \mu_i\} = \{\mu_i - \mu_i\} = \{0\} = \mathbf{0},$$

$$E(\boldsymbol{\epsilon\epsilon}') = E[(\mathbf{y} - \boldsymbol{\mu})(\mathbf{y} - \boldsymbol{\mu})'] = \{\sigma_{hi}\} = \boldsymbol{\Sigma} = V(\mathbf{y}) = V(\boldsymbol{\epsilon}).$$

The *covariance matrix of a pair of random vectors* is defined to be the matrix of covariances between the elements of one vector and the elements of the other vector. Thus if $\mathbf{z} = \{z_h\}$ is an $m \times 1$ random vector and $\mathbf{y} = \{y_i\}$ is an $n \times 1$ random vector, then

$$C(\mathbf{z}, \mathbf{y}) = E\{[\mathbf{z} - E(\mathbf{z})][\mathbf{y} - E(\mathbf{y})]'\}$$

is the $m \times n$ matrix whose (h, i)th element is $C(z_h, y_i)$, while $C(\mathbf{y}, \mathbf{z})$ is the $n \times m$ transpose of that matrix.

Here are a set of rules for calculating expectations, variances, and covariances of certain functions of $\mathbf{y}$. Throughout we suppose that the $n \times 1$ random vector $\mathbf{y}$ has expectation vector $E(\mathbf{y}) = \boldsymbol{\mu}$ and variance matrix $V(\mathbf{y}) = \boldsymbol{\Sigma}$, and write $\boldsymbol{\epsilon} = \mathbf{y} - \boldsymbol{\mu}$. The first two rules, which refer to linear functions, are straightforward generalizations of T5 and T6 in Section 5.1.

R1. SCALAR LINEAR FUNCTION. Let $z = g + \mathbf{h}'\mathbf{y}$, where the scalar g and the $n \times 1$ vector $\mathbf{h}$ are constants. Then the random variable z has

$$E(z) = g + \mathbf{h}'E(\mathbf{y}) = g + \mathbf{h}'\boldsymbol{\mu}.$$

Further, let $z^* = z - E(z)$. Then $z^* = \mathbf{h}'\mathbf{y} - \mathbf{h}'\boldsymbol{\mu} = \mathbf{h}'(\mathbf{y} - \boldsymbol{\mu}) = \mathbf{h}'\boldsymbol{\epsilon}$, and $z^{*2} = (\mathbf{h}'\boldsymbol{\epsilon})^2 = (\mathbf{h}'\boldsymbol{\epsilon})(\mathbf{h}'\boldsymbol{\epsilon}) = (\mathbf{h}'\boldsymbol{\epsilon})(\boldsymbol{\epsilon}'\mathbf{h}) = \mathbf{h}'\boldsymbol{\epsilon\epsilon}'\mathbf{h}$. So

$$V(z) = E(z^{*2}) = E(\mathbf{h}'\boldsymbol{\epsilon\epsilon}'\mathbf{h}) = \mathbf{h}'E(\boldsymbol{\epsilon\epsilon}')\mathbf{h} = \mathbf{h}'V(\boldsymbol{\epsilon})\mathbf{h} = \mathbf{h}'\boldsymbol{\Sigma}\mathbf{h}.$$

Incidentally, since the scalar variance $V(z)$ must be nonnegative, we see that every variance matrix $\boldsymbol{\Sigma}$ is nonnegative definite, and is positive definite iff it is nonsingular.

R2. VECTOR LINEAR FUNCTION. Let $\mathbf{z} = \mathbf{g} + \mathbf{H}\mathbf{y}$, where the $k \times 1$ vector $\mathbf{g}$ and the $k \times n$ matrix $\mathbf{H}$ are constants. Then the $k \times 1$ random vector $\mathbf{z}$ has

$$E(\mathbf{z}) = \mathbf{g} + \mathbf{H}\boldsymbol{\mu}.$$

Further, let $\mathbf{z}^* = \mathbf{z} - E(\mathbf{z})$. Then $\mathbf{z}^* = \mathbf{H}(\mathbf{y} - \boldsymbol{\mu}) = \mathbf{H}\boldsymbol{\epsilon}$, and $\mathbf{z}^*\mathbf{z}^{*\prime} = \mathbf{H}\boldsymbol{\epsilon}\boldsymbol{\epsilon}'\mathbf{H}'$. So

$$V(\mathbf{z}) = E(\mathbf{z}^*\mathbf{z}^{*\prime}) = E(\mathbf{H}\boldsymbol{\epsilon}\boldsymbol{\epsilon}'\mathbf{H}') = \mathbf{H}E(\boldsymbol{\epsilon}\boldsymbol{\epsilon}')\mathbf{H}' = \mathbf{H}\boldsymbol{\Sigma}\mathbf{H}'.$$

R3. MEAN SQUARES. Let $\mathbf{W} = \mathbf{y}\mathbf{y}'$. Then the $n \times n$ random matrix $\mathbf{W}$ has expectation $E(\mathbf{W}) = \boldsymbol{\Sigma} + \boldsymbol{\mu}\boldsymbol{\mu}'$.

Proof. Write

$$\mathbf{y}\mathbf{y}' = (\boldsymbol{\mu} + \boldsymbol{\epsilon})(\boldsymbol{\mu} + \boldsymbol{\epsilon})' = \boldsymbol{\mu}\boldsymbol{\mu}' + \boldsymbol{\mu}\boldsymbol{\epsilon}' + \boldsymbol{\epsilon}\boldsymbol{\mu}' + \boldsymbol{\epsilon}\boldsymbol{\epsilon}',$$

which, since $\boldsymbol{\mu}$ is constant and $E(\boldsymbol{\epsilon}) = \mathbf{0}$, implies

$$E(\mathbf{y}\mathbf{y}') = \boldsymbol{\mu}\boldsymbol{\mu}' + \boldsymbol{\Sigma}. \quad \blacksquare$$

R4. SUM OF SQUARES. Let $w = \mathbf{y}'\mathbf{y}$. Then the scalar random variable w has expectation $E(w) = \text{tr}(\boldsymbol{\Sigma}) + \boldsymbol{\mu}'\boldsymbol{\mu}$.

Proof. Write

$$\mathbf{y}'\mathbf{y} = \text{tr}(\mathbf{y}'\mathbf{y}) = \text{tr}(\mathbf{y}\mathbf{y}') = \text{tr}(\mathbf{W}),$$

so

$$\begin{aligned} E(\mathbf{y}'\mathbf{y}) &= E[\text{tr}(\mathbf{W})] = \text{tr}[E(\mathbf{W})] = \text{tr}(\boldsymbol{\Sigma} + \boldsymbol{\mu}\boldsymbol{\mu}') \\ &= \text{tr}(\boldsymbol{\Sigma}) + \text{tr}(\boldsymbol{\mu}\boldsymbol{\mu}') = \text{tr}(\boldsymbol{\Sigma}) + \text{tr}(\boldsymbol{\mu}'\boldsymbol{\mu}) = \text{tr}(\boldsymbol{\Sigma}) + \boldsymbol{\mu}'\boldsymbol{\mu}, \end{aligned}$$

using the facts that trace is a linear operator, and that if $\mathbf{AB}$ and $\mathbf{BA}$ are both square matrices, then $\text{tr}(\mathbf{AB}) = \text{tr}(\mathbf{BA})$. $\blacksquare$

R5. QUADRATIC FORM. Let $w = \mathbf{y}'\mathbf{T}\mathbf{y}$, where the $n \times n$ matrix $\mathbf{T}$ is constant. Then the random variable w has expectation $E(w) = \text{tr}(\mathbf{T}\boldsymbol{\Sigma}) + \boldsymbol{\mu}'\mathbf{T}\boldsymbol{\mu}$.

Proof. Write $\mathbf{y}'\mathbf{T}\mathbf{y} = \text{tr}(\mathbf{y}'\mathbf{T}\mathbf{y}) = \text{tr}(\mathbf{T}\mathbf{y}\mathbf{y}') = \text{tr}(\mathbf{T}\mathbf{W})$. Then

$$\begin{aligned} E(\mathbf{y}'\mathbf{T}\mathbf{y}) &= E[\text{tr}(\mathbf{T}\mathbf{W})] = \text{tr}[E(\mathbf{T}\mathbf{W})] = \text{tr}[\mathbf{T}E(\mathbf{W})] \\ &= \text{tr}[\mathbf{T}(\boldsymbol{\Sigma} + \boldsymbol{\mu}\boldsymbol{\mu}')] = \text{tr}(\mathbf{T}\boldsymbol{\Sigma}) + \text{tr}(\mathbf{T}\boldsymbol{\mu}\boldsymbol{\mu}') \\ &= \text{tr}(\mathbf{T}\boldsymbol{\Sigma}) + \boldsymbol{\mu}'\mathbf{T}\boldsymbol{\mu}. \quad \blacksquare \end{aligned}$$

R6. PAIR OF VECTOR LINEAR FUNCTIONS. Let $\mathbf{z}_1 = \mathbf{g}_1 + \mathbf{H}_1\mathbf{y}$, $\mathbf{z}_2 = \mathbf{g}_2 + \mathbf{H}_2\mathbf{y}$, where the $m_1 \times 1$ vector $\mathbf{g}_1$, the $m_2 \times 1$ vector $\mathbf{g}_2$, the $m_1 \times n$ matrix $\mathbf{H}_1$, and the $m_2 \times n$ matrix $\mathbf{H}_2$ are constants. Then $C(\mathbf{z}_1, \mathbf{z}_2) = \mathbf{H}_1\boldsymbol{\Sigma}\mathbf{H}_2'$.

Proof. Let $\mathbf{z}_1^* = \mathbf{z}_1 - E(\mathbf{z}_1) = \mathbf{H}_1\boldsymbol{\epsilon}$, and $\mathbf{z}_2^* = \mathbf{z}_2 - E(\mathbf{z}_2) = \mathbf{H}_2\boldsymbol{\epsilon}$. Then $\mathbf{z}_1^*\mathbf{z}_2^{*\prime} = \mathbf{H}_1\boldsymbol{\epsilon}\boldsymbol{\epsilon}'\mathbf{H}_2'$, so

$$C(\mathbf{z}_1, \mathbf{z}_2) = E(\mathbf{z}_1^*\mathbf{z}_2^{*\prime}) = \mathbf{H}_1E(\boldsymbol{\epsilon}\boldsymbol{\epsilon}')\,\mathbf{H}_2' = \mathbf{H}_1\boldsymbol{\Sigma}\mathbf{H}_2'. \quad \blacksquare$$

15.2. Classical Regression Model

We now set out the statistical model that is most commonly used to justify running a sample LS regression to estimate population parameters. That is, we provide a context for the data, one in which we observe a drawing on an $n \times 1$ random vector $\mathbf{y}$ and an $n \times k$ matrix $\mathbf{X} = (\mathbf{x}_1, \ldots, \mathbf{x}_k)$. The *classical regression*, or CR, *model* consists of these four assumptions:

(15.1) $E(\mathbf{y}) = \mathbf{X}\boldsymbol{\beta}$,

(15.2) $V(\mathbf{y}) = \sigma^2\mathbf{I}$,

(15.3) $\mathbf{X}$ nonstochastic,

(15.4) $\text{rank}(\mathbf{X}) = k$.

The understanding is that we observe $\mathbf{X}$ and $\mathbf{y}$, while $\boldsymbol{\beta}$ and σ^2 are unknown.

We interpret the assumptions briefly. In general, an $n \times 1$ random vector $\mathbf{y} = (y_1, \ldots, y_n)'$ will have expectation vector $\boldsymbol{\mu}$ and variance matrix $\boldsymbol{\Sigma}$, with

$$\boldsymbol{\mu} = \begin{pmatrix} \mu_1 \\ \cdot \\ \cdot \\ \cdot \\ \mu_n \end{pmatrix}, \qquad \boldsymbol{\Sigma} = \begin{pmatrix} \sigma_{11} & \cdot & \cdot & \cdot & \sigma_{1n} \\ \cdot & & & & \cdot \\ \cdot & \cdot & \sigma_{hi} & \cdot & \cdot \\ \cdot & & & & \cdot \\ \sigma_{n1} & \cdot & \cdot & \cdot & \sigma_{nn} \end{pmatrix}.$$

So in general, the elements of a random vector $\mathbf{y}$ will have different expectations, different variances, and free covariances. But in the CR model, we have

$$\text{(15.1)} \qquad \boldsymbol{\mu} = \mathbf{X}\boldsymbol{\beta},$$

which says that $\mu_i = \mathbf{x}_i'\boldsymbol{\beta}$, where $\mathbf{x}_i'$ is the ith row of $\mathbf{X}$. (Caution: Do not confuse $\mathbf{x}_i'$ with the transpose of the ith column of $\mathbf{X}$.) Consequently, all n of the unknown expectations, the μ_i's, are expressible in terms of k unknown parameters, the β_j's. The n expectations may well be different, but they all lie in the same k-dimensional plane in n-space. Further, in the CR model, we have

$$\text{(15.2)} \qquad \boldsymbol{\Sigma} = \sigma^2\mathbf{I},$$

which says that $\sigma_{ii} = \sigma^2$ for all i, and that $\sigma_{hi} = 0$ for all $h \neq i$. Thus the random variables $y_1, \ldots, y_n$ all have the same variance, and are uncorrelated. Further, we have

(15.3) $\mathbf{X}$ nonstochastic,

which says that the elements of $\mathbf{X}$ are constants, that is, degenerate random variables. Their values are fixed in repeated samples, unlike the elements of $\mathbf{y}$ which, being random variables, will vary from sample to sample. Finally, we have

$$\text{(15.4)} \qquad \text{rank}(\mathbf{X}) = k,$$

which says that the $n \times k$ matrix $\mathbf{X}$ has full column rank; its k columns are linearly independent in the matrix algebra sense.

In Chapter 16, we will return to the interpretation of the CR model, and to the population and sampling assumptions that underlie it.

15.3. Estimation of $\boldsymbol{\beta}$

We proceed to the estimation of the unknown parameters $\boldsymbol{\beta}$ and σ^2. We have a sample ($\mathbf{y}$, $\mathbf{X}$) produced by the CR model. How shall we process the sample data to obtain parameter estimates? The proposal is to use the sample LP, that is, to run the LS linear regression of $\mathbf{y}$ on $\mathbf{X}$. Because rank($\mathbf{X}$) = k, the normal equations of LS linear regression will have a unique solution, namely

$$\mathbf{b} = \mathbf{A}\mathbf{y}, \quad \text{where } \mathbf{A} = \mathbf{Q}^{-1}\mathbf{X}'.$$

This $k \times 1$ random vector $\mathbf{b}$ is our estimator of $\boldsymbol{\beta}$.

What properties does the estimator have? The matrix $\mathbf{A}$ is constant because it is a function of $\mathbf{X}$ alone. Hence $\mathbf{b}$ is a linear function of the random vector $\mathbf{y}$, and R2 of Section 15.1 applies. Recalling that $\mathbf{AX} = \mathbf{I}$ and that $\mathbf{AA}' = \mathbf{Q}^{-1}$, we calculate

$$E(\mathbf{b}) = \mathbf{A}E(\mathbf{y}) = \mathbf{A}\boldsymbol{\mu} = \mathbf{A}(\mathbf{X}\boldsymbol{\beta}) = (\mathbf{AX})\boldsymbol{\beta} = \mathbf{I}\boldsymbol{\beta} = \boldsymbol{\beta},$$

$$V(\mathbf{b}) = \mathbf{A}V(\mathbf{y})\mathbf{A}' = \mathbf{A}\boldsymbol{\Sigma}\mathbf{A}' = \mathbf{A}(\sigma^2\mathbf{I})\mathbf{A}' = \sigma^2\mathbf{AA}' = \sigma^2\mathbf{Q}^{-1}.$$

So the LS coefficient vector $\mathbf{b}$ is an unbiased estimator of the parameter vector $\boldsymbol{\beta}$, with $E(b_j) = \beta_j$ for $j = 1, \ldots, k$. And the variances and covariances of the k random variables in $\mathbf{b}$ are given by the appropriate elements of the $k \times k$ matrix $\sigma^2\mathbf{Q}^{-1}$:

$$V(b_j) = \sigma^2 q^{jj}, \qquad C(b_h, b_j) = \sigma^2 q^{hj},$$

where q^{hj} denotes the element in the hth row and jth column of $\mathbf{Q}^{-1}$.

15.4. Gauss-Markov Theorem

We now show that the LS estimator has an optimality property.

GAUSS-MARKOV THEOREM. In the CR model, the LS coefficient vector $\mathbf{b}$ is the minimum variance linear unbiased estimator of the parameter vector $\boldsymbol{\beta}$.

Proof. Let $\mathbf{b}^* = \mathbf{A}^*\mathbf{y}$, where $\mathbf{A}^*$ is any $k \times n$ nonstochastic matrix. Then $\mathbf{b}^*$ is a linear function of $\mathbf{y}$, that is a linear estimator. Rule R2 gives

$$E(\mathbf{b}^*) = \mathbf{A}^*E(\mathbf{y}) \quad = \mathbf{A}^*\mathbf{X}\boldsymbol{\beta},$$

$$V(\mathbf{b}^*) = \mathbf{A}^*V(\mathbf{y})\mathbf{A}^{*\prime} = \sigma^2\mathbf{A}^*\mathbf{A}^{*\prime}.$$

Clearly $\mathbf{b}^*$ will be unbiased for $\boldsymbol{\beta}$ iff $\mathbf{A}^*\mathbf{X} = \mathbf{I}$. Write $\mathbf{A}^* = \mathbf{A} + \mathbf{D}$, where $\mathbf{A} = \mathbf{Q}^{-1}\mathbf{X}'$ and $\mathbf{D} = \mathbf{A}^* - \mathbf{A}$. Observe that

$$\mathbf{A}^*\mathbf{X} = \mathbf{AX} + \mathbf{DX} = \mathbf{I} + \mathbf{DX},$$

$$\mathbf{A}^*\mathbf{A}^{*\prime} = (\mathbf{A} + \mathbf{D})(\mathbf{A} + \mathbf{D})' = \mathbf{AA}' + \mathbf{DD}' + \mathbf{AD}' + \mathbf{DA}'.$$

So the unbiasedness condition $\mathbf{A}^*\mathbf{X} = \mathbf{I}$ is equivalent to $\mathbf{DX} = \mathbf{O}$, that is, to $\mathbf{DXQ}^{-1} = \mathbf{O}$, that is, to $\mathbf{DA}' = \mathbf{O}$ (and hence to $\mathbf{AD}' = \mathbf{O}$). So if $\mathbf{b}^*$ is a linear unbiased estimator of $\boldsymbol{\beta}$, then

$$V(\mathbf{b}^*) = \sigma^2(\mathbf{AA}' + \mathbf{DD}') = V(\mathbf{b}) + \sigma^2\mathbf{DD}'.$$

The matrix $\mathbf{DD}'$ is nonnegative definite and the scalar σ^2 is positive, so $\sigma^2\mathbf{DD}'$ is nonnegative definite. Consequently, $V(\mathbf{b}^*) \geq V(\mathbf{b})$, with equality iff $\mathbf{DD}' = \mathbf{O}$, that is, iff $\mathbf{D} = \mathbf{O}$, that is, iff $\mathbf{A}^* = \mathbf{A}$, that is, iff $\mathbf{b}^* = \mathbf{b}$ in every sample. ■

Some explanations are in order:

- The matrix $\mathbf{DD}'$ is nonnegative definite because for any $k \times 1$ vector $\mathbf{h}$, the quadratic form $\mathbf{h}'\mathbf{DD}'\mathbf{h} = (\mathbf{D}'\mathbf{h})'(\mathbf{D}'\mathbf{h}) = \mathbf{v}'\mathbf{v} \geq 0$.
- If $\mathbf{t}^*$ and $\mathbf{t}$ are random vectors, we say that $V(\mathbf{t}^*) \geq V(\mathbf{t})$ iff $V(\mathbf{t}^*) - V(\mathbf{t})$ is nonnegative definite.

Observe the implications of the nonnegative definiteness of $V(\mathbf{b}^*) - V(\mathbf{b})$. Element by element, $\mathbf{b}$ is preferable to $\mathbf{b}^*$, any other linear unbiased estimator of $\boldsymbol{\beta}$, because its elements have smaller variances. But also consider a linear combination of the β_j's, say $\theta = \mathbf{h}'\boldsymbol{\beta}$, where $\mathbf{h}$ is a constant $k \times 1$ vector. Let $t = \mathbf{h}'\mathbf{b}$ and let $t^* = \mathbf{h}'\mathbf{b}^*$. Then both t and t^* are unbiased for θ, but $V(t^*) - V(t) = \mathbf{h}'[V(\mathbf{b}^*) - V(\mathbf{b})]\mathbf{h} \geq 0$. So $\mathbf{b}$ is also preferable to $\mathbf{b}^*$ for constructing estimators of linear combinations.

15.5. Estimation of σ^2 and $V(\mathbf{b})$

For estimation of the parameter σ^2, we draw on the LS residual vector $\mathbf{e} = \mathbf{My}$. The matrix $\mathbf{M}$ is constant because it is a function of $\mathbf{X}$ alone. Hence $\mathbf{e} = \mathbf{My}$ is a linear function of the random vector $\mathbf{y}$, and R2 applies. Recalling that $\mathbf{MX} = \mathbf{O}$ and $\mathbf{MM}' = \mathbf{MM} = \mathbf{M}$, we calculate

$$E(\mathbf{e}) = \mathbf{M}E(\mathbf{y}) = \mathbf{M}(\mathbf{X}\boldsymbol{\beta}) = (\mathbf{MX})\boldsymbol{\beta} = \mathbf{O}\boldsymbol{\beta} = \mathbf{0},$$

$$V(\mathbf{e}) = \mathbf{M}V(\mathbf{y})\mathbf{M}' = \mathbf{M}(\sigma^2\mathbf{I})\mathbf{M}' = \sigma^2\mathbf{MM}' = \sigma^2\mathbf{M}.$$

Thus, considered as random variables, the residuals $e_1, \ldots, e_n$ have zero expectations, generally different variances, and nonzero covariances. Now calculate the expectation of the random variable $\mathbf{e}'\mathbf{e}$, the sum of squared residuals. Apply R4, with $\mathbf{e}$ playing the role of $\mathbf{y}$:

$$E(\mathbf{e}'\mathbf{e}) = \text{tr}[V(\mathbf{e})] + [E(\mathbf{e})]'[E(\mathbf{e})] = \text{tr}(\sigma^2\mathbf{M}) + \mathbf{0}'\mathbf{0} = \sigma^2 \,\text{tr}(\mathbf{M}).$$

But $\mathbf{N} = \mathbf{XA}$ and $\mathbf{AX} = \mathbf{I}$, so

$$\text{tr}(\mathbf{N}) = \text{tr}(\mathbf{XA}) = \text{tr}(\mathbf{AX}) = \text{tr}(\mathbf{I}_k) = k.$$

Hence for $\mathbf{M} = \mathbf{I} - \mathbf{N}$, we have

$$\text{tr}(\mathbf{M}) = \text{tr}(\mathbf{I} - \mathbf{N}) = \text{tr}(\mathbf{I}_n) - \text{tr}(\mathbf{N}) = n - k.$$

So

$$E(\mathbf{e}'\mathbf{e}) = \sigma^2(n - k).$$

Defining the *adjusted mean squared residual,*

$$\hat{\sigma}^2 = \mathbf{e}'\mathbf{e}/(n - k),$$

we have $E(\hat{\sigma}^2) = E(\mathbf{e}'\mathbf{e})/(n - k) = \sigma^2$. So $\hat{\sigma}^2$ is an unbiased estimator of σ^2.

Finally, we estimate the variance matrix $V(\mathbf{b}) = \sigma^2\mathbf{Q}^{-1}$, by

$$\hat{V}(\mathbf{b}) = \hat{\sigma}^2\mathbf{Q}^{-1}.$$

Because $E(\hat{\sigma}^2) = \sigma^2$ and $\mathbf{Q}^{-1}$ is constant, it follows that

$$E[\hat{V}(\mathbf{b})] = \sigma^2\mathbf{Q}^{-1} = V(\mathbf{b}),$$

so that $\hat{V}(\mathbf{b})$ is an unbiased estimator of $V(\mathbf{b})$. In particular,

$$\hat{\sigma}^2_{b_j} = \hat{\sigma}^2 q^{jj}$$

is an unbiased estimator of $V(b_j) = \sigma^2_{b_j} = \sigma^2 q^{jj}$. The square root of the estimated variance,

$$\hat{\sigma}_{b_j} = \hat{\sigma}\sqrt{q^{jj}},$$

serves as the standard error of b_j.

Exercises

15.1 Suppose that the random vector $\mathbf{x}$ has $E(\mathbf{x}) = \boldsymbol{\mu}$, $V(\mathbf{x}) = \boldsymbol{\Sigma}$, and that $\mathbf{y} = \mathbf{g} + \mathbf{Hx}$, where

$$\boldsymbol{\mu} = \begin{pmatrix} 1 \\ 2 \\ 3 \end{pmatrix}, \quad \boldsymbol{\Sigma} = \begin{pmatrix} 2 & 1 & 2 \\ 1 & 3 & 1 \\ 2 & 1 & 4 \end{pmatrix}, \quad \mathbf{g} = \begin{pmatrix} 2 \\ 1 \end{pmatrix}, \quad \mathbf{H} = \begin{pmatrix} 1 & 2 & 1 \\ -1 & 2 & 1 \end{pmatrix}.$$

Calculate $E(\mathbf{y})$, $V(\mathbf{y})$, $E(\mathbf{yy}')$, $E(\mathbf{y}'\mathbf{y})$, $C(\mathbf{y}, \mathbf{x})$, and $C(\mathbf{x}, \mathbf{y})$.

15.2 Suppose the CR model applies with $n = 40$, $\sigma^2 = 4$, and

$$\mathbf{X}'\mathbf{X} = \begin{pmatrix} 40 & 10 \\ 10 & 5 \end{pmatrix}, \quad \boldsymbol{\beta} = \begin{pmatrix} 3 \\ 2 \end{pmatrix}.$$

Let $\mathbf{b}$ be the LS coefficient vector and $t = \mathbf{b}'\mathbf{b}$. Find $E(t)$.

15.3 The CR model applies with $\sigma^2 = 2$, and

$$\mathbf{X}'\mathbf{X} = \begin{pmatrix} 4 & 2 \\ 2 & 6 \end{pmatrix}, \quad \boldsymbol{\beta} = \begin{pmatrix} 3 \\ 5 \end{pmatrix}.$$

A sample is drawn and the LS coefficients b_1 and b_2 are calculated.

(a) Guess, as best you can, the value of b_2. Explain.
(b) Now you are told that $b_1 = 4$. Guess, as best you can, the value of b_2. Explain.

15.4 The CR model applies along with the usual notation. For each of the following statements, indicate whether it is true or false, and justify your answer.

(a) The random variable $t = \mathbf{b}'\mathbf{b}$ is an unbiased estimator of the parameter $\theta = \boldsymbol{\beta}'\boldsymbol{\beta}$.
(b) Since $\hat{\mathbf{y}} = \mathbf{Ny}$, it follows that $\mathbf{y} = \mathbf{N}^{-1}\hat{\mathbf{y}}$.
(c) Since $E(\hat{\mathbf{y}}) = E(\mathbf{y})$, it follows that the sum of the residuals is zero.
(d) If b_1 and b_2 are the first two elements of $\mathbf{b}$, $t_1 = b_1 + b_2$, and $t_2 = b_1 - b_2$, then $V(t_1) \geq V(t_2)$.

15.5 Show that the LS coefficients $\mathbf{b} = \mathbf{Ay}$ are uncorrelated with the residuals $\mathbf{e} = \mathbf{My}$. Hint: See R6, Section 15.1.

15.6 Suppose that the CR model applies to the data of Exercise 14.1. Report your estimates of the β_j parameters, with standard errors in parentheses beneath the coefficient estimates. Also report $\hat{\sigma}^2$.

16 Classical Regression: Interpretation and Application

16.1. Interpretation of the Classical Regression Model

It is instructive to compare our specification of the classical regression model to the more customary one. Our CR model is specified as

(16.1) $E(\mathbf{y}) = \mathbf{X}\boldsymbol{\beta}$,

(16.2) $V(\mathbf{y}) = \sigma^2\mathbf{I}$,

(16.3) $\mathbf{X}$ nonstochastic,

(16.4) $\text{rank}(\mathbf{X}) = k$.

Judge et al. (1988, pp. 178–183) specify a "General Linear Statistical Model" as follows (notation has been slightly changed):

(16.1*) $\mathbf{y} = \mathbf{X}\boldsymbol{\beta} + \boldsymbol{\epsilon}$,

(16.2*) $\mathbf{X}$ is a known nonstochastic matrix with linearly independent columns,

(16.3*) $E(\boldsymbol{\epsilon}) = \mathbf{0}$,

(16.4*) $E(\boldsymbol{\epsilon}\boldsymbol{\epsilon}') = \sigma^2\mathbf{I}$.

The two models are equivalent. Judge et al.'s $\boldsymbol{\epsilon}$ is simply the *disturbance vector*, the deviation of the random vector $\mathbf{y}$ from its expectation $\boldsymbol{\mu} = \mathbf{X}\boldsymbol{\beta}$. In that style, for a scalar random variable y with $E(y) = \mu$ and $V(y) = \sigma^2$, one might write $y = \mu + \epsilon$, $E(\epsilon) = 0$, $E(\epsilon^2) = \sigma^2$. There is no serious objection to doing so, except that it tends to give the disturbance a life of its own, rather than treating it as merely the deviation of a random variable from its expected value. Doing so may make one think of $\boldsymbol{\mu}$ as the "true value" of $\mathbf{y}$ and of $\boldsymbol{\epsilon}$ as an "error" or "mistake." For example, Judge et al. (1988, p. 179) say that the disturbance $\boldsymbol{\epsilon}$ "is a random vector representing the unpredictable or uncontrollable errors associated with the outcome of the experiment," and Johnston (1984, p. 169) says that "if the theorist has done a good job in specifying all the significant explanatory variables to be included in $\mathbf{X}$, it is reasonable to assume that both positive and negative discrepancies from the expected value will occur and that, on balance, they will average out at zero." Such language may overdramatize the primitive concept of the difference between the observed and the expected values of a random variable. In any event, we will want to distinguish between the *disturbance* vector $\boldsymbol{\epsilon} = \mathbf{y} - \boldsymbol{\mu}$, which is unobserved, and the *residual* vector $\mathbf{e} = \mathbf{y} - \hat{\mathbf{y}}$, which is observed.

In what situation would the CR model be justified? Suppose that there is a multivariate population for the random vector $(y, x_2, \ldots, x_k)'$, with pdf or pmf $f(y, x_2, \ldots, x_k)$. Expectations, variances, and covariances are defined in the usual manner:

$$E(y) = \mu_y, \qquad V(y) = \sigma_y^2, \qquad C(x_h, x_j) = \sigma_{hj}, \qquad C(x_j, y) = \sigma_{jy},$$

and so forth. Suppose further that the conditional expectation function of y given the x's is linear,

$$E(y|x_2, \ldots, x_k) = \beta_1 + \beta_2 x_2 + \cdots + \beta_k x_k,$$

and that the conditional variance function of y given the x's is constant,

$$V(y|x_2, \ldots, x_k) = \sigma^2,$$

say. We write these compactly as

$$(16.5) \qquad E(y|\mathbf{x}) = \mathbf{x}'\boldsymbol{\beta}, \qquad V(y|\mathbf{x}) = \sigma^2,$$

where $\mathbf{x} = (x_1, \ldots, x_k)'$ with $x_1 = 1$, and $\boldsymbol{\beta} = (\beta_1, \ldots, \beta_k)'$.

As for sampling schemes, the most natural one to consider would be:

Random Sampling from the Multivariate Population. Here n independent drawings, $(y_1, \mathbf{x}_1'), \ldots, (y_n, \mathbf{x}_n')$, are made, giving the observed sample data $(\mathbf{y}, \mathbf{X})$. In this scheme, the rows of the observed data matrix, namely the $(y_i, \mathbf{x}_i')$, are independent and identically distributed across i. So from Eq. (16.5), it follows that $E(y_i|\mathbf{x}_i) = \mathbf{x}_i'\boldsymbol{\beta}$ and $V(y_i|\mathbf{x}_i) = \sigma^2$. But also $E(y_i) = \mu_y$ for all i, $V(y_i) = \sigma_y^2$ for all i, and the $\mathbf{X}$ matrix is random. So this sampling scheme does not support the CR model, in which the expectations of the y_i differ and the $\mathbf{X}$ matrix is not random.

Instead of random sampling, we will rely on:

Stratified Sampling from the Multivariate Population. Here n values of the random vector $\mathbf{x}$ are specified. These values, the $\mathbf{x}_i$ ($i = 1, \ldots, n$), define n subpopulations, or strata. In the ith subpopulation, or stratum, the pdf or pmf of the dependent variable is $g(y|\mathbf{x}_i)$, with $E(y|\mathbf{x}_i) = \mathbf{x}_i'\boldsymbol{\beta} = \mu_i$, say, and $V(y|\mathbf{x}_i) = \sigma^2$. A random drawing is made from each subpopulation. That is, y_1 is drawn from subpopulation 1, y_2 is drawn from subpopulation 2, and so on. The successive drawings are independent. In this scheme, the sampled y's are not identically distributed; they are drawn from different subpopulations. The list of n selected $\mathbf{x}_i$ vectors is maintained in repeated sampling, so the expectations of the successive y's will depend only on i. We can then write $E(y_i)$ instead of $E(y_i|\mathbf{x}_i)$, and similarly we can write $V(y_i)$ instead of $V(y_i|\mathbf{x}_i)$. There is no need for all the $\mathbf{x}_i$'s to differ: the relevant requirement is that $\text{rank}(\mathbf{X}) = k$, so we need k linearly independent (in the matrix-algebra sense) $\mathbf{x}_i$'s. As discussed in Section 13.5, stratified sampling does not require that the researcher control the $\mathbf{x}$ values in the sense of imposing them on the subjects.

Under stratified sampling, it does not make sense to use the sample to estimate the population means and variances of the x's and y. The sample on $\mathbf{x}$ is not randomly drawn from the population joint distribution of $\mathbf{x}$, and consequently the sample on y is not randomly drawn from the population marginal distribution of y. Still, as in the bivariate case (Section 13.5), while stratification on $\mathbf{x}$ does induce a new marginal distribution for $\mathbf{x}$ and y, it preserves the conditional probability distributions of y given $\mathbf{x}$. That suffices when we are concerned with the conditional expectation of y given $\mathbf{x}$.

This stratified sampling scheme, also known as the nonstochastic explanatory variable scheme, will support the CR model. We adopt it now in order to simplify the theory. In Chapter 25 we will see how the conclusions carry over to the more natural scheme of random sampling.

Setting aside the sampling aspects, it is useful to compare this discussion of the underlying assumptions of the CR model with that in other textbooks. Johnston (1984, p. 169) seems to say that for the CR model to be correct, the theorist must have "done a good job in specifying all the significant explanatory variables." Judge et al. (1988, p. 186) say that "it is assumed that the $\mathbf{X}$ matrix contains the correct set of explanatory variables. In real-world situations we seldom, if ever, know the correct set of explanatory variables, and, consequently, certain relevant variables may be excluded or certain extraneous variables may be included." Here "the correct set of explanatory variables" seems to mean the variables that "could have or actually determined the outcomes that we have observed" (ibid., p. 178).

Such requirements are very stringent, and have a causal flavor that is not part of the explicit specification of the CR model. An alternative position is less stringent and is free of causal language. Nothing in the CR model itself requires an exhaustive list of the explanatory variables, nor any assumption about the direction of causality. We have in mind a joint probability distribution, in which any conditional expectation function is conceivably of interest. For example, suppose that the random vector (y, x_2, x_3) has a trivariate probability distribution. On the one hand, we might be interested in $E(y|x_2, x_3)$, but on the other hand we might be interested in $E(y|x_2)$ or, for that matter, in $E(x_2|x_3, y)$. It is possible that all of those CEF's are linear, and that none of them is causal. It may be true that causal relations are the most interesting ones, but that is a matter of economics rather than of statistics. More on this in Chapter 31.

16.2. Estimation of Linear Functions of $\boldsymbol{\beta}$

In the CR model we deal with an $n \times 1$ random vector $\mathbf{y}$. In general such a vector would have $E(\mathbf{y}) = \boldsymbol{\mu}$ and $V(\mathbf{y}) = \boldsymbol{\Sigma}$. One thing that makes the CR model special is the assumption that $\boldsymbol{\mu} = \mathbf{X}\boldsymbol{\beta}$, that is, $\mu_i = \mathbf{x}_i'\boldsymbol{\beta}$. The n unknown μ_i's may well be distinct, but all of them are expressible in terms of only k unknown β_j's.

In the CR model, we estimate $\boldsymbol{\beta}$ by $\mathbf{b}$, and thus estimate $\boldsymbol{\mu} = \mathbf{X}\boldsymbol{\beta}$ by $\hat{\boldsymbol{\mu}} = \mathbf{Xb} = \mathbf{Ny} = \hat{\mathbf{y}}$, rather than by $\mathbf{y}$ itself. Now $E(\hat{\mathbf{y}}) = \boldsymbol{\mu}$, and also $E(\mathbf{y}) = \boldsymbol{\mu}$, so both the fitted-value vector and the observed vector are unbiased estimators of $\boldsymbol{\mu}$. Why is it preferable to use $\hat{\mathbf{y}}$? An answer runs as follows. Because $V(\mathbf{y}) = \sigma^2\mathbf{I}$ and $V(\hat{\mathbf{y}}) = \sigma^2\mathbf{N}$, we have

$$V(\mathbf{y}) - V(\hat{\mathbf{y}}) = \sigma^2(\mathbf{I} - \mathbf{N}) = \sigma^2\mathbf{M}.$$

The matrix $\mathbf{M} = \mathbf{M}'\mathbf{M}$ is nonnegative definite, so $V(\mathbf{y}) \geq V(\hat{\mathbf{y}})$.

With respect to a single element of $\boldsymbol{\mu}$, say μ_i: the preferred estimator is $\hat{\mu}_i = \mathbf{x}_i'\mathbf{b} = \hat{y}_i = \mathbf{n}_i'\mathbf{y}$, where $\mathbf{n}_i'$ denotes the ith row of $\mathbf{N}$. A simpler unbiased estimator is $y_i = \mathbf{h}_i'\mathbf{y}$, where $\mathbf{h}_i$ is the $n \times 1$ vector with a 1 in its ith slot and zeroes elsewhere. Observe that $\hat{\mu}_i$ is a linear function of all n of the y's, while y_i is a function of only one of them. Evidently in the CR model it is desirable to combine information from all the observations in order to estimate the expectation of a single one. Such a preference is clear in random sampling from a univariate population, where all the observations have the same expectation. In the CR model, the preference persists even though the expectations are not the same. The reason is that the expectations are linked together, being functions of the same k β_j's.

Now, $\mu_i = \mathbf{x}_i'\boldsymbol{\beta}$ is a special case of a linear combination of the β's. The general case is $\theta = \mathbf{h}'\boldsymbol{\beta}$ where $\mathbf{h}$ is a nonrandom $k \times 1$ vector. Other special cases are of interest. For example, take $\mathbf{h} = (0, 0, 1, 0, \ldots, 0)'$, then $\theta = \beta_3$; or take $\mathbf{h} = (0, 1, -1, 0, \ldots, 0)'$, then $\theta = \beta_2 - \beta_3$. As indicated in Section 15.4, the preferred estimator of such a θ in the CR model is $t = \mathbf{h}'\mathbf{b}$. We elaborate on that point here.

By linear function rules, $E(t) = \mathbf{h}'E(\mathbf{b}) = \mathbf{h}'\boldsymbol{\beta} = \theta$, so that t is an unbiased estimator of θ. Further, $V(t) = \mathbf{h}'V(\mathbf{b})\mathbf{h} = \sigma^2\mathbf{h}'\mathbf{Q}^{-1}\mathbf{h}$. We can express t as a linear function of $\mathbf{y}$: $t = \mathbf{h}'\mathbf{b} = \mathbf{h}'\mathbf{A}\mathbf{y} = \mathbf{w}'\mathbf{y}$, where $\mathbf{w} = \mathbf{A}'\mathbf{h}$ is $n \times 1$ and nonstochastic. Consider all linear functions of $\mathbf{y}$ that might be used to estimate θ: $t^* = \mathbf{w}^{*\prime}\mathbf{y}$, where $\mathbf{w}^*$ is a nonstochastic $n \times 1$ vector. We have

$$E(t^*) = \mathbf{w}^{*\prime}\boldsymbol{\mu} = \mathbf{w}^{*\prime}\mathbf{X}\boldsymbol{\beta}, \qquad V(t^*) = \mathbf{w}^{*\prime}\boldsymbol{\Sigma}\mathbf{w}^* = \sigma^2\mathbf{w}^{*\prime}\mathbf{w}^*,$$

so t^* is unbiased iff $\mathbf{w}^{*\prime}\mathbf{X} = \mathbf{h}'$. In that event, we can write

$$\mathbf{h}'\mathbf{Q}^{-1}\mathbf{h} = \mathbf{w}^{*\prime}\mathbf{X}\mathbf{Q}^{-1}\mathbf{X}'\mathbf{w}^* = \mathbf{w}^{*\prime}\mathbf{N}\mathbf{w}^*,$$

and thus write

$$V(t) = \sigma^2\mathbf{h}'\mathbf{Q}^{-1}\mathbf{h} = \sigma^2\mathbf{w}^{*\prime}\mathbf{N}\mathbf{w}^*.$$

Observe that

$$V(t^*) - V(t) = \sigma^2\mathbf{w}^{*\prime}(\mathbf{I} - \mathbf{N})\mathbf{w}^* = \sigma^2\mathbf{w}^{*\prime}\mathbf{M}\mathbf{w}^* \geq 0,$$

because $\mathbf{M}$ is nonnegative definite. Thus the natural estimator of $\theta = \mathbf{h}'\boldsymbol{\beta}$, namely $t = \mathbf{h}'\mathbf{b}$, is in fact MVLUE in the CR model, where "linear" means linear in $\mathbf{y}$.

In practice, we will want to give some indication of the reliability of our estimate of θ. To estimate $V(t)$, replace σ^2 by $\hat{\sigma}^2$. The resulting standard error for $t = \mathbf{h}'\mathbf{b}$ is $\hat{\sigma}_t = \hat{\sigma}\sqrt{(\mathbf{h}'\mathbf{Q}^{-1}\mathbf{h})}$.

To recapitulate: in the CR model, whether our interest is in estimating the full vector $\boldsymbol{\beta}$, in estimating one of its elements β_j, or in estimating a linear combination θ of its elements, the preferred procedure is to use LS linear regression.

16.3. Estimation of Conditional Expectation, and Prediction

In the CR model, to estimate the parameter $\mu_i = E(y_i) = \mathbf{x}_i'\boldsymbol{\beta}$, our conclusion was to use $\hat{y}_i = \mathbf{x}_i'\mathbf{b}$. The expectation and variance of this estimator are

$$E(\hat{y}_i) = \mu_i, \qquad V(\hat{y}_i) = \sigma^2\mathbf{x}_i'\mathbf{Q}^{-1}\mathbf{x}_i = \sigma^2 n_{ii},$$

where n_{ii} is the ith diagonal element of $\mathbf{N}$. Now suppose that we are interested in estimating a point on the CEF, say $\mu_0 = \mathbf{x}_0'\boldsymbol{\beta}$, where $\mathbf{x}_0$ is some $k \times 1$ vector, not necessarily one of the points at which we have sampled. This parameter μ_0 is the expectation of y_0, where y_0 is a random drawing from the subpopulation defined by $\mathbf{x} = \mathbf{x}_0$. Because μ_0 is a linear combination of the elements of $\boldsymbol{\beta}$, the preferred estimator for it is $\hat{\mu}_0 = \mathbf{x}_0'\mathbf{b}$, which has expectation and variance

$$E(\hat{\mu}_0) = \mu_0, \qquad V(\hat{\mu}_0) = \sigma^2\mathbf{x}_0'\mathbf{Q}^{-1}\mathbf{x}_0.$$

The standard error for this estimator is $\hat{\sigma}\sqrt{(\mathbf{x}_0'\mathbf{Q}^{-1}\mathbf{x}_0)}$.

Prediction, or forecasting, is a distinct problem. There the objective is to predict the value of y_0, a single random drawing from the subpopulation defined by $\mathbf{x} = \mathbf{x}_0$. If we knew $\boldsymbol{\beta}$, our prediction would be $\mu_0 = \mathbf{x}_0'\boldsymbol{\beta}$. The prediction error would be $\epsilon_0 = y_0 - \mu_0$, with expectation $E(\epsilon_0) = 0$ and variance $E(\epsilon_0^2) = V(y_0) = \sigma^2$. In practice, we do not know $\boldsymbol{\beta}$, but we have a sample from the CR model, from which we have calculated $\mathbf{b}$. The natural predictor will be $\hat{\mu}_0 = \mathbf{x}_0'\mathbf{b}$. When that predictor is used, the prediction error will be $u = y_0 - \hat{\mu}_0$, with

$$E(u) = E(y_0) - E(\hat{\mu}_0) = 0,$$

$$V(u) = V(y_0) + V(\hat{\mu}_0) - 2C(y_0, \hat{\mu}_0) = \sigma^2 + \sigma^2\mathbf{x}_0'\mathbf{Q}^{-1}\mathbf{x}_0$$

$$= \sigma^2(1 + \mathbf{x}_0'\mathbf{Q}^{-1}\mathbf{x}_0),$$

taking the covariance to be zero on the understanding that the drawing on y_0 is independent of the sample observations $\mathbf{y}$. This predictor is unbiased, and the variance of the prediction error has two additive components: the variance of the prediction error that would be made were μ_0 known and used, and the variance of the estimator of μ_0. The "standard error of forecast," which is the square root of the estimate of $V(u)$, is given by $\hat{\sigma}\sqrt{(1 + \mathbf{x}_0'\mathbf{Q}^{-1}\mathbf{x}_0)}$.

16.4. Measuring Goodness of Fit

In empirical research that relies on the CR model, the objective is to estimate the population parameter vector $\boldsymbol{\beta}$, rather than to "fit the data" or to "explain the variation in the dependent variable." Nevertheless it is customary to report, along with the parameter estimates and their standard errors, a measure of goodness of fit.

To develop the measure, return to the algebra of least squares. Given the data $\mathbf{y}$, $\mathbf{X} = (\mathbf{x}_1, \ldots, \mathbf{x}_k)$, we have run the LS linear regression of $\mathbf{y}$ on $\mathbf{X}$, obtaining the coefficient vector $\mathbf{b}$, the fitted-value vector $\hat{\mathbf{y}}$, and the residual vector $\mathbf{e}$. Observe that $\mathbf{y} = \hat{\mathbf{y}} + \mathbf{e}$, and that $\hat{\mathbf{y}}'\mathbf{e} = (\mathbf{Xb})'\mathbf{e} = \mathbf{b}'(\mathbf{X}'\mathbf{e}) = \mathbf{b}'\mathbf{0} = 0$, by the FOC's. So

$$\mathbf{y}'\mathbf{y} = (\hat{\mathbf{y}} + \mathbf{e})'(\hat{\mathbf{y}} + \mathbf{e}) = \hat{\mathbf{y}}'\hat{\mathbf{y}} + \mathbf{e}'\mathbf{e},$$

which algebraically is

$$\sum_i y_i^2 = \sum_i \hat{y}_i^2 + \sum_i e_i^2. \tag{16.6}$$

This is an *analysis* (that is, decomposition) *of sum of squares*: the sum of squares of observed values is equal to the sum of squares of the fitted values plus the sum of squares of the residuals.

Further, $\Sigma_i y_i = \Sigma_i \hat{y}_i + \Sigma_i e_i$, so the mean of the observed values equals the mean of the fitted values plus the mean of the residuals:

$$\bar{y} = \bar{\hat{y}} + \bar{e}.$$

Now if $\bar{e} = 0$, then $\bar{\hat{y}} = \bar{y}$, so $n\bar{\hat{y}}^2 = n\bar{y}^2$, which subtracted from the decomposition in Eq. (16.6) gives

$$\sum_i (y_i - \bar{y})^2 = \sum_i (\hat{y}_i - \bar{y})^2 + \sum_i e_i^2. \tag{16.7}$$

This is an *analysis of variation*, where variation is defined to be the sum of squared deviations about the sample mean. Provided that the mean residual is zero, the variation of the observed values is equal to the variation of the fitted values plus the variation of the residuals.

Divide Eq. (16.7) through by $\Sigma_i(y_i - \bar{y})^2$ to get

$$(16.8) \qquad R^2 = \frac{\sum_i (\hat{y}_i - \bar{y})^2}{\sum_i (y_i - \bar{y})^2} = 1 - \frac{\sum_i e_i^2}{\sum_i (y_i - \bar{y})^2}.$$

The measure R^2, which will lie between zero and unity, is called the *coefficient of determination*, or squared multiple correlation coefficient. It measures, one says, the proportion of the variation of y that is accounted for (linearly) by variation in the x_j's; note that the fitted value $\hat{y}_i$ is an exact linear function of the x_{ij}'s. In this sense, R^2 measures the goodness of fit of the regression.

Consider an extreme case:

$$R^2 = 1 \quad \Leftrightarrow \quad \sum_i e_i^2 = 0 \quad \Leftrightarrow \quad \mathbf{e}'\mathbf{e} = 0 \quad \Leftrightarrow \quad \mathbf{e} = \mathbf{0} \quad \Leftrightarrow \quad \mathbf{y} = \mathbf{X}\mathbf{b},$$

in which case the observed y's fall on an exact linear function of the x's. The fit is perfect; all of the variation in y is accounted for by the variation in the x's. At the other extreme:

$$R^2 = 0 \quad \Leftrightarrow \quad \sum_i (\hat{y}_i - \bar{y})^2 = 0 \quad \Leftrightarrow \quad \hat{y}_i = \bar{y} \quad \text{for all } i,$$

in which case the best-fitting line is horizontal, and none of the variation in y is accounted for by variation in the x's.

From our perspective, R^2 has a very modest role in regression analysis, being a measure of the goodness of fit of a sample LS linear regression in a body of data. Nothing in the CR model requires that R^2 be high. Hence a high R^2 is not evidence in favor of the model, and a low R^2 is not evidence against it. Nevertheless, in empirical research reports, one often reads statements to the effect that "I have a high R^2, so my theory is good," or "My R^2 is higher than yours, so my theory is better than yours."

In fact the most important thing about R^2 is that it is not important in the CR model. The CR model is concerned with parameters in a population, not with goodness of fit in the sample. In Section 6.6 we did introduce the population coefficient of determination ρ^2, as a measure of strength of a relation in the population. But that measure will not be invariant when we sample selectively, as in the CR model, because it depends upon the marginal distribution of the explanatory variables. If one insists on a measure of predictive success (or rather failure), then $\hat{\sigma}^2$ might suffice: after all, the parameter σ^2 is the expected squared forecast error that would result if the population CEF were used as the predictor. Alternatively, the squared standard error of forecast (Section 16.3) at relevant values of $\mathbf{x}$ may be informative.

Some further remarks on the coefficient of determination follow.

• One should not calculate R^2 when $\bar{e} \neq 0$, for then the equivalence of the two versions of R^2 in Eq. (16.8) breaks down, and neither of them is bounded between 0 and 1. What guarantees that $\bar{e} = 0$? The only guarantee can come from the FOC's $\mathbf{X}'\mathbf{e} = \mathbf{0}$. It is customary to allow for an intercept in the regression, that is, to have, as one of the columns of $\mathbf{X}$, the $n \times 1$ vector $\mathbf{s} = (1, 1, \ldots, 1)'$. We refer to this $\mathbf{s}$ as the *summer vector*, because multiplying $\mathbf{s}'$ into any vector will sum up the elements in the latter. If $\mathbf{s}$ is one of the columns in $\mathbf{X}$, then $\mathbf{s}'\mathbf{e} = 0$ is one of the FOC's, so $\bar{e} = 0$. The same conclusion follows if there is a linear combination of the columns of $\mathbf{X}$ that equals the summer vector. Also if $\mathbf{y}$ and $\mathbf{x}_2, \ldots, \mathbf{x}_k$ all have zero column means in the sample, then $\bar{e} = 0$. But otherwise a zero mean residual is sheer coincidence.

• We can always find an $\mathbf{X}$ that makes $R^2 = 1$: take any n linearly independent $n \times 1$ vectors to form the $\mathbf{X}$ matrix. Because such a set of vectors forms a basis for n-space, any $n \times 1$ vector $\mathbf{y}$ will be expressible as an exact linear combination of the columns of that $\mathbf{X}$. But of course "fitting the data" is not a proper objective of research using the CR model.

• The fact that R^2 tends to increase as additional explanatory variables are included leads some researchers to report an *adjusted* (or "corrected") *coefficient of determination*, which discounts the fit when k is large relative to n. This measure, referred to as $\bar{R}^2$ (read as "R bar squared"), is defined via

$$1 - \bar{R}^2 = (n - 1)(1 - R^2)/(n - k),$$

which inflates the unexplained proportion and hence deflates the explained proportion. There is no strong argument for using this particular adjustment: for example, $(1 - k/n)R^2$ would have a similar effect. It may well be preferable to report R^2, n, and k, and let readers decide how to allow for n and k.

• The adjusted coefficient of determination may be written explicitly as

$$(16.9) \quad \bar{R}^2 = 1 - \left[\sum_i e_i^2/(n - k)\right] \Big/ \left[\sum_i (y_i - \bar{y})^2/(n - 1)\right].$$

It is sometimes said that in the CR model, the numerator $\Sigma_i e_i^2/(n - k)$ is an unbiased estimator of the disturbance variance, and that the denominator $\Sigma_i (y_i - \bar{y})^2/(n - 1)$ is an unbiased estimator of the variance of y. The first claim is correct, as we know. But the second claim is not correct: in the CR model the variance of the disturbance is the same thing as the common variance of the y_i, namely σ^2.

Exercises

16.1 Continuing the numerical example of Exercises 14.1 and 15.6, assume that the CR model applies. Let $\theta = \beta_1 + \beta_2$. Report your estimate of θ, along with its standard error.

16.2 The CR model applies to $E(\mathbf{y}) = \mathbf{X}\boldsymbol{\beta}$ with $\sigma^2 = 1$. Here $\mathbf{X}$ is an $n \times 2$ matrix with

$$\mathbf{X}'\mathbf{X} = \begin{pmatrix} 4 & -2 \\ -2 & 3 \end{pmatrix}.$$

You are offered the choice of two jobs: estimate $\beta_1 + \beta_2$, or estimate $\beta_1 - \beta_2$. You will be paid the dollar amount $10 - (t - \theta)^2$, where t is your estimate and θ is the parameter combination that you have chosen to estimate. To maximize your expected pay, which job should you take? What pay will you expect to receive?

16.3 In a regression analysis of the relation between earnings and various personal characteristics, a researcher includes these explanatory variables along with six others:

$$x_7 = \begin{cases} 1 & \text{if female} \\ 0 & \text{if male} \end{cases} \qquad x_8 = \begin{cases} 1 & \text{if male} \\ 0 & \text{if female} \end{cases}$$

but does not include a constant term.

(a) Does the sum of residuals from her LS regression equal zero?
(b) Why did she not also include a constant term?

16.4 Consider the customary situation, where the regression includes an intercept, and the first column of $\mathbf{X}$ is $\mathbf{x}_1 = \mathbf{s}$, the summer vector. Let $\mathbf{M}_1 = \mathbf{I} - \mathbf{x}_1(\mathbf{x}_1'\mathbf{x}_1)^{-1}\mathbf{x}_1'$.

(a) Show that $\mathbf{M}_1\mathbf{y}$ is the vector of residuals from a regression of $\mathbf{y}$ on the summer vector alone.
(b) Show that $\mathbf{y}'\mathbf{M}_1\mathbf{y} = \Sigma_i(y_i - \bar{y})^2$.
(c) Further suppose that the CR model applies to $E(\mathbf{y}) = \mathbf{X}\boldsymbol{\beta}$. Apply R5 (Section 15.1) to show that

$$E\left[\sum_i (y_i - \bar{y})^2\right] = (n - 1)\sigma^2 + \boldsymbol{\beta}_2'\mathbf{X}_2^{*\prime}\mathbf{X}_2^*\boldsymbol{\beta}_2,$$

where $\boldsymbol{\beta}_2$ is the $(k - 1) \times 1$ subvector that remains when the first element of $\boldsymbol{\beta}$ is deleted, $\mathbf{X}_2$ is the $n \times (k - 1)$ submatrix that remains when the first column of $\mathbf{X}$ is deleted, and $\mathbf{X}_2^* = \mathbf{M}_1\mathbf{X}_2$.
(d) Evaluate the claim that in Eq. (16.9), the denominator of the adjusted coefficient of determination is an unbiased estimator of the variance of the dependent variable.

16.5 GAUSS is a mathematical and statistical programming language, produced by Aptech Systems, Inc., Kent, Washington. We will rely on it frequently in the remainder of this book, presuming that it is installed on a computer available to you. Appendix B provides some introductory information about GAUSS; other information will be provided as hints in subsequent exercises. Version 1.49B of GAUSS is used here; modification to other versions should be straightforward.

Here is a GAUSS program to re-do Exercise 14.1. Enter it, run it, and print out the program file and the output file.

```
/* ASG1605 */ output file = ASG1605.OUT reset; format 8,4;
let x1 = 1 1 1 1 1; let x2 = 2 4 3 5 2; let y = 14 17 8 16 3;
X = x1~x2; Q = X'X; dq = det(Q); QI = invpd(Q);
A = QI*X'; N = X*A; I = eye(5); M = I - N;
trn = sumc(diag(N)); trm = sumc(diag(M)); b = A*y; yh = N*y; e = M*y;
"Q =          "        Q; ?; "det Q      =   "       dq; ?;
"Q inverse =     "   QI; ?;
"N =          "        N; ?; "  M        =   "       M; ?;
"tr(N)   =    "      trn;     "tr(M)      =   "    trm; ?;
"b' =       "          b'; ?;
"yhat' = "            yh'; ?; "  e'      = "         e'; end;
```

16.6 The algorithm used in Exercise 16.5 is not an efficient way to run LS linear regressions. Here is a more sensible way, which may serve as a starting point for your own future regression programs. The program also calculates the sum of squared residuals. Enter the program, personalizing it by completing names for the program and output files, and also entering your name. Run and print.

```
/* —1606 */ output file = —1606.OUT reset; format 8,4;
"Student name ———";
let x1 = 1 1 1 1 1; let x2 = 2 4 3 5 2; let y = 14 17 8 16 3;
X = x1~x2; Q= X'X; QI = invpd(Q); b = QI*X'y; sse = y'y - b'X'y;
"b'    =    " b'    ;  "         sse =   " sse;      end;
```

17 *Regression Algebra*

17.1. Regression Matrices

In this chapter we explore a variety of algebraic and interpretive results that may be relevant for empirical applications of multiple regression.

Associated with any $n \times k$ full-column-rank matrix $\mathbf{X}$ are the matrices

$$\mathbf{Q} = \mathbf{X}'\mathbf{X}, \qquad \mathbf{A} = \mathbf{Q}^{-1}\mathbf{X}', \qquad \mathbf{N} = \mathbf{X}\mathbf{A}, \qquad \mathbf{M} = \mathbf{I} - \mathbf{N}.$$

Multiplied into any $n \times 1$ vector, the matrices $\mathbf{A}$, $\mathbf{N}$, $\mathbf{M}$ will produce respectively the coefficients, fitted values, and residuals from the LS linear regression of that vector upon $\mathbf{X}$. With that in mind, we can interpret the results

$$\mathbf{AX} = \mathbf{I}, \qquad \mathbf{NX} = \mathbf{X}, \qquad \mathbf{MX} = \mathbf{O}.$$

Suppose we regress the jth column of $\mathbf{X}$, namely $\mathbf{x}_j$, upon all the columns of $\mathbf{X}$ (including the jth). Since LS linear regression chooses the linear combination of the columns of $\mathbf{X}$ that comes closest to $\mathbf{x}_j$, it is obvious that it will produce 1 as the coefficient on the jth explanatory variable, and 0's as the coefficients on all the other explanatory variables. It is equally obvious that the fit will be perfect: the fitted values will equal the observed values and the residuals will all be zero. That is,

$$\mathbf{A}\mathbf{x}_j = \mathbf{d}_j, \qquad \mathbf{N}\mathbf{x}_j = \mathbf{x}_j, \qquad \mathbf{M}\mathbf{x}_j = \mathbf{0},$$

where $\mathbf{d}_j$ is the jth column of the $k \times k$ identity matrix. Assembling those unusual regressions for $j = 1, \ldots, k$, we have indeed

$$\mathbf{AX} = \mathbf{I}, \qquad \mathbf{NX} = \mathbf{X}, \qquad \mathbf{MX} = \mathbf{O}.$$

17.2. Short and Long Regression Algebra

In empirical work, it is common to run a series of regressions, with successively shorter (or longer) lists of explanatory variables. We develop some algebra that is relevant to this practice.

Given as data the $n \times 1$ vector $\mathbf{y}$ and the $n \times k$ matrix $\mathbf{X} = (\mathbf{x}_1, \ldots, \mathbf{x}_k)$, whose rank is k, we regress $\mathbf{y}$ on $\mathbf{X}$. That is, we choose $\mathbf{c}$ to minimize $\mathbf{u}'\mathbf{u}$, where $\mathbf{u} = \mathbf{y} - \mathbf{Xc}$, producing the coefficient and residual vectors

$$\mathbf{b} = \mathbf{Ay}, \qquad \mathbf{e} = \mathbf{My},$$

where

$$\mathbf{A} = (\mathbf{X}'\mathbf{X})^{-1}\mathbf{X}', \qquad \mathbf{M} = \mathbf{I} - \mathbf{XA},$$

and

$$\mathbf{AX} = \mathbf{I}, \qquad \mathbf{MX} = \mathbf{O}, \qquad \mathbf{X}'\mathbf{e} = \mathbf{0}.$$

Partition $\mathbf{X}$ as $\mathbf{X} = (\mathbf{X}_1, \mathbf{X}_2)$, where $\mathbf{X}_1$ is $n \times k_1$ and $\mathbf{X}_2$ is $n \times k_2$. Correspondingly, partition $\mathbf{b}$ as $(\mathbf{b}_1', \mathbf{b}_2')'$, where $\mathbf{b}_1$ is $k_1 \times 1$ and $\mathbf{b}_2$ is $k_2 \times 1$. As a result of the fit, we have

$$(17.1) \qquad \mathbf{y} = \mathbf{Xb} + \mathbf{e} = (\mathbf{X}_1, \mathbf{X}_2)\begin{pmatrix}\mathbf{b}_1 \\ \mathbf{b}_2\end{pmatrix} + \mathbf{e} = \mathbf{X}_1\mathbf{b}_1 + \mathbf{X}_2\mathbf{b}_2 + \mathbf{e}.$$

Because $\mathbf{X}'\mathbf{e} = \mathbf{0}$, we know that $\mathbf{X}_1'\mathbf{e} = \mathbf{0}$ and $\mathbf{X}_2'\mathbf{e} = \mathbf{0}$.

Suppose that we shorten the list of explanatory variables and regress $\mathbf{y}$ on only the first k_1 columns of $\mathbf{X}$. Regressing $\mathbf{y}$ on $\mathbf{X}_1$, that is, choosing the $k_1 \times 1$ vector $\mathbf{c}_1$ to minimize $\mathbf{u}^{*\prime}\mathbf{u}^*$ where $\mathbf{u}^* = \mathbf{y} - \mathbf{X}_1\mathbf{c}_1$, is a full-rank problem. The resulting coefficient vector and residual vector are

$$\mathbf{b}_1^* = \mathbf{A}_1\mathbf{y}, \qquad \mathbf{e}^* = \mathbf{M}_1\mathbf{y},$$

where

$$\mathbf{A}_1 = (\mathbf{X}_1'\mathbf{X}_1)^{-1}\mathbf{X}_1', \qquad \mathbf{M}_1 = \mathbf{I} - \mathbf{X}_1\mathbf{A}_1,$$

and of course

$$\mathbf{A}_1\mathbf{X}_1 = \mathbf{I}, \qquad \mathbf{M}_1\mathbf{X}_1 = \mathbf{O}, \qquad \mathbf{X}_1'\mathbf{e}^* = \mathbf{0}.$$

As a result of this fit, we have

$$(17.2) \qquad \mathbf{y} = \mathbf{X}_1\mathbf{b}_1^* + \mathbf{e}^*.$$

How are $\mathbf{b}_1^*$ and $\mathbf{e}^*$ related to $\mathbf{b}_1$ and $\mathbf{e}$? That is, how is the *short regression* (Eq. 17.2) related to the *long regression* (Eq. 17.1)? Here "short" and "long" refer to the length of the list of explanatory variables.

To obtain the relations, consider the *auxiliary regressions*. Regress each column of $\mathbf{X}_2$ in turn upon $\mathbf{X}_1$ to obtain a set of auxiliary regressions:

$$\mathbf{x}_j = \mathbf{X}_1\mathbf{f}_j + \mathbf{x}_j^* \qquad (j = k_1 + 1, \ldots, k),$$

where $\mathbf{f}_j = \mathbf{A}_1\mathbf{x}_j$ and $\mathbf{x}_j^* = \mathbf{M}_1\mathbf{x}_j$. Assemble all these as

$$(17.3) \qquad \mathbf{X}_2 = \mathbf{X}_1\mathbf{F} + \mathbf{X}_2^*.$$

Here

$$\mathbf{F} = (\mathbf{f}_{k_1+1}, \ldots, \mathbf{f}_k) = \mathbf{A}_1\mathbf{X}_2$$

is a $k_1 \times k_2$ matrix, each column of which contains the coefficients from the regression of a column of $\mathbf{X}_2$ upon (all the columns of) $\mathbf{X}_1$, and

$$\mathbf{X}_2^* = (\mathbf{x}_{k_1+1}^*, \ldots, \mathbf{x}_k^*) = \mathbf{M}_1\mathbf{X}_2$$

is an $n \times k_2$ matrix, each column of which contains the residuals from the regression of a column of $\mathbf{X}_2$ upon (all the columns of) $\mathbf{X}_1$. Because $\mathbf{X}$ has full column rank, so does $\mathbf{X}_2^*$: see Exercise 17.3.

Use Eq. (17.1) to calculate

$$(17.4) \qquad \mathbf{b}_1^* = \mathbf{A}_1\mathbf{y} = \mathbf{A}_1(\mathbf{X}_1\mathbf{b}_1 + \mathbf{X}_2\mathbf{b}_2 + \mathbf{e}) = \mathbf{b}_1 + \mathbf{F}\mathbf{b}_2,$$

because $\mathbf{A}_1\mathbf{e} = (\mathbf{X}_1'\mathbf{X}_1)^{-1}\mathbf{X}_1'\mathbf{e}$ and $\mathbf{X}_1'\mathbf{e} = \mathbf{0}$. What Eq. (17.4) says is that the coefficients on $\mathbf{X}_1$ in the short regression are a mixture of the coefficients on $\mathbf{X}_1$ and on $\mathbf{X}_2$ in the long regression, with the auxiliary regression coefficients serving as weights in that mixture.

Use Eq. (17.1) again to calculate

$$(17.5) \qquad \mathbf{e}^* = \mathbf{M}_1\mathbf{y} = \mathbf{M}_1(\mathbf{X}_1\mathbf{b}_1 + \mathbf{X}_2\mathbf{b}_2 + \mathbf{e}) = \mathbf{X}_2^*\mathbf{b}_2 + \mathbf{e},$$

because $\mathbf{M}_1\mathbf{e} = (\mathbf{I} - \mathbf{X}_1\mathbf{A}_1)\mathbf{e}$ and $\mathbf{A}_1\mathbf{e} = \mathbf{0}$. What Eq. (17.5) says is that the residuals from the short regression equal the residuals from the long regression plus a mixture of the elements of $\mathbf{X}_2^*$, with the elements of $\mathbf{b}_2$ serving as weights in the mixture.

Use Eq. (17.5) along with the facts that $\mathbf{X}_2^{*\prime} = \mathbf{X}_2'\mathbf{M}_1$, $\mathbf{M}_1\mathbf{e} = \mathbf{e}$, and $\mathbf{X}_2'\mathbf{e} = \mathbf{0}$ to calculate

$$(17.6) \qquad \mathbf{e}^{*\prime}\mathbf{e}^* = \mathbf{e}'\mathbf{e} + \mathbf{b}_2'\mathbf{X}_2^{*\prime}\mathbf{X}_2^*\mathbf{b}_2.$$

Equation (17.6) says that the sum of squared residuals from the short regression exceeds the sum of squared residuals from the long regression by the nonnegative quantity $\mathbf{v}'\mathbf{v}$, where $\mathbf{v} = \mathbf{X}_2^*\mathbf{b}_2$. So shortening a regression cannot improve the fit. The two sums of squared residuals are equal iff $\mathbf{v} = \mathbf{0}$, that is, iff $\mathbf{X}_2^*\mathbf{b}_2 = \mathbf{0}$, that is (because $\text{rank}(\mathbf{X}_2^*) = k_2$), iff $\mathbf{b}_2 = \mathbf{0}$. A simpler argument leads to the same conclusion: running the short regression is equivalent to running the long regression subject to the constraint that the coefficients on $\mathbf{X}_2$ be zero; a constrained minimum cannot be less than an unconstrained one, and the minima are equal iff the constraint is not binding.

We have emphasized the contrast between the short and long regressions, but there are exceptional cases:

(1) If $\mathbf{b}_2 = \mathbf{0}$, then $\mathbf{b}_1^* = \mathbf{b}_1$, $\mathbf{e}^* = \mathbf{e}$, and $\mathbf{e}^{*\prime}\mathbf{e}^* = \mathbf{e}'\mathbf{e}$.

(2) If $\mathbf{X}_1'\mathbf{X}_2 = \mathbf{O}$ (each variable in $\mathbf{X}_1$ is *orthogonal*, in the matrix-algebra sense, to every variable in $\mathbf{X}_2$), then $\mathbf{F} = \mathbf{A}_1\mathbf{X}_2 = \mathbf{O}$ and $\mathbf{X}_2^* = \mathbf{X}_2$, so $\mathbf{b}_1^* = \mathbf{b}_1$, although $\mathbf{e}^* \neq \mathbf{e}$.

17.3. Residual Regression

When we run the long regression of $\mathbf{y}$ on $\mathbf{X}_1$ and $\mathbf{X}_2$ rather than a short regression of $\mathbf{y}$ on $\mathbf{X}_2$ alone, to get coefficients on $\mathbf{X}_2$, it is natural to say that we are getting the effect of $\mathbf{X}_2$ "after controlling for $\mathbf{X}_1$," or "after allowing for the effects of $\mathbf{X}_1$," or "after holding $\mathbf{X}_1$ constant." We can develop some algebra that gives content to that language.

Consider the *residual regression*. Regress $\mathbf{y}$ on $\mathbf{X}_2^* = \mathbf{M}_1\mathbf{X}_2$, the $n \times k_2$ matrix of residuals from the auxiliary regressions of $\mathbf{X}_2$ on $\mathbf{X}_1$. The coefficient vector will be

$$\mathbf{c}_2 = \mathbf{A}_2^*\mathbf{y},$$

where

$$\mathbf{A}_2^* = (\mathbf{X}_2^{*\prime}\mathbf{X}_2^*)^{-1}\mathbf{X}_2^{*\prime} = (\mathbf{X}_2^{*\prime}\mathbf{X}_2^*)^{-1}\mathbf{X}_2'\mathbf{M}_1.$$

Now

$$\mathbf{M}_1\mathbf{y} = \mathbf{M}_1(\mathbf{X}_1\mathbf{b}_1 + \mathbf{X}_2\mathbf{b}_2 + \mathbf{e}) = \mathbf{X}_2^*\mathbf{b}_2 + \mathbf{e},$$

$$\mathbf{X}_2'\mathbf{X}_2^* = \mathbf{X}_2'\mathbf{M}_1\mathbf{X}_2 = \mathbf{X}_2'\mathbf{M}_1'\mathbf{M}_1\mathbf{X}_2 = \mathbf{X}_2^{*\prime}\mathbf{X}_2^*,$$

$$\mathbf{X}_2'\mathbf{e} = \mathbf{0}.$$

So

$$\mathbf{c}_2 = \mathbf{A}_2^*\mathbf{y} = \mathbf{b}_2,$$

which is the $k_2 \times 1$ lower subvector of $\mathbf{b}$, the coefficient vector in the long regression of $\mathbf{y}$ on $\mathbf{X}$. To restate the result, the $\mathbf{b}_2$ subvector of $\mathbf{b}$ is obtainable by a two-step procedure:

(i) Regress $\mathbf{X}_2$ on $\mathbf{X}_1$ to get the residuals $\mathbf{X}_2^*$,
(ii) Regress $\mathbf{y}$ on $\mathbf{X}_2^*$ to get the coefficients $\mathbf{b}_2$.

Because $\mathbf{A}_2^*\mathbf{M}_1 = \mathbf{A}_2^*$, it is also clear that we obtain the same $\mathbf{b}_2$ by a *double residual regression*, regressing $\mathbf{y}^* = \mathbf{M}_1\mathbf{y}$ on $\mathbf{X}_2^*$. With $\mathbf{b}_2$ in hand, we can complete the calculation of the long-regression coefficient vector $\mathbf{b}$, by regressing $\mathbf{y}$ on $\mathbf{X}_1$ alone, obtaining $\mathbf{b}_1^* = \mathbf{A}_1\mathbf{y}$, and then recovering $\mathbf{b}_1$ as $\mathbf{b}_1 = \mathbf{b}_1^* - \mathbf{F}\mathbf{b}_2$, where $\mathbf{F} = \mathbf{A}_1\mathbf{X}_2$.

The residual regression result, namely $\mathbf{b}_2 = \mathbf{c}_2$, gives content to the language used above. For $\mathbf{c}_2$ indeed relates $\mathbf{y}$ to $\mathbf{X}_2$ "after controlling for the effects of $\mathbf{X}_1$" in the sense that only $\mathbf{X}_2^*$—the component of $\mathbf{X}_2$ that is not linearly related to $\mathbf{X}_1$—is used to account for $\mathbf{y}$. For example, in looking for the relation of earnings to experience in a regression that also includes education, we are in effect using not experience, but only the component of experience that is not linearly associated with education.

The situation here is quite reminiscent of the distinction between partial and total derivatives in calculus. Indeed $\mathbf{b}_1^* = \mathbf{b}_1 + \mathbf{F}\mathbf{b}_2$ has the same pattern as $dy/dx_1 = \partial y/\partial x_1 + (dx_2/dx_1)(\partial y/\partial x_2)$.

17.4. Applications of Residual Regression

The residual regression results are remarkably useful for theory and practice.

Trend Removal. A popular specification for economic time series takes the form $E(y_t) = \Sigma_{j=1}^k \beta_j x_{tj}$, where t indexes the observations, $x_1 = 1$, $x_2 = t$, and $x_3, \ldots, x_k$ are conventional explanatory variables. Here x_2 allows for a linear trend in the expected value of the dependent variable. The question arises: should the trend term be included in the regression, or should the variables first be "detrended" and then used without the trend terms included? In the first volume of *Econometrica*, Frisch and Waugh (1933) concluded that it does not matter. The residual regression results apply. Let $\mathbf{X}_1 = (\mathbf{x}_1, \mathbf{x}_2)$ and $\mathbf{X}_2 = (\mathbf{x}_3, \ldots, \mathbf{x}_k)$. The coefficients on the latter set will be the same whether we include $\mathbf{X}_1$ in the regression along with $\mathbf{X}_2$, or alternatively first detrend $\mathbf{X}_2$ (by calculating the residuals $\mathbf{X}_2^*$ from the regression of $\mathbf{X}_2$ on $\mathbf{X}_1$) and then use only those detrended explanatory variables.

Seasonal Adjustment. Some macroeconomic variables show a seasonal pattern. When tracking such a variable, it is useful for some purposes to "deseasonalize," or "seasonally adjust," the variable. Suppose that we have data on the variable y, quarter by quarter, for m years, with y_{th} being the value for year t, quarter h. Suppose that we are looking for a business-cycle pattern in this series, but we notice that first-quarter values are typically low and third-quarter values are typically high. It seems sensible to deseasonalize before judging, say, whether the series is now at a cyclical peak. A conventional way to deseasonalize is to calculate the seasonal means, $\bar{y}_1, \bar{y}_2, \bar{y}_3, \bar{y}_4$, say, and express each observation as a deviation from its seasonal mean: $y_{th}^* = y_{th} - \bar{y}_h$. These y^*'s form a seasonally adjusted series. (The grand mean $\bar{y}$ can be added back in to restore the level of the series.) The cyclical standing of the variable may be more apparent in the y^* series than in the y series.

This calculation can be performed by regression. We have $n = 4m$ observations on y, arranged by quarters within years. Define the four "seasonal dummy variables":

$$x_1 = \begin{cases} 1 \text{ in quarter 1} \\ 0 \text{ otherwise} \end{cases} \qquad x_2 = \begin{cases} 1 \text{ in quarter 2} \\ 0 \text{ otherwise} \end{cases}$$

$$x_3 = \begin{cases} 1 \text{ in quarter 3} \\ 0 \text{ otherwise} \end{cases} \qquad x_4 = \begin{cases} 1 \text{ in quarter 4} \\ 0 \text{ otherwise} \end{cases}$$

Let $\mathbf{X}_1 = (\mathbf{x}_1, \mathbf{x}_2, \mathbf{x}_3, \mathbf{x}_4)$. Then regress $\mathbf{y}$ on $\mathbf{X}_1$ to get coefficients $\mathbf{b}_1^* = \mathbf{A}_1\mathbf{y}$, and residuals $\mathbf{y}^* = \mathbf{M}_1\mathbf{y}$. These residuals form the seasonally adjusted series. (Again the grand mean can be added to restore the levels.)

To verify the assertion, observe that in view of the arrangement of the data (seasons within years), $\mathbf{X}_1 = (\mathbf{I}, \mathbf{I}, \ldots, \mathbf{I})'$, where each of the $\mathbf{I}$'s is 4×4. Then $\mathbf{X}_1'\mathbf{X}_1 = m\mathbf{I}_4$ and $\mathbf{X}_1'\mathbf{y} = m(\bar{y}_1, \bar{y}_2, \bar{y}_3, \bar{y}_4)'$. So the coefficient vector is $\mathbf{b}_1^* = (\bar{y}_1, \bar{y}_2, \bar{y}_3, \bar{y}_4)'$, and the residuals are $\mathbf{y}^* = \mathbf{M}_1\mathbf{y} = \{y_{th} - \bar{y}_h\}$, as asserted.

Linear Regression with Seasonal Data. Suppose that we are interested in the regression relation between $\mathbf{y}$ and a set of explanatory variables $\mathbf{X}_2$.

Ordinarily we would regress $\mathbf{y}$ on the summer vector $\mathbf{s} = (1, \ldots, 1)'$ and $\mathbf{X}_2$ to get an intercept and a set of slopes. But if $\mathbf{y}$ and $\mathbf{X}_2$ have seasonal components in them, we might want to remove the seasonals first. We have already done this for $\mathbf{y}$. To deseasonalize $\mathbf{X}_2$, regress $\mathbf{X}_2$ on the seasonal dummy variables $\mathbf{X}_1$, getting coefficients $\mathbf{F} = \mathbf{A}_1\mathbf{X}_2$ and residuals $\mathbf{X}_2^* = \mathbf{M}_1\mathbf{X}_2$. Then regress the seasonally-adjusted dependent variable $\mathbf{y}^* = \mathbf{M}_1\mathbf{y}$ on the seasonally adjusted explanatory variables $\mathbf{X}_2^*$ to get the slopes $\mathbf{c}_2$. Residual regression theory tells us that

$$\mathbf{c}_2 = \mathbf{A}_2^*\mathbf{y}^* = \mathbf{A}_2^*\mathbf{y} = \mathbf{b}_2,$$

which are the coefficients on $\mathbf{X}_2$ in the long regression of $\mathbf{y}$ on $(\mathbf{X}_1, \mathbf{X}_2)$. Running the regression on seasonally adjusted data in effect allows for parallel shifts in the relationship of $\mathbf{y}$ to $\mathbf{X}_2$—that is, separate intercepts for each of the quarters. To recover those intercepts, use $\mathbf{b}_1 = \mathbf{b}_1^* - \mathbf{F}\mathbf{b}_2$.

Deviations from Means. In simple regression, where there is only one explanatory variable along with the constant, $\hat{y}_i = a + bx_i$, the LS slope and intercept can be calculated as

$$b = \left[\sum_i (x_i - \bar{x})(y_i - \bar{y})\right] \bigg/ \left[\sum_i (x_i - \bar{x})^2\right], \qquad a = \bar{y} - b\bar{x}.$$

It is also true that

$$b = \left[\sum_i (x_i - \bar{x})y_i\right] \bigg/ \left[\sum_i (x_i - \bar{x})^2\right].$$

Deviations from the mean are residuals from regression on the summer vector, so these well-known formulas are applications of residual regression theory.

Turn to multiple regression of $\mathbf{y}$ on $\mathbf{X} = (\mathbf{x}_1, \mathbf{X}_2)$, where $\mathbf{x}_1$ is the summer vector. The fitted regression is

$$\hat{\mathbf{y}} = \mathbf{x}_1 b_1 + \mathbf{X}_2\mathbf{b}_2,$$

where b_1 is the intercept and $\mathbf{b}_2$ is the slope vector. Here

$$\mathbf{M}_1 = \mathbf{I} - \mathbf{x}_1(\mathbf{x}_1'\mathbf{x}_1)^{-1}\mathbf{x}_1' = \mathbf{I} - (1/n)\,\mathbf{x}_1\mathbf{x}_1'$$

is the idempotent matrix which, when multiplied into any column vector, produces deviations from the column mean. So $\mathbf{y}^* = \mathbf{M}_1\mathbf{y}$ and $\mathbf{X}_2^* = \mathbf{M}_1\mathbf{X}_2$ are the variables expressed as deviations from their respective means. Residual regression theory says that the slopes can be calculated as

$$\mathbf{b}_2 = (\mathbf{X}_2^{*\prime}\mathbf{X}_2^*)^{-1}\mathbf{X}_2^{*\prime}\mathbf{y} = (\mathbf{X}_2^{*\prime}\mathbf{X}_2^*)^{-1}\mathbf{X}_2^{*\prime}\mathbf{y}^*,$$

and the intercept recovered as

$$b_1 = b_1^* - \mathbf{F}\mathbf{b}_2 = \bar{y} - \bar{\mathbf{x}}_2'\mathbf{b}_2,$$

where

$$\bar{\mathbf{x}}_2' = \mathbf{A}_1\mathbf{X}_2 = (\mathbf{x}_1'\mathbf{x}_1)^{-1}\mathbf{x}_1'\mathbf{X}_2 = (1/n)\,\mathbf{x}_1'\mathbf{X}_2$$

is the row vector of means of the explanatory variables. Thus the familiar device that expresses variables as deviations about their means before running the regression carries over to the multiple regression case.

17.5. Short and Residual Regressions in the Classical Regression Model

We have considered the short and residual regressions from an algebraic perspective. We now reconsider them in a statistical context. Suppose that the CR model applies, so that $E(\mathbf{y}) = \mathbf{X}\boldsymbol{\beta}$, $V(\mathbf{y}) = \sigma^2\mathbf{I}$, $\mathbf{X}$ nonstochastic, rank($\mathbf{X}$) = k. Partition as follows:

$$\mathbf{X} = (\mathbf{X}_1, \mathbf{X}_2), \qquad \boldsymbol{\beta} = \begin{pmatrix} \boldsymbol{\beta}_1 \\ \boldsymbol{\beta}_2 \end{pmatrix},$$

where $\mathbf{X}_1$ is $n \times k_1$, $\mathbf{X}_2$ is $n \times k_2$, $\boldsymbol{\beta}_1$ is $k_1 \times 1$, and $\boldsymbol{\beta}_2$ is $k_2 \times 1$. We have

$$E(\mathbf{y}) = \mathbf{X}_1\boldsymbol{\beta}_1 + \mathbf{X}_2\boldsymbol{\beta}_2.$$

The long regression gives

$$\mathbf{y} = \mathbf{X}\mathbf{b} + \mathbf{e} = \mathbf{X}_1\mathbf{b}_1 + \mathbf{X}_2\mathbf{b}_2 + \mathbf{e},$$

and we know that

$$E\begin{pmatrix} \mathbf{b}_1 \\ \mathbf{b}_2 \end{pmatrix} = \begin{pmatrix} \boldsymbol{\beta}_1 \\ \boldsymbol{\beta}_2 \end{pmatrix}, \qquad V(\mathbf{b}) = \sigma^2\mathbf{Q}^{-1} = \sigma^2\begin{pmatrix} \mathbf{Q}^{11} & \mathbf{Q}^{12} \\ \mathbf{Q}^{21} & \mathbf{Q}^{22} \end{pmatrix},$$

where the $\mathbf{Q}$'s with superscripts denote submatrices in the inverse.

Short-Regression Coefficients. Consider the regression of $\mathbf{y}$ on $\mathbf{X}_1$ alone. If the CR model applies to $E(\mathbf{y}) = \mathbf{X}_1\boldsymbol{\beta}_1 + \mathbf{X}_2\boldsymbol{\beta}_2$, then in general $E(\mathbf{y}) \neq$

$\mathbf{X}_1\boldsymbol{\beta}_1$. Nevertheless we can evaluate the short regression results. The coefficient vector is $\mathbf{b}_1^* = \mathbf{A}_1\mathbf{y}$. Apply R2 (Section 15.1) to obtain

$$(17.7)\qquad E(\mathbf{b}_1^*) = \mathbf{A}_1 E(\mathbf{y}) = \mathbf{A}_1(\mathbf{X}_1\boldsymbol{\beta}_1 + \mathbf{X}_2\boldsymbol{\beta}_2) = \boldsymbol{\beta}_1 + \mathbf{F}\boldsymbol{\beta}_2,$$

$$(17.8)\qquad V(\mathbf{b}_1^*) = \mathbf{A}_1 V(\mathbf{y})\mathbf{A}_1' = \sigma^2\mathbf{A}_1\mathbf{A}_1' = \sigma^2(\mathbf{X}_1'\mathbf{X}_1)^{-1},$$

using $\mathbf{A}_1\mathbf{X}_1 = \mathbf{I}$, $\mathbf{A}_1\mathbf{X}_2 = \mathbf{F}$, and $\mathbf{A}_1\mathbf{A}_1' = (\mathbf{X}_1'\mathbf{X}_1)^{-1}$.

From Eq. (17.7), we conclude that in general $\mathbf{b}_1^*$ is a biased estimator of $\boldsymbol{\beta}_1$, a result known as *omitted-variable bias*. The exceptional cases are:

(a) *Irrelevant Omitted Variables*. If $\boldsymbol{\beta}_2 = \mathbf{0}$, then the CR model does apply to the short regression $E(\mathbf{y}) = \mathbf{X}_1\boldsymbol{\beta}_1$. Here $\mathbf{b}_1^*$ and $\mathbf{b}_1$ differ in any sample but have the same expectation.

(b) *Orthogonal Explanatory Variables*. If $\mathbf{F} = \mathbf{O}$, then $\mathbf{b}_1^*$ and $\mathbf{b}_1$ coincide in every sample, even though the CR model does not apply to the short regression.

From Eq. (17.8) it follows that $V(\mathbf{b}_1) \geq V(\mathbf{b}_1^*)$. Rewrite Eq. (17.4) as $\mathbf{b}_1 = \mathbf{b}_1^* - \mathbf{F}\mathbf{b}_2$. Now

$$C(\mathbf{b}_1^*, \mathbf{b}_2) = \mathbf{A}_1 V(\mathbf{y})\mathbf{A}_2^{*\prime} = \sigma^2\mathbf{A}_1\mathbf{A}_2^{*\prime} = \mathbf{O},$$

because $\mathbf{M}_1\mathbf{X}_1 = \mathbf{O}$ implies $\mathbf{A}_2^*\mathbf{A}_1' = \mathbf{O}$. Consequently,

$$(17.9)\qquad V(\mathbf{b}_1) = V(\mathbf{b}_1^*) + \mathbf{F}V(\mathbf{b}_2)\mathbf{F}'.$$

Because $\mathbf{F}V(\mathbf{b}_2)\mathbf{F}'$ is nonnegative definite, we have $V(\mathbf{b}_1) \geq V(\mathbf{b}_1^*)$.

Observe that the variance matrix of $\mathbf{b}_1^*$ does not depend on the true value of $\boldsymbol{\beta}_2$, although the expectation of $\mathbf{b}_1^*$ does. Whether or not $\boldsymbol{\beta}_2 = \mathbf{0}$, the short-regression coefficient estimator has smaller variance. This suggests that in practice there may be a bias-variance trade-off between short and long regressions when the target of interest is $\boldsymbol{\beta}_1$. Observe also that the short-regression coefficient vector $\mathbf{b}_1^*$ is an *unbiased* estimator of a certain mixture of parameters: $E(\mathbf{b}_1^*) = \boldsymbol{\beta}_1^*$, where $\boldsymbol{\beta}_1^* = \boldsymbol{\beta}_1 + \mathbf{F}\boldsymbol{\beta}_2$. We return to these two observations in Chapter 24.

Short-Regression Residuals. Next, turn to the short-regression residuals, $\mathbf{e}^* = \mathbf{M}_1\mathbf{y}$. We have

$$E(\mathbf{e}^*) = \mathbf{M}_1 E(\mathbf{y}) = \mathbf{M}_1(\mathbf{X}_1\boldsymbol{\beta}_1 + \mathbf{X}_2\boldsymbol{\beta}_2) = \mathbf{X}_2^*\boldsymbol{\beta}_2,$$

$$V(\mathbf{e}^*) = \mathbf{M}_1 V(\mathbf{y})\mathbf{M}_1' = \sigma^2\mathbf{M}_1,$$

using $\mathbf{M}_1\mathbf{X}_1 = \mathbf{O}$, $\mathbf{M}_1\mathbf{X}_2 = \mathbf{X}_2^*$, $\mathbf{M}_1\mathbf{M}_1' = \mathbf{M}_1$. So in general the short-regression residual vector has a nonzero expectation. Because $\text{rank}(\mathbf{X}_2^*) = k_2$, the only exceptional case is $\boldsymbol{\beta}_2 = \mathbf{0}$. For the sum of squared residuals, we have $\mathbf{e}^{*\prime}\mathbf{e}^{*} = \mathbf{y}'\mathbf{M}_1\mathbf{y}$, and R5 (Section 15.1) applies with $\mathbf{M}_1$ taking the role of $\mathbf{T}$:

$$E(\mathbf{e}^{*\prime}\mathbf{e}^{*}) = \text{tr}[\mathbf{M}_1 V(\mathbf{y})] + E(\mathbf{y})'\mathbf{M}_1 E(\mathbf{y}).$$

But $\mathbf{M}_1 V(\mathbf{y}) = \sigma^2\mathbf{M}_1$, $\mathbf{M}_1 E(\mathbf{y}) = \mathbf{X}_2^*\boldsymbol{\beta}_2$, and $\text{tr}(\mathbf{M}_1) = n - k_1$, so

$$E(\mathbf{e}^{*\prime}\mathbf{e}^{*}) = \sigma^2\,\text{tr}(\mathbf{M}_1) + \boldsymbol{\beta}_2'\mathbf{X}_2^{*\prime}\mathbf{X}_2^*\boldsymbol{\beta}_2 = \sigma^2(n - k_1) + \boldsymbol{\beta}_2'\mathbf{X}_2^{*\prime}\mathbf{X}_2^*\boldsymbol{\beta}_2.$$

And $E(\mathbf{e}'\mathbf{e}) = \sigma^2(n - k)$, so

$$E(\mathbf{e}^{*\prime}\mathbf{e}^{*}) - E(\mathbf{e}'\mathbf{e}) = \sigma^2 k_2 + \boldsymbol{\beta}_2'\mathbf{X}_2^{*\prime}\mathbf{X}_2^*\boldsymbol{\beta}_2.$$

On the right-hand side, the first term is positive and the second term is nonnegative, and is zero iff $\boldsymbol{\beta}_2 = \mathbf{0}$.

We conclude that omission of explanatory variables leads to an increase in the expected sum of squared residuals, even if $\boldsymbol{\beta}_2 = \mathbf{0}$. The increase in expectation should come as no surprise because we know from Eq. (17.6) that $\mathbf{e}^{*\prime}\mathbf{e}^{*} - \mathbf{e}'\mathbf{e} = \mathbf{b}_2'\mathbf{X}_2^{*\prime}\mathbf{X}_2^*\mathbf{b}_2 \geq 0$ in every sample, with equality iff $\mathbf{b}_2 = \mathbf{0}$. Omitting $\mathbf{X}_2$ never reduces the sum of squared residuals, and almost always increases it, so on average it must increase it.

Residual Regression. Finally, consider the residual regression of $\mathbf{y}$ on $\mathbf{X}_2^* = \mathbf{M}_1\mathbf{X}_2$. The coefficient vector is $\mathbf{c}_2 = \mathbf{A}_2^*\mathbf{y} = \mathbf{b}_2$. Apply R2 to find

$$E(\mathbf{b}_2) = \mathbf{A}_2^* E(\mathbf{y}) \quad = \mathbf{A}_2^*(\mathbf{X}_1\boldsymbol{\beta}_1 + \mathbf{X}_2\boldsymbol{\beta}_2) = \boldsymbol{\beta}_2,$$

$$V(\mathbf{b}_2) = \mathbf{A}_2^* V(\mathbf{y})\mathbf{A}_2^{*\prime} = \sigma^2\mathbf{A}_2^*\mathbf{A}_2^{*\prime} = \sigma^2(\mathbf{X}_2^{*\prime}\mathbf{X}_2^*)^{-1} = \sigma^2(\mathbf{Q}_{22}^*)^{-1},$$

using

$$\mathbf{A}_2^*\mathbf{X}_1 = \mathbf{O}, \qquad \mathbf{A}_2^*\mathbf{X}_2 = \mathbf{I}, \qquad \mathbf{A}_2^*\mathbf{A}_2^{*\prime} = (\mathbf{X}_2^{*\prime}\mathbf{X}_2^*)^{-1},$$

and defining

$$\mathbf{Q}_{22}^* = \mathbf{X}_2^{*\prime}\mathbf{X}_2^* = \mathbf{X}_2'\mathbf{M}_1\mathbf{X}_2 = \mathbf{X}_2'\mathbf{X}_2 - \mathbf{X}_2'\mathbf{X}_1(\mathbf{X}_1'\mathbf{X}_1)^{-1}\mathbf{X}_1'\mathbf{X}_2.$$

But $\mathbf{b}_2$ is the lower $k_2 \times 1$ subvector of $\mathbf{b}$, so $V(\mathbf{b}_2)$ is also given by the southeast $(k_2 \times k_2)$ block of the matrix $V(\mathbf{b}) = \sigma^2\mathbf{Q}^{-1}$. Thus we have proved an algebraic result:

SUBMATRIX OF INVERSE THEOREM. Suppose that a positive definite matrix $\mathbf{Q}$ and its inverse $\mathbf{Q}^{-1}$ are partitioned conformably as

$$\mathbf{Q} = \begin{pmatrix} \mathbf{Q}_{11} & \mathbf{Q}_{12} \\ \mathbf{Q}_{21} & \mathbf{Q}_{22} \end{pmatrix}, \qquad \mathbf{Q}^{-1} = \begin{pmatrix} \mathbf{Q}^{11} & \mathbf{Q}^{12} \\ \mathbf{Q}^{21} & \mathbf{Q}^{22} \end{pmatrix},$$

where the diagonal blocks are square. Then $\mathbf{Q}^{22} = (\mathbf{Q}_{22}^*)^{-1}$, where

$$\mathbf{Q}_{22}^* = \mathbf{Q}_{22} - \mathbf{Q}_{21}(\mathbf{Q}_{11})^{-1}\mathbf{Q}_{12}.$$

Exercises

17.1 Using national time series data, I run the linear regression of $\mathbf{y}_1$ (consumption) on $\mathbf{x}_1$ (the summer vector) and $\mathbf{x}_2$ (disposable income), obtaining $\mathbf{b}_1$, $\hat{\mathbf{y}}_1$, and $\mathbf{e}_1$ as the vectors of coefficients, fitted values, and residuals. I also run the linear regression of $\mathbf{y}_2$ (saving) on the same two explanatory variables, obtaining $\mathbf{b}_2$, $\hat{\mathbf{y}}_2$, and $\mathbf{e}_2$. In the data, of course, consumption + saving = disposable income, so $\mathbf{y}_1 + \mathbf{y}_2 = \mathbf{x}_2$.

(a) Use the **A**, **N**, **M** matrices to show as concisely as possible that

$$\mathbf{b}_1 + \mathbf{b}_2 = \begin{pmatrix} 0 \\ 1 \end{pmatrix}, \qquad \hat{\mathbf{y}}_1 + \hat{\mathbf{y}}_2 = \mathbf{x}_2, \qquad \mathbf{e}_1 + \mathbf{e}_2 = \mathbf{0}.$$

(b) Show that the sum of squared residuals for the savings regression is identical to the sum of squared residuals for the consumption regression.

(c) True or false? (Explain briefly.) The coefficients of determination, the R^2's, are the same for the two regressions.

17.2 Let $\mathbf{Z} = \mathbf{XT}$, where $\mathbf{X}$ is an $n \times k$ matrix with rank k, and $\mathbf{T}$ is a $k \times k$ matrix with rank k. Let $\mathbf{b}$ and $\mathbf{e}$ be the LS coefficient and residual vectors for regression of an $n \times 1$ vector $\mathbf{y}$ on $\mathbf{X}$. Show that regression of $\mathbf{y}$ on $\mathbf{Z}$ gives coefficient vector $\mathbf{c} = \mathbf{T}^{-1}\mathbf{b}$ and residual vector $\mathbf{e}^* = \mathbf{e}$.

17.3 Suppose that the $n \times k$ matrix $\mathbf{X} = (\mathbf{X}_1, \mathbf{X}_2)$ has full column rank. Let $\mathbf{X}_2^* = \mathbf{M}_1\mathbf{X}_2$ be the $n \times k_2$ matrix of residuals from the auxiliary regression of $\mathbf{X}_2$ on $\mathbf{X}_1$. Show that rank$(\mathbf{X}_2^*) = k_2$. Hint: Use proof by contradiction.

17.4 Table A.3 contains a cross-section data set on $n = 100$ family heads from the 1963 Survey of Consumer Finances, as taken from Mirer (1988, pp. 18–22). The variables are:

V1 = Identification number (1, . . . , 100)
V2 = Family size
V3 = Education (years)

V4 = Age (years)
V5 = Experience (years)
V6 = Months worked
V7 = Race (coded as 1 = white, 2 = black)
V8 = Region (coded as 1 = northeast, 2 = northcentral, 3 = south, 4 = west)
V9 = Earnings ($1000)
V10 = Income ($1000)
V11 = Wealth ($1000)
V12 = Savings ($1000)

Experience is defined as age − education − 5. We will use this data set, presumed to be available as an ASCII file labeled SCF, frequently.

Run these four linear earnings regressions:

(a) y on x_1, x_2.
(b) y on x_1, x_2, x_3, x_4.
(c) y on x_1, x_2, x_3, x_4, x_5.
(d) y on x_1, x_2, x_3, x_4, x_5, x_6, x_7, x_8.

Here y = earnings; $x_1 = 1$; x_2 = education; x_3 = experience; $x_4 = x_3^2$; $x_5 = 1$ if black, 0 if white; $x_6 = 1$ if northcentral, 0 otherwise; $x_7 = 1$ if south, 0 otherwise; $x_8 = 1$ if west, 0 otherwise.

For each regression in turn, assume that the CR model holds. Report coefficient estimates, their standard errors, the estimate of σ, and the R^2. Also report the means and standard deviations of all variables used in (d).

GAUSS Hints:

(1) n = 100; load D[n,12] = scf; [This will read in the data set as a 100 × 12 matrix called D.]
(2) y = D[.,9]; [The vector y is equal to the 9th column of D.]
(3) If x is an n × 1 vector, then z = x .* x is the n × 1 vector of squares of the elements of x.
(4) If x and w are n × 1 vectors, then z = x .== w is the n × 1 vector with 1's where the elements in x and w are equal, and 0's where they differ.
(5) If D is a square matrix, then c = diag(D) is the column vector whose elements are the diagonal elements of D.

17.5 Comment on any aspect of the regression results in Exercise 17.4 that puzzles you.

17.6 Continuing Exercise 17.4, let $\mathbf{X}_1 = (\mathbf{x}_1, \mathbf{x}_2, \mathbf{x}_3, \mathbf{x}_4)$ and $\mathbf{X}_2 = (\mathbf{x}_5, \mathbf{x}_6, \mathbf{x}_7, \mathbf{x}_8)$. Write and run a program to:

(a) Calculate the auxiliary regressions of $\mathbf{X}_2$ on $\mathbf{X}_1$, obtaining the coefficient matrix $\mathbf{F} = (\mathbf{X}_1'\mathbf{X}_1)^{-1}\mathbf{X}_1'\mathbf{X}_2$ and the residual matrix $\mathbf{X}_2^* = \mathbf{X}_2 - \mathbf{X}_1\mathbf{F}$.

(b) Calculate the residual regression of $\mathbf{y}$ on $\mathbf{X}_2^*$, obtaining the coefficient vector $\mathbf{c}_2$.

(c) Use those results, along with those found in Exercise 17.4, to verify numerically the relations $\mathbf{b}_1^* = \mathbf{b}_1 + \mathbf{F}\mathbf{b}_2$ and $\mathbf{c}_2 = \mathbf{b}_2$.

18 *Multivariate Normal Distribution*

18.1. Introduction

For the classical regression model we now have considerable information on the sampling distributions of the LS statistics $\mathbf{b}$ and $\mathbf{e}'\mathbf{e}$. That information, which concerns expectations, variances, and covariances, suffices to justify the use of certain sample statistics as estimates of the population parameters, and to provide estimates of their precision as well. We have seen why the sample LS coefficient b_j serves as an estimate of β_j, and why its standard error, $\hat{\sigma}_{b_j} = \hat{\sigma}\sqrt{q^{jj}}$, serves as an estimate of its standard deviation, $\sqrt{V(b_j)} = \sigma_{b_j} = \sigma\sqrt{q^{jj}}$.

But we need more information to undertake further exact statistical inference for the regression parameters, that is, to construct exact confidence intervals and to test hypotheses at exact significance levels. Following the traditional practice, we will specialize the CR model to the case where the y's are normally distributed, and then deduce the exact sampling distributions of the LS statistics. As a preliminary, in this chapter we set aside the regression context in order to develop some general theory on the multivariate normal distribution.

18.2. Multivariate Normality

Suppose that the joint pdf of the $n \times 1$ random vector $\mathbf{y}$ is

$$f(\mathbf{y}) = (2\pi)^{-n/2}|\boldsymbol{\Sigma}|^{-1/2}e^{-w/2}$$

where

$$w = (\mathbf{y} - \boldsymbol{\mu})'\boldsymbol{\Sigma}^{-1}(\mathbf{y} - \boldsymbol{\mu}) \text{ is a scalar,}$$

$\boldsymbol{\mu}$ is an $n \times 1$ parameter vector,

$\boldsymbol{\Sigma}$ is an $n \times n$ positive definite parameter matrix,

$|\boldsymbol{\Sigma}|^{-1/2} = 1/\sqrt{\det(\boldsymbol{\Sigma})}$.

Then we say that the distribution of $\mathbf{y}$ is *multivariate normal,* or *multinormal,* with parameters $\boldsymbol{\mu}$ and $\boldsymbol{\Sigma}$, and write $\mathbf{y} \sim \mathcal{N}(\boldsymbol{\mu}, \boldsymbol{\Sigma})$.

Consider a couple of special cases. If $n = 1$, then $\mathbf{y} = y$, $\boldsymbol{\mu} = \mu$, $\boldsymbol{\Sigma} = \sigma^2$, $w = z^2$, with $z = (y - \mu)/\sigma$. So $f(y)$ is the familiar univariate normal density $f(y) = \exp(-z^2/2)/\sqrt{(2\pi\sigma^2)}$.

If $n = 2$, then

$$\mathbf{y} = \begin{pmatrix} y_1 \\ y_2 \end{pmatrix}, \qquad \boldsymbol{\mu} = \begin{pmatrix} \mu_1 \\ \mu_2 \end{pmatrix}, \qquad \boldsymbol{\Sigma} = \begin{pmatrix} \sigma_{11} & \sigma_{12} \\ \sigma_{21} & \sigma_{22} \end{pmatrix}.$$

So

$$|\boldsymbol{\Sigma}| = \sigma_{11}\sigma_{22} - \sigma_{12}^2 = \sigma_1^2\sigma_2^2(1 - \rho^2),$$

where

$$\rho = \sigma_{12}/(\sigma_1\sigma_2), \qquad \sigma_1^2 = \sigma_{11}, \qquad \sigma_2^2 = \sigma_{22},$$

and

$$\boldsymbol{\Sigma}^{-1} = (1 - \rho^2)^{-1} \begin{pmatrix} 1/\sigma_1^2 & -\rho/(\sigma_1\sigma_2) \\ -\rho/(\sigma_1\sigma_2) & 1/\sigma_2^2 \end{pmatrix}.$$

Also

$$w = (z_1^2 + z_2^2 - 2\rho z_1 z_2)/(1 - \rho^2),$$

with $z_1 = (y_1 - \mu_1)/\sigma_1$, $z_2 = (y_2 - \mu_2)/\sigma_2$. So $f(y_1, y_2)$ is the bivariate normal density of Section 7.3.

Returning to the general multivariate case, partition the $n \times 1$ vector $\mathbf{y}$ into the $n_1 \times 1$ vector $\mathbf{y}_1$ and the $n_2 \times 1$ vector $\mathbf{y}_2$, and correspondingly partition $\boldsymbol{\mu}$ and $\boldsymbol{\Sigma}$:

$$\mathbf{y} = \begin{pmatrix} \mathbf{y}_1 \\ \mathbf{y}_2 \end{pmatrix}, \qquad \boldsymbol{\mu} = \begin{pmatrix} \boldsymbol{\mu}_1 \\ \boldsymbol{\mu}_2 \end{pmatrix}, \qquad \boldsymbol{\Sigma} = \begin{pmatrix} \boldsymbol{\Sigma}_{11} & \boldsymbol{\Sigma}_{12} \\ \boldsymbol{\Sigma}_{21} & \boldsymbol{\Sigma}_{22} \end{pmatrix}.$$

We can now state these generalizations of the properties shown for the bivariate normal in Section 7.4.

If $\mathbf{y} \sim \mathcal{N}(\boldsymbol{\mu}, \boldsymbol{\Sigma})$, *then:*

P1. The expectation vector and variance matrix are $E(\mathbf{y}) = \boldsymbol{\mu}$ and $V(\mathbf{y}) = \boldsymbol{\Sigma}$, thus justifying the symbols used for the parameters.

P2. The marginal distribution of $\mathbf{y}_1$ is multinormal:

$$\mathbf{y}_1 \sim \mathcal{N}(\boldsymbol{\mu}_1, \boldsymbol{\Sigma}_{11}).$$

P3. The conditional distribution of $\mathbf{y}_2$ given $\mathbf{y}_1$ is multinormal:

$$\mathbf{y}_2|\mathbf{y}_1 \sim \mathcal{N}(\boldsymbol{\mu}_2^*, \boldsymbol{\Sigma}_{22}^*),$$

where

$$\boldsymbol{\mu}_2^* = E(\mathbf{y}_2|\mathbf{y}_1) = \boldsymbol{\alpha} + \mathbf{B}'\mathbf{y}_1,$$

$$\mathbf{B} = (\boldsymbol{\Sigma}_{11})^{-1}\boldsymbol{\Sigma}_{12},$$

$$\boldsymbol{\alpha} = \boldsymbol{\mu}_2 - \mathbf{B}'\boldsymbol{\mu}_1,$$

$$\boldsymbol{\Sigma}_{22}^* = V(\mathbf{y}_2|\mathbf{y}_1) = \boldsymbol{\Sigma}_{22} - \mathbf{B}'\boldsymbol{\Sigma}_{11}\mathbf{B}.$$

Observe that the CEF vector is linear in $\mathbf{y}_1$, and that the conditional variance matrix is constant across $\mathbf{y}_1$.

P4. Uncorrelatedness implies independence: If $\boldsymbol{\Sigma}_{12} = \mathbf{O}$, then $\mathbf{y}_1$ and $\mathbf{y}_2$ are independent random vectors.

Proof. If $\boldsymbol{\Sigma}_{12} = \mathbf{O}$, then $\mathbf{B} = \mathbf{O}$, so $\boldsymbol{\mu}_2^* = \boldsymbol{\alpha} = \boldsymbol{\mu}_2$, and $\boldsymbol{\Sigma}_{22}^* = \boldsymbol{\Sigma}_{22}$. Consequently, $\mathbf{y}_2|\mathbf{y}_1 \sim \mathcal{N}(\boldsymbol{\mu}_2, \boldsymbol{\Sigma}_{22})$ for all $\mathbf{y}_1$. These conditional distributions are all the same—they all coincide with the marginal distribution $\mathbf{y}_2 \sim \mathcal{N}(\boldsymbol{\mu}_2, \boldsymbol{\Sigma}_{22})$—so the vectors are stochastically independent. ∎

Of course, the roles of $\mathbf{y}_1$ and $\mathbf{y}_2$ can be reversed throughout.

Consider a special case. Take the bivariate normal distribution, by setting $n_1 = n_2 = 1$. Then P2 and P3 specialize to

$$y_1 \sim \mathcal{N}(\mu_1, \sigma_{11}), \qquad y_2|y_1 \sim \mathcal{N}(\alpha + \beta y_1, \sigma^2),$$

where

$$\beta = \sigma_{12}/\sigma_{11}, \qquad \alpha = \mu_2 - \beta\mu_1, \qquad \sigma^2 = \sigma_{22} - \beta^2\sigma_{11},$$

as in Section 7.4.

For a second special case, set $n_1 = n - 1$, $n_2 = 1$. Here y_2 is scalar, while $\mathbf{y}_1$ is a vector. The variables and parameters partition as

$$\mathbf{y} = \begin{pmatrix} \mathbf{y}_1 \\ y_2 \end{pmatrix}, \qquad \boldsymbol{\mu} = \begin{pmatrix} \boldsymbol{\mu}_1 \\ \mu_2 \end{pmatrix}, \qquad \boldsymbol{\Sigma} = \begin{pmatrix} \boldsymbol{\Sigma}_{11} & \boldsymbol{\sigma}_{12} \\ \boldsymbol{\sigma}'_{12} & \sigma_{22} \end{pmatrix},$$

and we have

$$y_2|\mathbf{y}_1 \sim \mathcal{N}(\alpha + \boldsymbol{\beta}'\mathbf{y}_1, \sigma^2),$$

where

$$\boldsymbol{\beta} = (\boldsymbol{\Sigma}_{11})^{-1}\boldsymbol{\sigma}_{12}, \qquad \alpha = \mu_2 - \boldsymbol{\beta}'\boldsymbol{\mu}_1, \qquad \sigma^2 = \sigma_{22} - \boldsymbol{\beta}'\boldsymbol{\Sigma}_{11}\boldsymbol{\beta}.$$

Evidently this case specifies a population that could support the CR model: the conditional expectation function $E(y_2|\mathbf{y}_1) = \alpha + \boldsymbol{\beta}'\mathbf{y}_1$ is linear, and the conditional variance $V(y_2|\mathbf{y}_1) = \sigma^2$ is constant.

Concluding the properties of the $\mathcal{N}(\boldsymbol{\mu}, \boldsymbol{\Sigma})$ distribution, we have:

P5. Linear functions of a multinormal vector are multinormal: If $\mathbf{z} = \mathbf{g} + \mathbf{H}\mathbf{y}$, where $\mathbf{g}$ and $\mathbf{H}$ are nonrandom, and $\mathbf{H}$ has full row rank, then $\mathbf{z} \sim \mathcal{N}(\mathbf{g} + \mathbf{H}\boldsymbol{\mu}, \mathbf{H}\boldsymbol{\Sigma}\mathbf{H}')$.

As in Section 7.4, the full-row-rank condition is required to rule out degeneracies, by ensuring that $V(\mathbf{z})$ is nonsingular, a prerequisite for multinormality. To see the problem, suppose that $\mathbf{y} = (y_1, y_2)'$, and consider $\mathbf{z} = \mathbf{H}\mathbf{y}$, with

$$\mathbf{H} = \begin{pmatrix} 1 & 1 \\ 1 & 1 \end{pmatrix}.$$

Then $z_1 = y_1 + y_2 = z_2$. So z_1 and z_2, being linear functions of the bivariate normal vector $\mathbf{y}$, are each univariate normal. But with $z_1 = z_2$, their joint density lies entirely over the 45° line, rather than having the characteristic bell shape of the bivariate normal. Also, if $\mathbf{y}$ is $n \times 1$ and $\mathbf{H}$ is $m \times n$ with $m > n$, then $\mathbf{H}$ cannot have full row rank, so $\mathbf{z} = \mathbf{H}\mathbf{y}$ will not be multinormal. Nevertheless, any subvector of $\mathbf{z}$ that is expressible as a full-row-rank linear function of $\mathbf{y}$ will be multinormal. Some texts speak of a degenerate multinormal distribution when $\text{rank}(\mathbf{H}) < m$.

18.3. Functions of a Standard Normal Vector

If the $n \times 1$ random vector $\mathbf{z}$ is distributed $\mathcal{N}(\mathbf{0}, \mathbf{I})$, then we say that $\mathbf{z}$ is a *standard normal vector*. In that event, $z_1, \ldots, z_n$ are independent standard normal variables. The multinormal pdf specializes to

$$f(\mathbf{z}) = (2\pi)^{-n/2} \exp(-\mathbf{z}'\mathbf{z}/2) = \prod_{i=1}^{n} [\exp(-z_i^2/2)/\sqrt{(2\pi)}],$$

which is indeed the product of n standard normal densities.

We restate and extend the theory of Section 8.5 for functions of independent standard normal variables:

(18.1) If $w = \mathbf{z}'\mathbf{z}$, where the $n \times 1$ vector $\mathbf{z} \sim \mathcal{N}(\mathbf{0}, \mathbf{I})$, then $w \sim \chi^2(n)$.

That is, the sum of squares of n independent standard normal variables has the chi-square distribution with parameter n.

(18.2) If $v = (w_1/m)/(w_2/n)$, where $w_1 \sim \chi^2(m)$ and $w_2 \sim \chi^2(n)$ are independent, then $v \sim F(m, n)$.

That is, the ratio of two independent chi-square variables, each divided by its degrees of freedom, has the *Snedecor F distribution*, with numerator and denominator degrees of freedom equal to those of the respective chi-squares. The cdf of this distribution is tabulated in many textbooks.

(18.3) If $u = z/\sqrt{(w/n)}$ where $z \sim \mathcal{N}(0, 1)$ and $w \sim \chi^2(n)$ are independent, then $u \sim t(n)$.

That is, the ratio of a standard normal variable to the square root of an independent chi-square variable divided by its degrees of freedom, has the Student's t-distribution. The parameter is the same as the parameter of the chi-square. Observe that if $u \sim t(n)$, then $u^2 \sim F(1, n)$, which parallels the result that if $z \sim \mathcal{N}(0, 1)$, then $z^2 \sim \chi^2(1)$.

All three arguments reverse. For example, if $w \sim \chi^2(n)$, then w is expressible as the sum of squares of n independent standard normal variables.

Consider a sequence of random variables indexed by n. Two convenient asymptotic (in n) results follow:

(18.4) If $v \sim F(m, n)$ then $mv \xrightarrow{D} \chi^2(m)$.

Proof. Write $v = (w_1/m)/(w_2/n)$, so $mv = w_1/(w_2/n)$. Since $w_2/n = (1/n)\Sigma_i z_i^2$ can be interpreted as a sample mean in random sampling (sample size n) on the random variable z^2, whose expectation is 1, the Law of Large Numbers says $w_2/n \xrightarrow{P} 1$. So by S4 (Section 9.5), the variable mv has the same limiting distribution as $w_1/1$, namely $\chi^2(m)$. ■

(18.5) If $u \sim t(n)$ then $u \xrightarrow{D} \mathcal{N}(0, 1)$.

Proof. Write $u = z/\sqrt{(w/n)}$. Since $w/n \xrightarrow{P} 1$, it follows that $\sqrt{(w/n)} \xrightarrow{P} \sqrt{1} = 1$. So u has the same limiting distribution as $z/1$, namely $\mathcal{N}(0, 1)$. ■

Consequently, if n is large, then one can rely on the chi-square and normal tables for approximate probabilities, rather than referring to the Snedecor F and Student's t tables.

18.4. Quadratic Forms in Normal Vectors

We now establish the distributions of certain functions of a general multinormal vector, by reducing them to the functions of a standard normal vector introduced above.

Q1. Suppose that the $n \times 1$ vector $\mathbf{y} \sim \mathcal{N}(\boldsymbol{\mu}, \boldsymbol{\Sigma})$. Let $w = (\mathbf{y} - \boldsymbol{\mu})'\boldsymbol{\Sigma}^{-1}(\mathbf{y} - \boldsymbol{\mu})$. Then $w \sim \chi^2(n)$.

Proof. It suffices to show that $w = \mathbf{z}'\mathbf{z}$, where the $n \times 1$ vector $\mathbf{z}$ is distributed $\mathcal{N}(\mathbf{0}, \mathbf{I})$. The steps follow.

(i) Since $\boldsymbol{\Sigma}$ is positive definite, we can write $\boldsymbol{\Sigma} = \mathbf{C}\boldsymbol{\Lambda}\mathbf{C}'$ where $\mathbf{C}$ is *orthonormal* (that is, $\mathbf{CC}' = \mathbf{I} = \mathbf{C}'\mathbf{C}$) and $\boldsymbol{\Lambda}$ is diagonal with all diagonal elements positive. The diagonal elements of $\boldsymbol{\Lambda}$ are the characteristic roots (eigenvalues) of $\boldsymbol{\Sigma}$, and the columns of $\mathbf{C}$ are the corresponding characteristic vectors (eigenvectors) of $\boldsymbol{\Sigma}$.

(ii) Let $\boldsymbol{\Lambda}^*$ be the diagonal matrix whose diagonal elements are the reciprocal square roots of the corresponding diagonal elements of $\boldsymbol{\Lambda}$.

(iii) Let $\mathbf{H} = \mathbf{C}\boldsymbol{\Lambda}^*\mathbf{C}'$. Then $\mathbf{H}' = \mathbf{H}$, $\mathbf{H}'\mathbf{H} = \mathbf{C}\boldsymbol{\Lambda}^{-1}\mathbf{C}' = \boldsymbol{\Sigma}^{-1}$, and $\mathbf{H}\boldsymbol{\Sigma}\mathbf{H}' = \mathbf{I}$.

(iv) Let $\boldsymbol{\epsilon} = \mathbf{y} - \boldsymbol{\mu}$. Then $\boldsymbol{\epsilon} \sim \mathcal{N}(\mathbf{0}, \boldsymbol{\Sigma})$.

(v) Let $\mathbf{z} = \mathbf{H}\boldsymbol{\epsilon}$. Then $\mathbf{z} \sim \mathcal{N}(\mathbf{0}, \mathbf{I})$.
(vi) $w = \boldsymbol{\epsilon}'\boldsymbol{\Sigma}^{-1}\boldsymbol{\epsilon} = \boldsymbol{\epsilon}'\mathbf{H}'\mathbf{H}\boldsymbol{\epsilon} = (\mathbf{H}\boldsymbol{\epsilon})'(\mathbf{H}\boldsymbol{\epsilon}) = \mathbf{z}'\mathbf{z}$. ■

Q2. Suppose that the $n \times 1$ vector $\mathbf{u} \sim \mathcal{N}(\mathbf{0}, \mathbf{I})$. Let $\mathbf{M}$ be a nonrandom $n \times n$ idempotent matrix with rank($\mathbf{M}$) $= r \leq n$. Let $w = \mathbf{u}'\mathbf{M}\mathbf{u}$. Then $w \sim \chi^2(r)$.

Proof. It suffices to show that $w = \mathbf{z}_1'\mathbf{z}_1$, where the $r \times 1$ vector $\mathbf{z}_1$ is distributed $\mathcal{N}(\mathbf{0}, \mathbf{I})$. The steps follow.

(i) Since $\mathbf{M}$ is symmetric, $\mathbf{M} = \mathbf{C}\boldsymbol{\Lambda}\mathbf{C}'$, where $\mathbf{C}$ is orthonormal and $\boldsymbol{\Lambda}$ is diagonal.
(ii) Since $\mathbf{M}$ is idempotent, its characteristic roots are either zeroes or ones. Since its rank is r, there are r unit roots and $n - r$ zero roots. These roots are displayed on the diagonal of $\boldsymbol{\Lambda}$, which without loss of generality can be arranged as

$$\boldsymbol{\Lambda} = \begin{pmatrix} \mathbf{I}_{r\times r} & \mathbf{O}_{r\times(n-r)} \\ \mathbf{O}_{(n-r)\times r} & \mathbf{O}_{(n-r)\times(n-r)} \end{pmatrix}.$$

(iii) Partition $\mathbf{C}$ correspondingly as $\mathbf{C} = (\mathbf{C}_1, \mathbf{C}_2)$, where $\mathbf{C}_1$ is $n \times r$, and $\mathbf{C}_2$ is $n \times (n - r)$.
(iv) Because $\mathbf{C}$ is orthonormal, we have

$$\mathbf{C}\mathbf{C}' = (\mathbf{C}_1, \mathbf{C}_2)\begin{pmatrix} \mathbf{C}_1' \\ \mathbf{C}_2' \end{pmatrix} = \mathbf{C}_1\mathbf{C}_1' + \mathbf{C}_2\mathbf{C}_2' = \mathbf{I}_n,$$

$$\mathbf{C}'\mathbf{C} = \begin{pmatrix} \mathbf{C}_1' \\ \mathbf{C}_2' \end{pmatrix}(\mathbf{C}_1, \mathbf{C}_2) = \begin{pmatrix} \mathbf{C}_1'\mathbf{C}_1 & \mathbf{C}_1'\mathbf{C}_2 \\ \mathbf{C}_2'\mathbf{C}_1 & \mathbf{C}_2'\mathbf{C}_2 \end{pmatrix} = \begin{pmatrix} \mathbf{I}_r & \mathbf{O} \\ \mathbf{O} & \mathbf{I}_{n-r} \end{pmatrix}.$$

(v) So

$$\mathbf{C}\boldsymbol{\Lambda} = (\mathbf{C}_1, \mathbf{C}_2)\begin{pmatrix} \mathbf{I} & \mathbf{O} \\ \mathbf{O} & \mathbf{O} \end{pmatrix} = (\mathbf{C}_1, \mathbf{O}),$$

$$\mathbf{M} = \mathbf{C}\boldsymbol{\Lambda}\mathbf{C}' = (\mathbf{C}_1, \mathbf{O})\begin{pmatrix} \mathbf{C}_1' \\ \mathbf{C}_2' \end{pmatrix} = \mathbf{C}_1\mathbf{C}_1'.$$

(vi) Because the $r \times n$ matrix $\mathbf{C}_1'$ is nonrandom with rank r, the $r \times 1$ random vector $\mathbf{z}_1 = \mathbf{C}_1'\mathbf{u}$ is multinormal with

$$E(\mathbf{z}_1) = \mathbf{C}_1'E(\mathbf{u}) = \mathbf{C}_1'\mathbf{0} = \mathbf{0},$$

$$V(\mathbf{z}_1) = \mathbf{C}_1'V(\mathbf{u})\mathbf{C}_1 = \mathbf{C}_1'\mathbf{I}\mathbf{C}_1 = \mathbf{C}_1'\mathbf{C}_1 = \mathbf{I}_r.$$

That is, $\mathbf{z}_1 \sim \mathcal{N}(\mathbf{0}, \mathbf{I}_r)$.

(vii) $w = \mathbf{u}'\mathbf{M}\mathbf{u} = \mathbf{u}'(\mathbf{C}_1\mathbf{C}_1')\mathbf{u} = (\mathbf{u}'\mathbf{C}_1)(\mathbf{C}_1'\mathbf{u}) = \mathbf{z}_1'\mathbf{z}_1$. ∎

Q3. Suppose that the $n \times 1$ vector $\mathbf{u} \sim \mathcal{N}(\mathbf{0}, \mathbf{I})$. Let $\mathbf{M}$ be a nonrandom $n \times n$ idempotent matrix with $\text{rank}(\mathbf{M}) = r \leq n$. Let $\mathbf{L}$ be a nonrandom matrix such that $\mathbf{LM} = \mathbf{O}$. Let $\mathbf{t}_1 = \mathbf{Mu}$ and $\mathbf{t}_2 = \mathbf{Lu}$. Then $\mathbf{t}_1$ and $\mathbf{t}_2$ are independent random vectors.

Proof. It suffices to show that $\mathbf{t}_1$ and $\mathbf{t}_2$ are respectively functions of two independent random vectors. The steps follow.

(i) Using the construction of Q2, again let $\mathbf{z}_1 = \mathbf{C}_1'\mathbf{u}$, and also let $\mathbf{z}_2 = \mathbf{C}_2'\mathbf{u}$. The $n \times 1$ random vector $\mathbf{z} = (\mathbf{z}_1', \mathbf{z}_2')' = \mathbf{C}'\mathbf{u}$ is standard normal, with $\mathbf{z}_1$ and $\mathbf{z}_2$ being independent.
(ii) Now $\mathbf{M} = \mathbf{C}_1\mathbf{C}_1'$, so $\mathbf{t}_1 = \mathbf{Mu} = (\mathbf{C}_1\mathbf{C}_1')\mathbf{u} = \mathbf{C}_1(\mathbf{C}_1'\mathbf{u}) = \mathbf{C}_1\mathbf{z}_1$.
(iii) Let $\mathbf{N} = \mathbf{I} - \mathbf{M} = \mathbf{C}_2\mathbf{C}_2'$. Then $\mathbf{LN} = \mathbf{L}(\mathbf{I} - \mathbf{M}) = \mathbf{L} - \mathbf{LM} = \mathbf{L}$.
(iv) So $\mathbf{t}_2 = \mathbf{Lu} = (\mathbf{LN})\mathbf{u} = \mathbf{L}(\mathbf{C}_2\mathbf{C}_2')\mathbf{u} = \mathbf{LC}_2(\mathbf{C}_2'\mathbf{u}) = \mathbf{LC}_2\mathbf{z}_2$.
(v) Thus $\mathbf{t}_1$ is a function of $\mathbf{z}_1$, and $\mathbf{t}_2$ is a function of $\mathbf{z}_2$. ∎

Exercises

18.1 Suppose that $\mathbf{y} \sim \mathcal{N}(\boldsymbol{\mu}, \boldsymbol{\Sigma})$, with

$$\boldsymbol{\mu} = \begin{pmatrix} 1 \\ 2 \\ 3 \end{pmatrix}, \qquad \boldsymbol{\Sigma} = \begin{pmatrix} 2 & -1 & 1 \\ -1 & 5 & 1 \\ 1 & 1 & 3 \end{pmatrix}.$$

(a) Calculate $E(y_3|y_1, y_2)$ and $V(y_3|y_1, y_2)$.
(b) Find the best prediction of y_3 given that $y_1 = 1 = y_2$.
(c) Calculate $E(y_3|y_1)$ and $V(y_3|y_1)$.
(d) Find the best prediction of y_3 given that $y_1 = 1$.
(e) Find $\Pr(-1 \leq y_3 \leq 2)$.

18.2 For the $\boldsymbol{\Sigma}$ matrix in Exercise 18.1:

(a) Find the characteristic roots $\lambda_1, \lambda_2, \lambda_3$, and a corresponding set of orthonormal characteristic vectors $\mathbf{c}_1, \mathbf{c}_2, \mathbf{c}_3$.
(b) Verify that $\mathbf{C}\boldsymbol{\Lambda}\mathbf{C}' = \boldsymbol{\Sigma}$.

GAUSS Hints:

(1) If S is an n × n symmetric matrix, then the commands

```
C = 0; r = eigsym(S,"C");
```

will produce the n × 1 vector r of characteristic roots of S, and the n × n matrix C of orthonormal characteristic vectors of S.

(2) If T is an n × n matrix and r is an n × 1 vector, then the command

```
T = diagrv(T,r);
```

will replace the diagonal elements of T by the elements of r.

19 *Classical Normal Regression*

19.1. Classical Normal Regression Model

We now strengthen the CR model by assuming that the random vector $\mathbf{y}$ is multivariate normally distributed. What results is the *classical normal regression*, or CNR, *model*, which consists of the assumptions

$$\text{(19.1)} \quad \mathbf{y} \sim \mathcal{N}(\mathbf{X}\boldsymbol{\beta}, \sigma^2\mathbf{I}),$$

$$\text{(19.2)} \quad \mathbf{X} \text{ nonstochastic},$$

$$\text{(19.3)} \quad \text{rank}(\mathbf{X}) = k.$$

Recalling the properties of multinormality, the interpretation is that the random variables $y_1, \ldots, y_n$ are independent, with $y_i \sim \mathcal{N}(\mu_i, \sigma^2)$, where $\mu_i = \mathbf{x}_i'\boldsymbol{\beta}$. (Caution: $\mathbf{x}_i'$ denotes the ith row of $\mathbf{X}'$, not the transpose of the ith column of $\mathbf{X}$.) So the y_i's are independent normal variables that differ in their means, but have the same variance. With respect to the underlying population, we have in mind a joint probability distribution for the random vector $(y, x_2, \ldots, x_k)'$ in which the conditional distribution of y given the x's is

$$y|x_2, \ldots, x_k \sim \mathcal{N}(\beta_1 + \beta_2 x_2 + \cdots + \beta_k x_k, \sigma^2).$$

The normality refers to the conditional distribution of y given the x's. No normality assumption for the x's is being made, although it is true that if the joint distribution of $(y, x_2, \ldots, x_k)'$ is multinormal, then the conditional distribution above will automatically hold. With respect to the sampling, we continue to rely on the classical, stratified-on-x, scheme set out in Section 16.1.

19.2. Maximum Likelihood Estimation

For estimation in the CNR model, we might consider the maximum likelihood approach. In our initial application of this approach (Section 12.4), the data were randomly sampled from a single population. In the CNR model that feature is lacking, but ML estimation is still possible, because the pdf for the vector $\mathbf{y}$ is fully specified up to the parameters $\boldsymbol{\beta}$ and σ^2.

Recall that if an $n \times 1$ random vector $\mathbf{y}$ is distributed $\mathcal{N}(\boldsymbol{\mu}, \boldsymbol{\Sigma})$, then its pdf is

$$f(\mathbf{y}) = (2\pi)^{-n/2}|\boldsymbol{\Sigma}|^{-1/2} \exp(-w/2),$$

where

$$w = \boldsymbol{\epsilon}'\boldsymbol{\Sigma}^{-1}\boldsymbol{\epsilon}, \qquad \boldsymbol{\epsilon} = \mathbf{y} - \boldsymbol{\mu}, \qquad |\boldsymbol{\Sigma}|^{-1/2} = 1/\sqrt{\det(\boldsymbol{\Sigma})}.$$

In the CNR model, $\mathbf{y}$ is distributed $\mathcal{N}(\boldsymbol{\mu}, \boldsymbol{\Sigma})$ with $\boldsymbol{\mu} = \mathbf{X}\boldsymbol{\beta}$ and $\boldsymbol{\Sigma} = \sigma^2\mathbf{I}$. So

$$\boldsymbol{\Sigma}^{-1} = (1/\sigma^2)\mathbf{I}, \qquad |\boldsymbol{\Sigma}| = (\sigma^2)^n, \qquad \boldsymbol{\epsilon} = \mathbf{y} - \mathbf{X}\boldsymbol{\beta},$$

and the pdf simplifies to

$$f(\mathbf{y}) = (2\pi)^{-n/2}(\sigma^2)^{-n/2} \exp[-\boldsymbol{\epsilon}'\boldsymbol{\epsilon}/(2\sigma^2)].$$

As a pdf this is viewed as a function whose argument is $\mathbf{y}$, with (the true values of) $\boldsymbol{\beta}$ and σ^2, and the observed nonrandom $\mathbf{X}$, as givens. But we can also read $f(\mathbf{y})$ as a function $\mathcal{L}(\boldsymbol{\beta}, \sigma^2)$ whose arguments are $\boldsymbol{\beta}$ and σ^2 with $\mathbf{y}$ and $\mathbf{X}$ as givens. Doing so, we have the likelihood function for our sample. The ML estimates of $\boldsymbol{\beta}$ and σ^2 are the values that maximize $\mathcal{L}$, or equivalently maximize its logarithm,

$$L = L(\boldsymbol{\beta}, \sigma^2) = \log \mathcal{L} = -(n/2)\log(2\pi) - (n/2)\log(\sigma^2) - (1/2)\boldsymbol{\epsilon}'\boldsymbol{\epsilon}/\sigma^2.$$

With $\boldsymbol{\epsilon} = \mathbf{y} - \mathbf{X}\boldsymbol{\beta}$, it is immediate that L is maximized with respect to $\boldsymbol{\beta}$ by minimizing $\boldsymbol{\epsilon}'\boldsymbol{\epsilon}$ with respect to $\boldsymbol{\beta}$. But that is just the least-squares criterion, so in the CNR model, the ML estimator of $\boldsymbol{\beta}$ is identical to the LS estimator $\mathbf{b}$.

Inserting the solution value for $\boldsymbol{\beta}$ makes $\boldsymbol{\epsilon}'\boldsymbol{\epsilon} = \mathbf{e}'\mathbf{e}$, with $\mathbf{e} = \mathbf{y} - \mathbf{X}\mathbf{b}$, which leaves the "concentrated log-likelihood function,"

$$L^*(\sigma^2) = L(\mathbf{b}, \sigma^2) = -(n/2)\log(2\pi) - (n/2)\log(\sigma^2) - (1/2)\mathbf{e}'\mathbf{e}/\sigma^2,$$

to be maximized with respect to σ^2. The first derivative is

$$\partial L^*/\partial\sigma^2 = -(n/2)/\sigma^2 + (1/2)\mathbf{e}'\mathbf{e}/\sigma^4.$$

Equating $\partial L^*/\partial\sigma^2$ to zero and solving gives the ML estimator of σ^2 as $\mathbf{e}'\mathbf{e}/n$, which differs only slightly from our previous estimator $\hat{\sigma}^2$. So under the CNR model, ML estimation essentially coincides with LS estimation.

Further, with the pdf specified up to parameters, one may evaluate the Cramér-Rao lower bound for the variance of unbiased estimators. Doing so would show that the LS coefficient vector $\mathbf{b}$ is the minimum variance unbiased estimator of $\boldsymbol{\beta}$ in the CNR model: see Judge et al. (1988, pp. 227–229) or Amemiya (1985, pp. 17–19).

19.3. Sampling Distributions

From a practical point of view, the relevant implications of the CNR model are those that refer to the sampling distributions of the LS statistics $\mathbf{b}$ and $\mathbf{e}'\mathbf{e}$. We defer discussion of $\mathbf{e}'\mathbf{e}$ until Section 21.1.

The key distribution result in the CNR model is that the random vector $\mathbf{b}$ is multinormally distributed:

D1. $\mathbf{b} \sim \mathcal{N}(\boldsymbol{\beta}, \sigma^2\mathbf{Q}^{-1})$.

Proof. Recall that $\mathbf{b} = \mathbf{Ay}$, where $\mathbf{A} = \mathbf{Q}^{-1}\mathbf{X}'$ is a constant $k \times n$ matrix. Multiplication by a nonsingular matrix preserves rank, so $\text{rank}(\mathbf{A}) = \text{rank}(\mathbf{X}') = \text{rank}(\mathbf{X}) = k$. So $\mathbf{b}$ is a full-row-rank linear function of the multinormal vector $\mathbf{y}$, and hence $\mathbf{b}$ is multinormal by P5 (Section 18.2). ■

Any full-row-rank linear function of the multinormal vector $\mathbf{b}$ will also be multinormal. Thus:

D2. Let $\mathbf{t} = \mathbf{Hb}$ and $\boldsymbol{\theta} = \mathbf{H}\boldsymbol{\beta}$, where the $p \times k$ matrix $\mathbf{H}$ is nonrandom with $\text{rank}(\mathbf{H}) = p$. Then $\mathbf{t} \sim \mathcal{N}(\boldsymbol{\theta}, \sigma^2\mathbf{D}^{-1})$, where $\mathbf{D} = (\mathbf{HQ}^{-1}\mathbf{H}')^{-1}$.

As a special case we have:

D3. Let b_j be the jth element of $\mathbf{b}$. Then $b_j \sim \mathcal{N}(\beta_j, \sigma^2 q^{jj})$.

Proof. Take $\mathbf{H} = \mathbf{h}'$, where $\mathbf{h}$ is the $k \times 1$ vector with a 1 in the jth slot and 0's elsewhere. Then $\mathbf{H}\boldsymbol{\beta} = \beta_j$, $\mathbf{Hb} = b_j$, $\mathbf{HQ}^{-1}\mathbf{H}' = q^{jj}$. ■

The result D2 also subsumes other linear functions of the elements of $\mathbf{b}$, such as $\hat{\mu}_i = \mathbf{x}_i'\mathbf{b}$. Indeed it subsumes D1: just take $\mathbf{H} = \mathbf{I}_k$.

Now proceed to quadratic functions of $\mathbf{b}$. Recall from Q1 (Section 18.4) that if the $n \times 1$ random vector $\mathbf{y}$ is distributed $\mathcal{N}(\boldsymbol{\mu}, \boldsymbol{\Sigma})$, then the random variable $w = (\mathbf{y} - \boldsymbol{\mu})'\boldsymbol{\Sigma}^{-1}(\mathbf{y} - \boldsymbol{\mu})$ is distributed $\chi^2(n)$. Applying this to D2 and D3 yields

D4. $w = (\mathbf{t} - \boldsymbol{\theta})'\mathbf{D}(\mathbf{t} - \boldsymbol{\theta})/\sigma^2 \sim \chi^2(p)$,

D5. $w_j = (b_j - \beta_j)^2/(\sigma^2 q^{jj}) \sim \chi^2(1)$.

19.4. Confidence Intervals

We use the distribution results to construct exact confidence intervals and regions, supposing that σ^2 is known. The more practical results, those that are operational when σ^2 is unknown, are deferred until Chapter 21.

Rewrite D3 as

D3A. $z_j = (b_j - \beta_j)/\sigma_{b_j} \sim \mathcal{N}(0, 1)$,

where $\sigma_{b_j}^2 = \sigma^2 q^{jj}$. Then by the logic of Section 11.5, $b_j \pm 1.96\sigma_{b_j}$ is a 95% confidence interval for the unknown parameter β_j. Intervals for different confidence levels can be constructed: for example, in place of 1.96, use 1.645 to get a 90% confidence interval, or 2.576 to get a 99% confidence interval. The higher the confidence level requested, the wider the interval.

For a given confidence level, say 95%, the interval will be wide if $\sigma_{b_j}^2$ is large, that is, if σ^2 is large and/or q^{jj} is large. Focus on the latter component, $q^{jj} = 1/q_{jj}^*$, where

$$q_{jj}^* = \mathbf{x}_j^{*\prime}\mathbf{x}_j^* = \sum_i (x_{ij}^*)^2$$

is the sum of squared residuals in the auxiliary regression of x_j on all the other x's: see the Submatrix of Inverse Theorem (Section 17.5). The confidence interval will be wide (ceteris paribus) if that sum of squared residuals is small. Confine attention to the leading case where β_j is a slope coefficient in a regression that includes a constant. Then the auxiliary regression will also include a constant, so its coefficient of determination, R_j^2 say, is well defined. Hence

$$q_{jj}^* = (1 - R_j^2) \sum_i (x_{ij} - \bar{x}_j)^2.$$

So the interval will be wide (ceteris paribus) if $\Sigma_i(x_{ij} - \bar{x}_j)^2$ is small, and/or if R_j^2 is large. The latter says that *collinearity* of x_j with the other x's tends to produce wide confidence intervals, a topic to which we return in Chapter 23.

To summarize, for an individual slope coefficient β_j, the confidence interval for a given level will be wide—the estimate of β_j will be imprecise—if the population conditional variance of y is large, the variation of x_j about its mean is small, and/or the auxiliary R_j^2 is large.

The procedure developed here also applies to constructing a confidence interval for a single linear combination of the elements of $\boldsymbol{\beta}$. Let $\theta = \mathbf{h}'\boldsymbol{\beta}$ and $t = \mathbf{h}'\mathbf{b}$, where $\mathbf{h}$ is a nonrandom $k \times 1$ vector. With $p = 1$, we can rewrite D2 as $z = (t - \theta)/\sigma_t \sim \mathcal{N}(0, 1)$, where $\sigma_t^2 = \sigma^2\mathbf{h}'\mathbf{Q}^{-1}\mathbf{h} = \mathbf{h}'V(\mathbf{b})\mathbf{h}$. A 95% confidence interval for the scalar parameter θ is given by $t \pm 1.96\sigma_t$.

19.5. Confidence Regions

Preliminaries

Suppose that we are concerned with the parameter pair (θ_1, θ_2). From our sample, we have constructed the two 95% confidence intervals

$$t_1 \pm 1.96\sigma_{t_1}, \qquad t_2 \pm 1.96\sigma_{t_2}.$$

The probability that the first interval covers the true θ_1 is 0.95, and the probability that the second interval covers the true θ_2 is 0.95. Can we combine these two *intervals* to get a 95% confidence *region* for the pair (θ_1, θ_2)? The intersection of the two intervals in the (θ_1, θ_2) plane is a rectangular region, a box. What is the probability that this random box covers the true parameter point (θ_1, θ_2)? Let A_1 be the event that the true θ_1 lies in the first interval, and let A_2 be the event that the true θ_2 lies in the second interval. Then $A = A_1 \cap A_2$ is the event that the true point (θ_1, θ_2) lies in the box. While $\Pr(A_1) = 0.95 = \Pr(A_2)$, it does not follow that $\Pr(A) = 0.95$. So the box is not a 95% joint confidence region.

Indeed if A_1 and A_2 are independent events—that is, if t_1 and t_2 are independent random variables—then $\Pr(A) = \Pr(A_1)\Pr(A_2) = (0.95)^2 = 0.9025$, and the box will be a 90.25% joint confidence region. In general, $C(t_1, t_2)$ is nonzero, so in general A_1 and A_2 are not independent events, so $\Pr(A) \neq 0.9025$. It is possible to calculate $\Pr(A)$ from the BVN distribution of t_1 and t_2, and thus to ascertain the exact confidence level of the box. By the same token it is possible to get an exact 95% box by using an appropriate critical value in place of 1.96. Let A_1^* be the event that the true θ_1 lies in the interval $t_1 \pm c^*\sigma_{t_1}$, let A_2^* be the event that the true θ_2 lies in the interval $t_2 \pm c^*\sigma_{t_2}$, and let $A^* = A_1^* \cap A_2^*$. Then from the BVN distribution of t_1 and t_2, one can find c^* such that $\Pr(A^*) = 0.95$, and thus obtain a box whose exact confidence level is 95%.

But an alternative approach to constructing joint confidence regions is simpler, and conventional as well.

Joint Confidence Region

Suppose that we are concerned with the $p \times 1$ parameter vector $\boldsymbol{\theta} = \mathbf{H}\boldsymbol{\beta}$. Given a sample value of $\mathbf{t}$, and knowledge of σ^2, we rely on D4 to propose

$$(\boldsymbol{\theta} - \mathbf{t})'\mathbf{D}(\boldsymbol{\theta} - \mathbf{t})/\sigma^2 \leq c_p \tag{19.4}$$

as the 95% confidence region for the unknown parameter vector $\boldsymbol{\theta}$. Here c_p is the 5% critical value from the $\chi^2(p)$ table, that is, $G_p(c_p) = 0.95$, where $G_p(\cdot)$ is the cdf of the $\chi^2(p)$ distribution. For example, from Table A.2, $c_1 = 3.84$, $c_2 = 5.99$. The region consists of all $p \times 1$ vectors $\boldsymbol{\theta}$ that satisfy the inequality. Centered at the point $\mathbf{t}$, the region is an ellipsoid, because the matrix $\mathbf{D}/\sigma^2$ is positive definite. Observe that here $\boldsymbol{\theta}$ denotes the argument of a function, not necessarily the true parameter point.

The rationale for the proposal is clear. For arbitrary $\boldsymbol{\theta}$, the left-hand side of the inequality (19.4) is a random ellipsoid, with center that varies from sample to sample as $\mathbf{t}$ varies. For the true $\boldsymbol{\theta}$, the left-hand side of the inequality is the random variable

$$w = (\mathbf{t} - \boldsymbol{\theta})'\mathbf{D}(\mathbf{t} - \boldsymbol{\theta})/\sigma^2,$$

which, according to D4, is distributed $\chi^2(p)$. The true parameter point $\boldsymbol{\theta}$ will lie in or on the random ellipsoid iff $w \leq c_p$. Let A be the event that $w \leq c_p$. We conclude that $\Pr(A) = 0.95$. Thus the probability that A occurs, that is, that the random ellipsoid covers the true $\boldsymbol{\theta}$, is 0.95. That justifies saying that inequality (19.4) provides a 95% confidence region for the parameter vector $\boldsymbol{\theta}$.

Regions for different confidence levels are constructed by using appropriate critical values from the $\chi^2(p)$ table. The higher the confidence level requested, the larger the critical value, and hence the larger the resulting ellipsoid.

If we apply D4 for a single parameter θ, we get the 95% confidence region $w \leq c_1$, where $w = (t - \theta)^2/\sigma_t^2$. Now, $w \leq c_1$ defines an interval on the real line, centered at t. But $w = z^2$, where $z = (t - \theta)/\sigma_t$, and $c_1 = 3.84 = (1.96)^2$. So $w \leq c_1$ is equivalent to $|z| \leq \sqrt{c_1} = 1.96$. That is, the region $w \leq c_1$ coincides with the interval $t \pm 1.96\sigma_t$ of Section 19.4.

It is worth noting that the w of D4 can be written as

$$w = (\mathbf{t} - \boldsymbol{\theta})'[V(\mathbf{t})]^{-1}(\mathbf{t} - \boldsymbol{\theta}),$$

because $(\mathbf{D}/\sigma^2) = (\sigma^2\mathbf{D}^{-1})^{-1} = [V(\mathbf{t})]^{-1}$. So w is a natural generalization of the scalar $z^2 = (t - \theta)^2/\sigma_t^2$.

19.6. Shape of the Joint Confidence Region

To study the joint confidence region and its relation to univariate confidence intervals, it is convenient to take the case $p = 2$, and further specialize as follows. Suppose that

$$\begin{pmatrix} t_1 \\ t_2 \end{pmatrix} \sim \mathcal{N}\left\{\begin{pmatrix} \theta_1 \\ \theta_2 \end{pmatrix}, \begin{pmatrix} 1 & r \\ r & 1 \end{pmatrix}\right\},$$

where r lies between -1 and 1 to ensure positive definiteness.

Here $z_1 = t_1 - \theta_1$ and $z_2 = t_2 - \theta_2$ are each distributed $\mathcal{N}(0, 1)$, while $w_1 = z_1^2$ and $w_2 = z_2^2$ are each distributed $\chi^2(1)$. The relevant 95% critical values are $c_1 = 3.84$ and $c = \sqrt{c_1} = 1.96$. The box, centered at the sample point (t_1, t_2), which is defined by intersecting the intervals $|z_1| \leq c$ and $|z_2| \leq c$, is not a 95% confidence region. However, consider the variable

$$w = (\boldsymbol{\theta} - \mathbf{t})'\mathbf{D}(\boldsymbol{\theta} - \mathbf{t})/\sigma^2 = \mathbf{z}'(\mathbf{D}/\sigma^2)\mathbf{z},$$

where $\mathbf{z} = (z_1, z_2)'$. With

$$V(\mathbf{t}) = \sigma^2\mathbf{D}^{-1} = \begin{pmatrix} 1 & r \\ r & 1 \end{pmatrix},$$

we have

$$\mathbf{D}/\sigma^2 = (\sigma^2\mathbf{D}^{-1})^{-1} = (1 - r^2)^{-1}\begin{pmatrix} 1 & -r \\ -r & 1 \end{pmatrix},$$

so $w = (z_1^2 + z_2^2 - 2rz_1z_2)/(1 - r^2)$. For the true $\boldsymbol{\theta}$, the theory says that $w \sim \chi^2(2)$, so a 95% confidence region for $\boldsymbol{\theta}$ is $w \leq c_2$, where $c_2 = 5.99$, the 5% critical value from the $\chi^2(2)$ table. We can write the region as

$$z_1^2 + z_2^2 - 2rz_1z_2 \leq c_2(1 - r^2).$$

In the θ_1, θ_2 plane, this is an ellipse centered at (t_1, t_2). For convenience, let us translate the axes so the origin is now located at the point (t_1, t_2). The ellipse is centered at the origin in the z_1, z_2 plane. We illustrate the possibilities with two figures.

Figure 19.1 refers to the case $r = 0$, which arises when the estimators are uncorrelated. The ellipse is just a circle,

$$z_1^2 + z_2^2 \leq c_2,$$

centered at the origin with radius $\sqrt{c_2} = \sqrt{5.99} = 2.45$. The box is a square centered at the origin, with each half-side equal to $\sqrt{c_1}$. Look along a coordinate axis: because $\sqrt{c_2} > \sqrt{c_1}$, it is evident that there are points in the circle that are not in the square. Look along the 45° line emanating from the center, that is, along the ray $z_1 = z_2$: the circle passes through the point $z_1 = z_2 = \sqrt{(c_2/2)}$, while the northeast corner of the square is located at the point $z_1 = z_2 = \sqrt{c_1}$. Since $\sqrt{(c_2/2)} < \sqrt{c_1}$, it is evident that there are points in the square that are not in the circle. This exhibits the distinction between intersecting two univariate 95% confidence intervals and constructing a 95% joint confidence region.

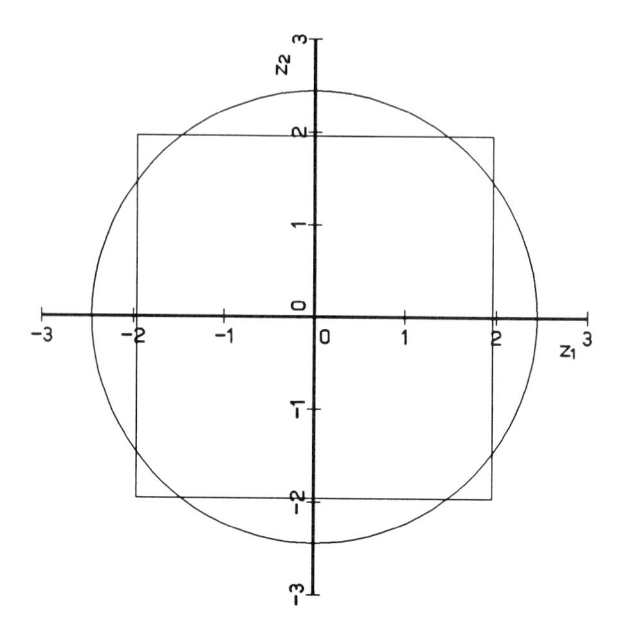

Figure 19.1 Confidence ellipse and box: $r = 0$.

Figure 19.2 refers to the case $r = 0.6$. Here we have a proper ellipse,

$$z_1^2 + z_2^2 - 2rz_1z_2 \leq c_2(1 - r^2),$$

centered at the origin. The major axis of the ellipse runs along the 45° line, so its vertices can be located by setting $z_1 = z_2 = z$, say, and solving the equation $z^2 + z^2 - 2rz^2 = c_2(1 - r^2)$ to get $z^2 = c_2(1 + r)/2$. So the northeast vertex is located at the point $z_1 = z_2 = \sqrt{[c_2(1 + r)/2]}$. The minor axis of the ellipse runs along the $-45°$ line, so its vertices can be located by setting $z_1 = z$, and $z_2 = -z$, say, and solving the equation $z^2 + z^2 - 2rz(-z) = c_2(1 - r^2)$. The solution is $z^2 = c_2(1 - r)/2$, so the southeast vertex is located at the point $z_1 = \sqrt{[c_2(1 - r)/2]}$, $z_2 = -\sqrt{[c_2(1 - r)/2]}$. Specifically, with $r = 0.6$, the northeast vertex is located at (2.19, 2.19) and the southeast vertex is located at (1.09, −1.09). Here again we see points in the ellipse that are not in the square, and points in the square that are not in the ellipse. As compared with the circle that prevailed when $r = 0$, which intersected the 45° line at the point $[\sqrt{(c_2/2)}, \sqrt{(c_2/2)}] = (1.73, 1.73)$, the ellipse has been stretched out in one direction and pulled in somewhat in the other.

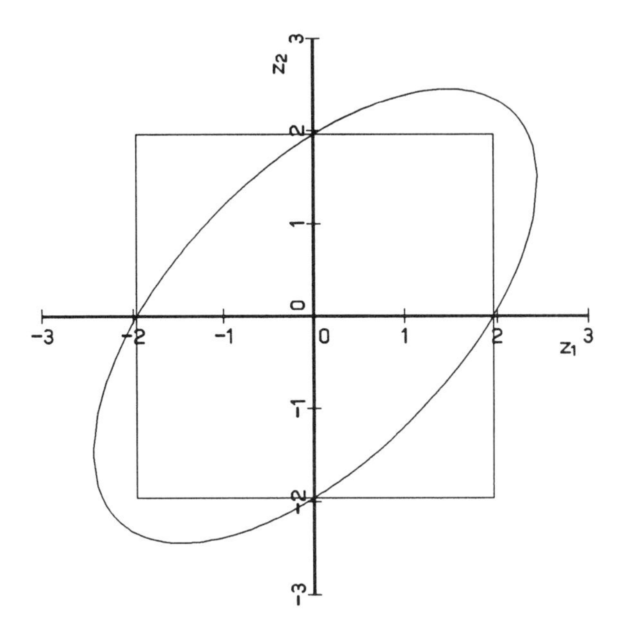

Figure 19.2 Confidence ellipse and box: $r = 0.6$.

Exercises

19.1 The CNR model applies with $k = 4$, $\mathbf{X'X} = \mathbf{I}$, $\sigma^2 = 2$, and $\boldsymbol{\beta} = \mathbf{0}$. Let $t = \mathbf{b'b}$. Find the number c such that $\Pr(t > c) = 0.10$.

19.2 The CNR model applies to $E(\mathbf{y}) = \mathbf{X}\boldsymbol{\beta}$, with $\sigma^2 = 7$ and

$$\mathbf{X'X} = \begin{pmatrix} 4 & 1 \\ 1 & 2 \end{pmatrix}.$$

A sample gives these LS estimates: $b_1 = 3$, $b_2 = 2$. Determine whether the point $\beta_1 = 2$, $\beta_2 = -2$ lies within the 95% confidence region for $\boldsymbol{\beta}$.

19.3 The CNR model applies to $E(y) = \beta_1 x_1 + \beta_2 x_2$, with $\sigma^2 = 2$ and

$$\mathbf{X'X} = \begin{pmatrix} 5 & 2 \\ 2 & 4 \end{pmatrix}.$$

Your sample has $b_1 = 3$, $b_2 = 2$.

(a) Construct a 95% confidence interval for $\theta = \beta_1 + \beta_2$.
(b) Construct a 90% confidence region for the pair (β_1, β_2).

20 CNR Model: Hypothesis Testing

20.1. Introduction

We proceed to another type of statistical inference, the testing of hypotheses about the population parameter vector $\boldsymbol{\beta}$. We suppose that the CNR model holds, so that

$$\mathbf{y} \sim \mathcal{N}(\mathbf{X}\boldsymbol{\beta}, \sigma^2\mathbf{I}), \qquad \mathbf{X}\ (n \times k)\ \text{nonstochastic}, \qquad \text{rank}(\mathbf{X}) = k.$$

We continue under the assumption that σ^2 is known, so that the distribution results of Chapter 19 are applicable.

20.2. Test on a Single Parameter

Suppose that we have a hypothesis about the jth regression coefficient, specifically the *null hypothesis* that $\beta_j = \beta_j^\circ$, where β_j° is a specific number. We propose the following *5%-significance-level* test against the *alternative hypothesis* that $\beta_j \neq \beta_j^\circ$. With a sample in hand, *accept* the null hypothesis if β_j° lies within the 95% confidence interval for β_j, namely $b_j \pm 1.96\sigma_{b_j}$; *reject* the null hypothesis otherwise. Equivalently, calculate the *test statistic*

$$z_j^\circ = (b_j - \beta_j^\circ)/\sigma_{b_j},$$

and compare its absolute value with the *critical value* 1.96.

If $|z_j^\circ| > 1.96$, then reject the null hypothesis $\beta_j = \beta_j^\circ$.
If $|z_j^\circ| \leq 1.96$, then accept the null hypothesis $\beta_j = \beta_j^\circ$.

The rationale is as follows. Think of b_j, and hence z_j°, as a random variable rather than the value obtained in a particular sample. Let A be the event $\{|z_j^\circ| > 1.96\}$. The probability that A occurs depends on what the true value of β_j is. If the true value is β_j°, so that the null hypothesis is true, then the random variable z_j° is identical to the random variable z_j defined in

D3A. $z_j = (b_j - \beta_j)/\sigma_{b_j} \sim \mathcal{N}(0, 1)$.

Consequently, $\Pr(A|\beta_j = \beta_j^\circ) = 0.05$. So the significance level, namely the probability of rejecting the null hypothesis when it is true, is 5%.

Suppose a sample has $|z_j^\circ| > 1.96$. If the null is true, then a low-probability event has occurred. The probability of the event is so low that its occurrence is taken to be evidence against the null; so the decision is to reject the null. Heuristically, the point estimate b_j is so far from the hypothesized parameter value β_j° that it is implausible that b_j has in fact been drawn from a distribution with expected value β_j°. However, finding a sample with $|z_j^\circ| \leq 1.96$ is not surprising when the null is true, so then the decision is to accept the null.

When $|z_j^\circ| > 1.96$, one says that *b_j is significantly different from* β_j° at the 5% level; when $|z_j^\circ| \leq 1.96$, one says that *b_j is not significantly different from* β_j° at the 5% level.

Several lessons are immediate:

• Rejection of the null is not proof that the null is false. After all, there is a nonzero probability of rejecting the null if it is true: $\Pr(|z_j^\circ| > 1.96|\beta_j = \beta_j^\circ) = 0.05$. Loosely speaking, when the null is true, in 5% of the samples drawn from the population, the decision will be "reject the null."

• Acceptance of the null is not proof that the null is true. After all, different null hypotheses would also have been acceptable. Indeed if the null had been $\beta_j = \beta_j^{\circ\circ}$, where $\beta_j^{\circ\circ}$ is any other point that happens to lie in the confidence interval $b_j \pm 1.96\sigma_{b_j}$, it too would have been accepted as a null hypothesis.

• If σ_{b_j} is large, then the 95% confidence interval is wide, and widely diverse null hypotheses about β_j are all acceptable at the 5% level. In that situation, the sample contains little information about the true value of β_j. The LS estimator b_j may well be the best estimator, but it need not be a precise estimator.

The test procedure adapts to handle a null hypothesis about β_j at different significance levels. Further, to test a null hypothesis about a single linear combination of the elements of $\boldsymbol{\beta}$: accept iff the null θ° lies in the confidence interval for θ, equivalently, iff the test statistic $z^\circ = (t - \theta^\circ)/\sigma_t$ is in absolute value less than or equal to the critical value.

It is good practice to use correct wording in reporting the outcome of a test: it is the estimate b_j, not the parameter β_j, whose significance is being assessed; and the test is being conducted at a 5% significance level, say, not at a 95% confidence level or at a 95% significance level.

20.3. Test on a Set of Parameters

Next suppose we have a *joint null hypothesis* about $\boldsymbol{\beta}$, specifically about p linear functions of $\boldsymbol{\beta}$. Let $\boldsymbol{\theta} = \mathbf{H}\boldsymbol{\beta}$, where the $p \times k$ nonrandom matrix $\mathbf{H}$ has rank p. Our null hypothesis is

$$\boldsymbol{\theta} = \boldsymbol{\theta}^\circ,$$

where $\boldsymbol{\theta}^\circ$ is a specific numerical $p \times 1$ vector. By appropriate choice of $\mathbf{H}$, this subsumes the situations where the hypothesis concerns the full vector $\boldsymbol{\beta}$, or a $k_2 \times 1$ subvector $\boldsymbol{\beta}_2$, or a single element β_j.

We propose the following 5%-significance-level test against the alternative hypothesis that $\boldsymbol{\theta} \neq \boldsymbol{\theta}^\circ$. With a sample in hand, accept the null hypothesis if $\boldsymbol{\theta}^\circ$ lies within the 95% confidence region for $\boldsymbol{\theta}$ given by

$$w = (\boldsymbol{\theta} - \mathbf{t})'\mathbf{D}(\boldsymbol{\theta} - \mathbf{t})/\sigma^2 \leq c_p;$$

reject the null hypothesis otherwise. Here $\mathbf{t} = \mathbf{Hb}$, while c_p is the 5% critical value from the $\chi^2(p)$ table; that is, $G_p(c_p) = 0.95$ where $G_p(\cdot)$ is the cdf of the $\chi^2(p)$ distribution. Equivalently, calculate the test statistic

$$w^\circ = (\mathbf{t} - \boldsymbol{\theta}^\circ)'\mathbf{D}(\mathbf{t} - \boldsymbol{\theta}^\circ)/\sigma^2.$$

If $w^\circ > c_p$, then reject the null hypothesis $\boldsymbol{\theta} = \boldsymbol{\theta}^\circ$.
If $w^\circ \leq c_p$, then accept the null hypothesis $\boldsymbol{\theta} = \boldsymbol{\theta}^\circ$.

The rationale is as follows. Think of w° as a random variable rather than as the value obtained in a particular sample. Let $A = \{w^\circ > c_p\}$. The probability that A occurs depends on the true value of $\boldsymbol{\theta}$. If the null hypothesis is true, so that $\boldsymbol{\theta} = \boldsymbol{\theta}^\circ$, then the random variable w° is identical to the random variable w defined in

D4. $w = (\mathbf{t} - \boldsymbol{\theta})'\mathbf{D}(\mathbf{t} - \boldsymbol{\theta})/\sigma^2 \sim \chi^2(p)$.

Because $\Pr(w > c_p) = 1 - G_p(c_p) = 0.05$, we have $\Pr(A|\boldsymbol{\theta} = \boldsymbol{\theta}^\circ) = 0.05$, so the significance level, namely the probability of rejecting the null hypothesis when it is true, is 5%. Suppose that a sample has $w^\circ > c_p$. If the null hypothesis is true, then a low-probability event has occurred. The probability of the event is so low that its occurrence is taken to be evidence against the null. Heuristically, the point estimate $\mathbf{t}$ is so far from the hypothesized parameter value $\boldsymbol{\theta}^\circ$ that it is implausible that $\mathbf{t}$ has in fact been drawn from a distribution with expectation $\boldsymbol{\theta}^\circ$. But finding a sample with $w^\circ \leq c_p$ is not surprising when the null is true.

Because $V(\mathbf{t}) = \sigma^2\mathbf{D}^{-1}$, we can write

$$w^\circ = (\mathbf{t} - \boldsymbol{\theta}^\circ)'[V(\mathbf{t})]^{-1}(\mathbf{t} - \boldsymbol{\theta}^\circ),$$

which shows that w° measures the deviation $\mathbf{t} - \boldsymbol{\theta}^\circ$ in the same way that z_j° measures the deviation $b_j - \beta_j^\circ$, that is, relative to the variability of the estimator.

The discussion of ellipses and boxes in Section 19.6 implies that one cannot tell the outcome of the test of a joint hypothesis from the outcomes of univariate tests of its separate components. It is quite possible to accept each of the separate null hypotheses $\theta_j = \theta_j^\circ$ tested one by one, while rejecting the joint null hypothesis $\theta_j = \theta_j^\circ$ $(j = 1, \ldots, p)$. There is no paradox here: the conjunction of two hypotheses may be untenable, even though either hypothesis by itself is tenable.

20.4. Power of the Test

To test (at the 5% significance level) the null hypothesis $\boldsymbol{\theta} = \boldsymbol{\theta}^\circ$ against the alternative $\boldsymbol{\theta} \neq \boldsymbol{\theta}^\circ$, we have proposed the test statistic

$$w^\circ = (\mathbf{t} - \boldsymbol{\theta}^\circ)'\mathbf{D}(\mathbf{t} - \boldsymbol{\theta}^\circ)/\sigma^2,$$

and the decision rule: reject the null iff $w^\circ > c$, where c is the 5% critical value in the $\chi^2(p)$ table. Our rationale was that the event $\{w^\circ > c\}$ is rare when the null hypothesis is true: $\Pr[(w^\circ > c)|\boldsymbol{\theta} = \boldsymbol{\theta}^\circ] = 0.05$. Why choose this particular rejection region, $w^\circ > c$? Taking as the rejection region any other interval for w° whose probability is 0.05 when the null is true would also provide a test at the 5% significance level. Indeed, one might just toss a fair 20-sided die and reject the null iff the "1" turns up; that would also provide a 5% significance level test of the null $\boldsymbol{\theta} = \boldsymbol{\theta}^\circ$.

To understand our choice, one must consider the *power function* for the test, namely the probability of rejecting the null $\boldsymbol{\theta} = \boldsymbol{\theta}^\circ$, as a function of the true parameter value $\boldsymbol{\theta}$. The "rare if null true" rationale for rejecting the null when the event $w^\circ > c$ occurs would lose its force if it turned out that the event $w^\circ > c$ was even rarer when the null is false, that is, if it turned out that $\Pr[(w^\circ > c)|\boldsymbol{\theta} \neq \boldsymbol{\theta}^\circ] < 0.05$. Our claim is that

$$\Pr[(w^\circ > c)|\boldsymbol{\theta}] \geq 0.05,$$

with equality iff $\boldsymbol{\theta} = \boldsymbol{\theta}^\circ$. That is, the *power of the test*, namely the probability of rejecting the null, is everywhere greater than the significance level, except at $\boldsymbol{\theta} = \boldsymbol{\theta}^\circ$. Further, the power is increasing in a sensible measure of the distance between the hypothesized and true value of $\boldsymbol{\theta}$. Tests that use a different rejection region may not have those desirable properties. For example, the power of the 20-sided-die test is everywhere equal to its significance level.

The argument rests on the distribution of w° as a function of $\boldsymbol{\theta}$. We first focus on the expectation, showing that $E(w^\circ)$ increases as $\boldsymbol{\theta}$ departs from $\boldsymbol{\theta}^\circ$. Define the random *miss vector*,

$$\mathbf{m} = \mathbf{t} - \boldsymbol{\theta}^\circ,$$

and rewrite the test statistic as

$$w^\circ = \mathbf{m}'(\mathbf{D}/\sigma^2)\mathbf{m}.$$

Now $E(\mathbf{m}) = E(\mathbf{t}) - \boldsymbol{\theta}^\circ = \mathbf{H}\boldsymbol{\beta} - \boldsymbol{\theta}^\circ = \boldsymbol{\theta} - \boldsymbol{\theta}^\circ = \boldsymbol{\mu}$, say, and $V(\mathbf{m}) = V(\mathbf{t}) = \sigma^2\mathbf{D}^{-1}$. (Caution: Do not confuse $\boldsymbol{\mu} = E(\mathbf{m})$ with $\boldsymbol{\mu} = E(\mathbf{y}) = \mathbf{X}\boldsymbol{\beta}$.) Use R5 (Section 15.1, with $\mathbf{D}/\sigma^2$ playing the role of $\mathbf{T}$) to calculate

$$\begin{aligned} E(w^\circ) &= \operatorname{tr}[(\mathbf{D}/\sigma^2)\sigma^2\mathbf{D}^{-1}] + \boldsymbol{\mu}'(\mathbf{D}/\sigma^2)\boldsymbol{\mu} \\ &= \operatorname{tr}(\mathbf{I}_p) + \boldsymbol{\mu}'(\mathbf{D}/\sigma^2)\boldsymbol{\mu} \\ &= p + \boldsymbol{\mu}'(\mathbf{D}/\sigma^2)\boldsymbol{\mu}. \end{aligned}$$

Because $\mathbf{D}/\sigma^2$ is positive definite, we conclude that $E(w^\circ) \geq p$, with equality iff $\boldsymbol{\mu} = \mathbf{0}$, that is, iff $\boldsymbol{\theta} = \boldsymbol{\theta}^\circ$. The farther the null is from the truth, that is, the larger the magnitude of the $p \times 1$ vector $\boldsymbol{\mu}$, as measured by the nonnegative scalar $\boldsymbol{\mu}'(\mathbf{D}/\sigma^2)\boldsymbol{\mu}$, the larger is $E(w^\circ)$. Incidentally, the calculation so far does not rely on normality.

Finding that $E(w^\circ)$ rises as $\boldsymbol{\theta}$ departs from $\boldsymbol{\theta}^\circ$ makes it plausible that $\Pr(w^\circ > c)$ also rises as $\boldsymbol{\theta}$ departs from $\boldsymbol{\theta}^\circ$. But to establish the latter requires examination of the probability distribution of w° as a function of the parameter $\boldsymbol{\theta}$.

20.5. Noncentral Chi-square Distribution

The relevant distribution theory starts at the level of Section 18.3:

Suppose that the $n \times 1$ random vector $\mathbf{z}$ is distributed $\mathcal{N}(\boldsymbol{\alpha}, \mathbf{I})$. Let $w = \mathbf{z}'\mathbf{z}$. Then $w \sim \chi^{2*}(n, \lambda^2)$, where $\lambda^2 = \boldsymbol{\alpha}'\boldsymbol{\alpha}$.

That is, the sum of squares of n independent $\mathcal{N}(\alpha_i, 1)$ variables has the *noncentral chi-square* distribution with degrees of freedom parameter n and *noncentrality* parameter $\lambda^2 = \Sigma_i \alpha_i^2$. The familiar (central) chi-square distribution is the special case that arises when $\boldsymbol{\alpha} = \mathbf{0}$. Table A.4 gives a small display of $1 - G_k^*(c_k; \lambda^2)$, where $G_k^*(\cdot; \lambda^2)$ denotes the cdf of the $\chi^{2*}(k, \lambda^2)$ distribution, and c_k is the critical value relevant for testing at the 5% significance level. That is, c_k is defined by $G_k^*(c_k; 0) = G_k(c_k) = 0.95$; thus $c_1 = 3.84$, $c_2 = 5.99$, $c_3 = 7.81$. In each column of the table, we see that the probability of exceeding c_k increases with λ^2. (Caution: This table records the complement of the cdf, not a cdf itself as Table A.2 did.)

Retracing the steps used in the proof of Q1 (Section 18.4), it follows immediately that:

Q4. Suppose that the $n \times 1$ vector $\mathbf{y}$ is distributed $\mathcal{N}(\boldsymbol{\mu}, \boldsymbol{\Sigma})$. Let $w = \mathbf{y}'\boldsymbol{\Sigma}^{-1}\mathbf{y}$. Then $w \sim \chi^{2*}(n, \lambda^2)$, where $\lambda^2 = \boldsymbol{\mu}'\boldsymbol{\Sigma}^{-1}\boldsymbol{\mu}$.

Now return to the CNR model. As a linear function of $\mathbf{b}$, the $p \times 1$ miss vector, $\mathbf{m} = \mathbf{t} - \boldsymbol{\theta}^\circ$, is distributed $\mathcal{N}(\boldsymbol{\mu}, \sigma^2\mathbf{D}^{-1})$, with $\boldsymbol{\mu} = \boldsymbol{\theta} - \boldsymbol{\theta}^\circ$. Applying Q4 to our test statistic $w^\circ = \mathbf{m}'(\mathbf{D}/\sigma^2)\mathbf{m}$, gives a new distribution result for the CNR model:

$$w^\circ \sim \chi^{2*}(p, \lambda^2), \quad \text{with } \lambda^2 = (\boldsymbol{\theta} - \boldsymbol{\theta}^\circ)'\mathbf{D}(\boldsymbol{\theta} - \boldsymbol{\theta}^\circ)/\sigma^2.$$

So the power of our test is given by

$$\Pr[(w^\circ > c_p) | \boldsymbol{\theta}] = 1 - G_p^*(c_p; \lambda^2), \quad \text{with } \lambda^2 = (\boldsymbol{\theta} - \boldsymbol{\theta}^\circ)'\mathbf{D}(\boldsymbol{\theta} - \boldsymbol{\theta}^\circ)/\sigma^2.$$

Observe that the probability distribution of the test statistic w° depends on $\boldsymbol{\theta}$ only through the scalar λ^2. For a given significance level, that is for a given c_p, the power in $\boldsymbol{\theta}$-space will be constant along ellipsoids centered at $\boldsymbol{\theta}^\circ$. As $\boldsymbol{\theta}$ departs from $\boldsymbol{\theta}^\circ$, that is, as we move farther out along a ray through $\boldsymbol{\theta} - \boldsymbol{\theta}^\circ$, the scalar λ^2 increases, and so the power increases. Observe also that the power depends on the direction, not merely upon the magnitude, of $\boldsymbol{\theta} - \boldsymbol{\theta}^\circ$.

This argument completes the rationale that supports the joint hypothesis test procedure of Section 20.3. It also supports the single hypothesis test procedure of Section 20.2, because that, as we have seen, is equivalent to a joint test with $p = 1$.

Actually, for $p = 1$, one can deduce the noncentral chi-square probabilities from the $\mathcal{N}(0, 1)$ cdf. The calculation runs as follows. Suppose $w^\circ \sim \chi^{2*}(1, \lambda^2)$. Then $w^\circ = z^{\circ 2}$, where $z^\circ \sim \mathcal{N}(\lambda, 1)$, that is, $(z^\circ - \lambda) \sim \mathcal{N}(0, 1)$. Let $A = \{w^\circ > c\} = A_1 \cup A_2$, where

$$A_1 = \{z^\circ > \sqrt{c}\} = \{(z^\circ - \lambda) > (\sqrt{c} - \lambda)\},$$

$$A_2 = \{z^\circ < -\sqrt{c}\} = \{(z^\circ - \lambda) < -(\sqrt{c} + \lambda)\}.$$

Now $\Pr(A_1) = 1 - F(\sqrt{c} - \lambda) = F(\lambda - \sqrt{c})$, and $\Pr(A_2) = F[-(\lambda + \sqrt{c})]$, where $F(\cdot)$ denotes the $\mathcal{N}(0, 1)$ cdf. Because A_1 and A_2 are disjoint, we have

$$\Pr(w^\circ > c) = \Pr(A) = \Pr(A_1) + \Pr(A_2) = F(\lambda - \sqrt{c}) + F[-(\lambda + \sqrt{c})].$$

For example, suppose $\lambda^2 = 1$, and $c = 3.84$. Then $\lambda = 1$ and $\sqrt{c} = 1.96$, so

$$\Pr(w^\circ > c) = F(-0.96) + F(-2.96) = 0.168 + 0.002 = 0.170,$$

as in Table A.4.

Exercises

20.1 The CNR model applies to $E(\mathbf{y}) = \mathbf{X}\boldsymbol{\beta}$. You know that $\sigma^2 = 2$ and that

$$\mathbf{X}'\mathbf{X} = \begin{pmatrix} 3 & 1 \\ 1 & 2 \end{pmatrix}.$$

In a sample of 32 observations, the LS coefficients are $b_1 = 2$, $b_2 = 2$.

(a) Test at the 5% significance level the joint null hypothesis that $\beta_1 = 3 = \beta_2$.
(b) State the alternative hypothesis against which you are testing.

20.2 The CNR model applies to $E(\mathbf{y}) = \mathbf{X}\boldsymbol{\beta}$, with $\sigma^2 = 7$ and

$$\mathbf{X}'\mathbf{X} = \begin{pmatrix} 4 & 1 \\ 1 & 2 \end{pmatrix}.$$

The null hypothesis $\beta_2 = 1$ will be tested at the 10% significance level, against the alternative that $\beta_2 \neq 1$. What is the probability of rejecting that null hypothesis, if the true value of β_2 is 3?

20.3 The regression slope b in a CNR model is distributed $\mathcal{N}(\beta, 1)$. The null hypothesis $\beta = 0$ will be tested at the 10% significance level by using the statistic $z° = b/\sigma_b$. That is, the null will be rejected if and only if $|z°| > 1.645$.

(a) Write and run a program that tabulates the power of the test at these 9 values of the true parameter β:

$$-2 \quad -1.5 \quad -1 \quad -0.5 \quad 0 \quad 0.5 \quad 1 \quad 1.5 \quad 2$$

(b) Redo (a) for the situation where $b \sim \mathcal{N}(\beta, 4)$.
(c) What do your two tables tell you about the effect of σ_b^2 on the power of the test?

GAUSS Hint:
The command cdfn(c) evaluates the standard normal cdf at the point c.

20.4 The pair of regression slopes b_1, b_2 in a CNR model is distributed BVN$(\beta_1, \beta_2, 1, 1, r)$ with $r = 0.6$. The joint null hypothesis $\beta_1 = 0$, $\beta_2 = 0$ will be tested at the 5% significance level by using the statistic

$$w° = (b_1^2 + b_2^2 - 2rb_1b_2)/(1 - r^2).$$

That is, the null will be rejected iff $w° > 5.99$.

(a) Write and run a program that tabulates the power of the test at these 9 values of the true parameter pair (β_1, β_2):

$-1, 1$	$0, 1$	$1, 1$
$-1, 0$	$0, 0$	$1, 0$
$-1, -1$	$0, -1$	$1, -1$

(b) Redo (a) for the situation where $r = -0.6$.

(c) What do your two tables tell you about the effect of the sign of the correlation r on the power of the test?

GAUSS Hint:

The command cdfchinc(c,n,m) gives the cdf of the noncentral chi-square distribution with degrees of freedom n and noncentrality parameter m^2, at the point c. For the tests in (a) and (b), the noncentrality parameter is

$$\lambda^2 = (\beta_1^2 + \beta_2^2 - 2r\beta_1\beta_2)/(1 - r^2).$$

21 CNR Model: Inference with σ^2 Unknown

21.1. Distribution Theory

Thus far, the procedures for constructing confidence intervals and regions for $\boldsymbol{\beta}$, and for testing hypotheses about its elements, have required knowledge of σ^2. We now extend the theory to obtain procedures that are operational in practice, where σ^2 is unknown. It is natural to use

$$\hat{\sigma}^2 = \mathbf{e}'\mathbf{e}/(n - k)$$

in place of σ^2, and thus to use

$$\hat{z}_j = (b_j - \beta_j)/\hat{\sigma}_{b_j}$$

in place of $z_j = (b_j - \beta_j)/\sigma_{b_j}$, and

$$\hat{w} = (\mathbf{t} - \boldsymbol{\theta})'\mathbf{D}(\mathbf{t} - \boldsymbol{\theta})/\hat{\sigma}^2$$

in place of $w = (\mathbf{t} - \boldsymbol{\theta})'\mathbf{D}(\mathbf{t} - \boldsymbol{\theta})/\sigma^2$. To assess the distribution of the new statistics, we draw on some additional implications of the CNR model.

For the CNR model in which $\mathbf{y} \sim \mathcal{N}(\mathbf{X}\boldsymbol{\beta}, \sigma^2\mathbf{I})$, the relevant theory resumes with

D6. $w_o = \mathbf{e}'\mathbf{e}/\sigma^2 \sim \chi^2(n - k)$.

Proof. Recall the theory of Chapter 18 on functions of normal vectors, and take these steps:

(i) Let $\boldsymbol{\epsilon} = \mathbf{y} - \mathbf{X}\boldsymbol{\beta}$. So $\boldsymbol{\epsilon} \sim \mathcal{N}(\mathbf{0}, \sigma^2\mathbf{I})$ by P5 (Section 18.2).
(ii) Let $\mathbf{u} = (1/\sigma)\boldsymbol{\epsilon}$. So the $n \times 1$ vector $\mathbf{u} \sim \mathcal{N}(\mathbf{0}, \mathbf{I})$.

(iii) Rewrite $\mathbf{y} = \mathbf{X}\boldsymbol{\beta} + \boldsymbol{\epsilon}$ as $\mathbf{y} = \mathbf{X}\boldsymbol{\beta} + \sigma\mathbf{u}$.
(iv) Then $\mathbf{e} = \mathbf{My} = \mathbf{M}(\mathbf{X}\boldsymbol{\beta} + \sigma\mathbf{u}) = \sigma\mathbf{Mu}$, using $\mathbf{MX} = \mathbf{O}$.
(v) So $\mathbf{e}'\mathbf{e} = \sigma^2\mathbf{u}'\mathbf{Mu}$, and $w_o = \mathbf{e}'\mathbf{e}/\sigma^2 = \mathbf{u}'\mathbf{Mu}$.
(vi) The nonrandom matrix $\mathbf{M}$ is idempotent with rank $n - k$.
(vii) So by Q2 (Section 18.4), $w_o \sim \chi^2(n - k)$. ∎

Next,

D7. The random vectors $\mathbf{b}$ and $\mathbf{e}$ are independent.

Proof. Continue the construction above. The steps:

(i) $\mathbf{b} = \mathbf{Ay} = \mathbf{A}(\mathbf{X}\boldsymbol{\beta} + \sigma\mathbf{u}) = \boldsymbol{\beta} + \sigma\mathbf{Au}$, using $\mathbf{AX} = \mathbf{I}$.
(ii) Let $\mathbf{t}_1 = (1/\sigma)\mathbf{e} = \mathbf{Mu}$ and $\mathbf{t}_2 = (1/\sigma)(\mathbf{b} - \boldsymbol{\beta}) = \mathbf{Au}$.
(iii) Since $\mathbf{AM} = \mathbf{O}$, the conditions of Q3 (Section 18.4) are met, so $\mathbf{t}_1$ and $\mathbf{t}_2$ are independent.
(iv) So $\mathbf{e} = \sigma\mathbf{t}_1$ and $\mathbf{b} = \boldsymbol{\beta} + \sigma\mathbf{t}_2$ are independent. ∎

As a corollary, we have that any function of $\mathbf{e}$ is independent of any function of $\mathbf{b}$. Specifically,

D8. Each of the statistics $\mathbf{e}$, $\mathbf{e}'\mathbf{e}$, w_o, $\hat{\sigma}^2$, $\hat{V}(\mathbf{b})$, is independent of each of the statistics $\mathbf{b}$, b_j, $\mathbf{t} = \mathbf{Hb}$, w, w_j.

Now turn to the statistics that use the estimator $\hat{\sigma}^2$ instead of the population parameter σ^2. Let $v = \hat{w}/p$. Then

D9. $v = (\mathbf{t} - \boldsymbol{\theta})'\mathbf{D}(\mathbf{t} - \boldsymbol{\theta})/(p\hat{\sigma}^2) \sim F(p, n - k)$.

Proof. Recall from Eq. (18.2) the requirement for a random variable to have the Snedecor F-distribution. It suffices to show that v is the ratio of two independent chi-square variables, each divided by its degrees of freedom parameter. The steps:

(i) $\hat{\sigma}^2/\sigma^2 = [\mathbf{e}'\mathbf{e}/(n - k)]/\sigma^2 = (\mathbf{e}'\mathbf{e}/\sigma^2)/(n - k) = w_o/(n - k)$.
(ii) $v = \hat{w}/p = (w/p)/(\hat{\sigma}^2/\sigma^2) = (w/p)/[w_o/(n - k)]$.
(iii) But $w \sim \chi^2(p)$ is independent of $w_o \sim \chi^2(n - k)$. ∎

Continuing, let $u_j = \hat{z}_j$. Then

D10. $u_j = (b_j - \beta_j)/\hat{\sigma}_{b_j} \sim t(n - k)$.

Proof. Recall from Eq. (18.3) the requirement for a random variable to have Student's t-distribution. It suffices to show that u_j is the ratio of a standard normal variable to the square root of an independent chi-square variable divided by its parameter. The steps:

(i) $(b_j - \beta_j)/\sigma_{b_j} = z_j$.

(ii) $\hat{\sigma}_{b_j}/\sigma_{b_j} = \sqrt{[\hat{\sigma}^2 q^{jj}/(\sigma^2 q^{jj})]} = \sqrt{(\hat{\sigma}^2/\sigma^2)} = \sqrt{[w_o/(n - k)]}$.

(iii) $u_j = [(b_j - \beta_j)/\sigma_{b_j}]/[\hat{\sigma}_{b_j}/\sigma_{b_j}] = z_j/\sqrt{[w_o/(n - k)]}$.

(iv) But $z_j \sim \mathcal{N}(0, 1)$ is independent of $w_o \sim \chi^2(n - k)$. ∎

21.2. Confidence Intervals and Regions

Under the CNR model, to construct confidence intervals and regions when σ^2 is unknown, one uses the t and F distribution results in the same way that the $\mathcal{N}(0, 1)$ and χ^2 distribution results would be used were σ^2 known. So the following discussion can be concise.

Confidence Intervals

For a single regression coefficient, draw on D10. Let c be the two-tail 5% critical value in the $t(n - k)$ table; that is, $G(c) = 0.975$, where $G(\cdot)$ is the cdf of the $t(n - k)$ distribution. Then $b_j \pm c\hat{\sigma}_{b_j}$ provides a 95% confidence interval for β_j. The rationale: the event that the random variable b_j lies within $c\hat{\sigma}_{b_j}$ of the fixed parameter β_j is

$$A = \{\beta_j - c\hat{\sigma}_{b_j} \leq b_j \leq \beta_j + c\hat{\sigma}_{b_j}\} = \{|(b_j - \beta_j)/\hat{\sigma}_{b_j}| \leq c\} = \{|u_j| \leq c\}.$$

Since D10 says that $u_j \sim t(n - k)$, we conclude that $\Pr(A) = G(c) - G(-c) = 0.975 - (1 - 0.975) = 0.95$.

Alternatively, we might draw on D9, specialized to $p = 1$. Let d be the 5% critical value in the $F(1, n - k)$ table; that is, $G_1(d) = 0.95$, where $G_1(\cdot)$ is the cdf of the $F(1, n - k)$ distribution. Then a 95% confidence interval will consist of all values β_j satisfying the inequality $v_j \leq d$, where

$$v_j = (b_j - \beta_j)^2/\hat{\sigma}^2_{b_j}.$$

Let $B = \{v_j \leq d\}$ be the event that the true β_j lies in this interval. Now, B is identical to $A = \{|u_j| \leq c\}$ because $v_j = |u_j|^2$ and $d = c^2$, so this is the same interval we got from D10.

In the same manner, to construct a confidence interval for a single linear function of coefficients $\theta = \mathbf{h}'\boldsymbol{\beta}$, draw on D9 (with $p = 1$), or its $t(n - k)$ equivalent.

Joint Confidence Regions

Similarly for joint confidence regions: suppose we are concerned with the $p \times 1$ parameter vector $\boldsymbol{\theta} = \mathbf{H}\boldsymbol{\beta}$. Let d_p be the 5% critical value of the $F(p, n - k)$ distribution; that is $G_p(d_p) = 0.95$, where $G_p(\cdot)$ is the cdf of the $F(p, n - k)$ distribution. We rely on D9 to propose

$$(\boldsymbol{\theta} - \mathbf{t})'\mathbf{D}(\boldsymbol{\theta} - \mathbf{t})/(p\hat{\sigma}^2) \leq d_p$$

as the 95% confidence region for the unknown parameter vector $\boldsymbol{\theta}$. The region consists of all $p \times 1$ vectors $\boldsymbol{\theta}$ that satisfy the inequality. Observe that here $\boldsymbol{\theta}$ denotes the argument of a function, not necessarily the true parameter vector. The rationale for the proposal is that the true parameter vector $\boldsymbol{\theta}$ lies in that random region iff $v \leq d_p$, an event whose probability is 0.95.

It is instructive to rewrite this operational confidence region $v \leq d_p$ as

$$\hat{w} = (\boldsymbol{\theta} - \mathbf{t})'\mathbf{D}(\boldsymbol{\theta} - \mathbf{t})/\hat{\sigma}^2 \leq pd_p.$$

Observe its similarity to the ellipsoidal region $w \leq c_p$ that would be used were σ^2 known, namely

$$w = (\boldsymbol{\theta} - \mathbf{t})'\mathbf{D}(\boldsymbol{\theta} - \mathbf{t})/\sigma^2 \leq c_p.$$

The region $v \leq d_p$ is also an ellipsoid centered at $\mathbf{t}$, and indeed has precisely the same shape as the region $w \leq c_p$, being merely expanded. So our previous analysis (Section 19.6) of the shape of the region, and of its relation to the rectangular region obtained by intersecting single-parameter confidence intervals, carries over directly.

21.3. Hypothesis Tests

The theory just developed for constructing confidence intervals and regions for $\boldsymbol{\beta}$ adapts to testing hypotheses about $\boldsymbol{\beta}$ when σ^2 is unknown.

Single Parameter

Suppose that we have a null hypothesis about the jth regression coefficient, namely $\beta_j = \beta_j^\circ$, where β_j° is a specific number. At the 5% significance level, accept the null iff β_j° lies in the 95% confidence interval for β_j. Equivalently, calculate the test statistic

$$u_j^\circ = (b_j - \beta_j^\circ)/\hat{\sigma}_{b_j},$$

and compare it with c, the two-tail 5% critical value in the $t(n - k)$ table. If $|u_j^\circ| > c$, then reject the null hypothesis $\beta_j = \beta_j^\circ$. If $|u_j^\circ| \leq c$, then accept the null hypothesis $\beta_j = \beta_j^\circ$.

In presenting the results of an empirical study, a correct practice is to report the regression coefficients b_j along with their standard errors $\hat{\sigma}_{b_j}$. This gives readers the information they need to construct a confidence interval for each regression coefficient, and to test a hypothesis about any one of them. It is common practice to report the regression coefficients along with their "t-ratios" or "t-statistics," the $u_j^\circ = b_j/\hat{\sigma}_{b_j}$, and to say, if u_j° is large, that b_j is "significant," meaning "significantly different from zero." This common practice is not a good one, because it encourages readers to consider only "zero null hypotheses" $\beta_j = 0$, which are not necessarily the interesting ones. Of course, a knowledgeable reader can always unscramble b_j and u_j° to recover $\hat{\sigma}_{b_j}$.

Set of Parameters

Suppose we have a joint hypothesis, one that concerns several parameters, specifically p linear functions of $\boldsymbol{\beta}$. Let $\boldsymbol{\theta} = \mathbf{H}\boldsymbol{\beta}$, where the $p \times k$ nonrandom matrix $\mathbf{H}$ has rank p. Our null hypothesis is $\boldsymbol{\theta} = \boldsymbol{\theta}^\circ$, where $\boldsymbol{\theta}^\circ$ is a numerical $p \times 1$ vector. By appropriate choice of $\mathbf{H}$, this setup subsumes the situation where the hypothesis is about the full vector $\boldsymbol{\beta}$, or about a $k_2 \times 1$ subvector $\boldsymbol{\beta}_2$, or about a single element β_j.

To test at the 5% significance level, against the alternative $\boldsymbol{\theta} \neq \boldsymbol{\theta}^\circ$, accept the null if $\boldsymbol{\theta}^\circ$ lies in the 95% confidence region for $\boldsymbol{\theta}$, reject otherwise. Equivalently, calculate the test statistic

$$v^\circ = (\mathbf{t} - \boldsymbol{\theta}^\circ)'\mathbf{D}(\mathbf{t} - \boldsymbol{\theta}^\circ)/(p\hat{\sigma}^2),$$

and compare it with d_p, the 5% critical value from the $F(p, n - k)$ table. If $v^\circ > d_p$, then reject the null hypothesis $\boldsymbol{\theta} = \boldsymbol{\theta}^\circ$. If $v^\circ \leq d_p$, then accept the null hypothesis $\boldsymbol{\theta} = \boldsymbol{\theta}^\circ$.

The power function for the F-test is very similar to that for the chi-square test discussed in Sections 20.4 and 20.5, so we need not discuss it explicitly.

21.4. Zero Null Subvector Hypothesis

A leading special case of a joint hypothesis arises when the null says that several of the β_j's are zero. We refer to this as a *zero null subvector* hypothesis. Without loss of generality, let those β_j's be the last k_2 elements of $\boldsymbol{\beta}$. Partitioning as

$$E(\mathbf{y}) = \mathbf{X}\boldsymbol{\beta} = (\mathbf{X}_1, \mathbf{X}_2)\begin{pmatrix}\boldsymbol{\beta}_1\\ \boldsymbol{\beta}_2\end{pmatrix} = \mathbf{X}_1\boldsymbol{\beta}_1 + \mathbf{X}_2\boldsymbol{\beta}_2,$$

we state the null as $\boldsymbol{\beta}_2 = \mathbf{0}$. To fit it into the framework of Section 21.3, set $p = k_2$, $\mathbf{H} = (\mathbf{O}, \mathbf{I})$ where the $\mathbf{O}$ is $k_2 \times k_1$, the $\mathbf{I}$ is $k_2 \times k_2$, and $\boldsymbol{\theta}^\circ = \mathbf{0}$. Then

$$\boldsymbol{\theta} = \mathbf{H}\boldsymbol{\beta} = \boldsymbol{\beta}_2, \qquad \mathbf{t} = \mathbf{H}\mathbf{b} = \mathbf{b}_2,$$

$$\mathbf{H}\mathbf{Q}^{-1}\mathbf{H}' = (\mathbf{O}, \mathbf{I})\begin{pmatrix}\mathbf{Q}^{11} & \mathbf{Q}^{12}\\ \mathbf{Q}^{21} & \mathbf{Q}^{22}\end{pmatrix}\begin{pmatrix}\mathbf{O}\\ \mathbf{I}\end{pmatrix} = \mathbf{Q}^{22},$$

$$\mathbf{D} = (\mathbf{H}\mathbf{Q}^{-1}\mathbf{H}')^{-1} = (\mathbf{Q}^{22})^{-1} = \mathbf{Q}_{22}^* = \mathbf{X}_2^{*\prime}\mathbf{X}_2^*,$$

with $\mathbf{X}_2^* = \mathbf{M}_1\mathbf{X}_2$. So the test statistic v° becomes

$$v^\circ = \mathbf{b}_2'\mathbf{Q}_{22}^*\mathbf{b}_2/(k_2\hat{\sigma}^2).$$

Using Residual Sums of Squares

There is another way to calculate this test statistic. Recall from Eq. (17.6) that

$$\mathbf{b}_2'\mathbf{Q}_{22}^*\mathbf{b}_2 = \mathbf{b}_2'\mathbf{X}_2^{*\prime}\mathbf{X}_2^*\mathbf{b}_2 = \mathbf{e}^{*\prime}\mathbf{e}^* - \mathbf{e}'\mathbf{e},$$

where $\mathbf{e}^{*\prime}\mathbf{e}^*$ is the sum of squared residuals from the short regression of $\mathbf{y}$ on $\mathbf{X}_1$ alone, while $\mathbf{e}'\mathbf{e}$ is the sum of squared residuals from the long regression of $\mathbf{y}$ on $\mathbf{X} = (\mathbf{X}_1, \mathbf{X}_2)$. Recall also that $\hat{\sigma}^2 = \mathbf{e}'\mathbf{e}/(n - k)$. So the test statistic can be written as

$$v^\circ = \frac{(n - k)}{k_2}\left(\frac{\mathbf{e}^{*\prime}\mathbf{e}^* - \mathbf{e}'\mathbf{e}}{\mathbf{e}'\mathbf{e}}\right).$$

This version may be computationally convenient. To get the test statistic for a zero null hypothesis $\boldsymbol{\beta}_2 = \mathbf{0}$, there is no need to extract the $k_2 \times k_2$ submatrix $\hat{V}(\mathbf{b}_2) = \hat{\sigma}^2\mathbf{Q}^{22}$ from $\hat{V}(\mathbf{b})$, and invert it. Instead, just run the appropriate long and short regressions and use their sums of squared residuals.

The result is also analytically instructive. Large values of v° lead to rejection of the null, and v° will be large (ceteris paribus) when the ratio $(\mathbf{e}^{*\prime}\mathbf{e}^* - \mathbf{e}'\mathbf{e})/\mathbf{e}'\mathbf{e}$ is large. That is to say, the null hypothesis $\boldsymbol{\beta}_2 = \mathbf{0}$ is rejected when dropping $\mathbf{X}_2$ from the regression leads to a large proportional increase in the sum of squared residuals, that is, to a substantial worsening of the fit. This is quite natural. In terms of the underlying population, suppose that the model has

$$E(y \mid x_2, \ldots, x_8) = \beta_1 + \beta_2 x_2 + \cdots + \beta_8 x_8,$$

while the null hypothesis is $\beta_6 = \beta_7 = \beta_8 = 0$, that is,

$$E(y \mid x_2, \ldots, x_8) = \beta_1 + \beta_2 x_2 + \cdots + \beta_5 x_5.$$

This null says that the conditional expectation of the dependent variable y given all the x's does not in fact vary with x_6, x_7, x_8. So the null imposes a restriction on the CEF. To test the null, impose its restriction in estimating the CEF, by running the short regression, and see whether the fit is much worse than the fit obtained when the restriction was not imposed. If the fit is not much worse, then accept the hypothesis that the conditional expectation of y given all the x's does not in fact vary with x_6, x_7, x_8.

Using R^2's

Continuing, suppose that $\mathbf{X}_1$ contains the summer vector (and perhaps more columns). Then both the short and long regressions contain an intercept, and so their R^2's are well defined. For the long regression we have

$$\mathbf{e}'\mathbf{e} = (1 - R^2) \sum_i (y_i - \bar{y})^2,$$

and for the short regression we have

$$\mathbf{e}^{*\prime}\mathbf{e}^{*} = (1 - R^{2*}) \sum_i (y_i - \bar{y})^2.$$

So

$$(\mathbf{e}^{*\prime}\mathbf{e}^{*} - \mathbf{e}'\mathbf{e})/\mathbf{e}'\mathbf{e} = (R^2 - R^{2*})/(1 - R^2) = \Delta R^2/(1 - R^2),$$

where ΔR^2 is the reduction in R^2 that occurs when $\mathbf{X}_2$ is dropped from the list of explanatory variables. For this standard situation, then, the test statistic can be written as

$$v^{\circ} = \frac{(n - k)}{k_2} \left(\frac{\Delta R^2}{1 - R^2} \right).$$

We now see that $\boldsymbol{\beta}_2 = \mathbf{0}$ is rejected when dropping $\mathbf{X}_2$ from the regression leads to a large decrease in R^2, relative to $1 - R^2$ (especially when $n - k$ is large and/or k_2 is small). This version may be computationally convenient: one needs to record only the R^2's of the short and long regressions.

All Slopes Zero

Finally suppose that $\mathbf{X}_1$ contains *only* the summer vector, so $\boldsymbol{\beta}_1$ contains only the intercept. Then the null $\boldsymbol{\beta}_2 = \mathbf{0}$ asserts that all of the slopes $\beta_2, \ldots, \beta_k$ are zero. Here the short-regression sum of squared residuals is $\mathbf{e}^{*\prime}\mathbf{e}^{*} = \Sigma_i(y_i - \bar{y})^2$, so the short-regression R^{2*} is in effect zero. So $\Delta R^2 = R^2$, and the test statistic simplifies to

$$v^{\circ} = \frac{(n - k)}{(k - 1)} \left(\frac{R^2}{1 - R^2} \right).$$

Evidently, an "all slopes zero" null hypothesis will be rejected when R^2 is large, especially when n is large and/or k is small.

The all-slopes-zero test is sometimes referred to as the "test of significance of the complete regression" or the "overall significance test of the regression." Many packaged computer programs routinely calculate this statistic and report it as the "regression F-statistic." As a result, in many journal articles and textbooks it is routinely reported along with R^2. Typically the regression F-statistic is large, and one sees dramatic statements. To take a textbook example, Intriligator (1978, pp. 138–141), with $n = 12$ annual observations, regresses GNP on $k = 3$ explanatory variables (the constant, lagged GNP, and government expenditures). The R^2 is 0.9958, so $v^\circ = 1072$, while the 5% critical value from the $F(2,9)$ table is $d = 4.26$. He writes "it is clear that the overall regression is highly significant and that the hypothesis that all [slope] coefficients are zero is overwhelmingly rejected" (p. 140). A less dramatic description of the situation would run as follows. The economic model, which allows $\mu_i = E(Y_i)$ to vary linearly with Y_{i-1} and G, fits much better than a model that insists that $\mu_i = \mu$ for all i. In short, the economic model is "much better than nothing."

Exercises

21.1 Consider the special case of the CNR model in which the only explanatory variable is the constant. Use the present distribution results to derive F1–F4 of Section 8.6, the theorems on the distribution of the sample mean and variance in random sampling from a univariate normal distribution.

21.2 The CNR model applies to $E(\mathbf{y}) = \mathbf{X}\boldsymbol{\beta}$, with $\sigma^2 = 1$ and

$$\boldsymbol{\beta} = \begin{pmatrix} 2 \\ 4 \end{pmatrix}, \qquad \mathbf{X}'\mathbf{X} = \begin{pmatrix} 2 & 1 \\ 1 & 3 \end{pmatrix}.$$

For each of the following two samples, determine the best guess of the random variable b_2, justifying your answer.

(a) A sample has $\mathbf{e}'\mathbf{e} = 10$.
(b) A sample has $b_1 = 3$.

21.3 Here are the results of two regressions run on annual time series for the years 1935–1978:

(i) $\hat{y} = 50 + 0.2x_2 + 0.5x_3 - 2.0x_4, \quad R^2 = 0.80.$
(ii) $\hat{y} = 100 + 0.3x_2 + 0.4x_3, \quad R^2 = 0.76.$

Determine whether the following is true or false: in equation (i), the standard error for b_4 is $1/\sqrt{2}$.

21.4 The CNR model applies to $E(\mathbf{y}) = \mathbf{x}_1\beta_1 + \mathbf{x}_2\beta_2$. A sample of size $n = 102$ gives these statistics:

$$\mathbf{b} = \begin{pmatrix} 5 \\ 2 \end{pmatrix}, \qquad \mathbf{Q} = \begin{pmatrix} 2 & 1 \\ 1 & 3 \end{pmatrix}, \qquad \mathbf{e}'\mathbf{e} = 80.$$

Let $\theta = \beta_1 - \beta_2$. Test at the 5% significance level, the null hypothesis that $\theta = 1$.

21.5 The CNR model applies to $E(\mathbf{y}) = \mathbf{X}\boldsymbol{\beta}$ with

$$\mathbf{X}'\mathbf{X} = \begin{pmatrix} 4 & 2 \\ 2 & 5 \end{pmatrix}.$$

A sample of size 32 gives $b_1 = 2$, $b_2 = 5$, $\mathbf{e}'\mathbf{e} = 60$. Construct a 90% confidence interval for $\beta_2 - \beta_1$.

21.6 Suppose that the CNR model applies to the earnings function

$$E(y \mid x_1, \ldots, x_8) = \beta_1 + \beta_2 x_2 + \beta_3 x_3 + \beta_4 x_4 + \beta_5 x_5 + \beta_6 x_6 + \beta_7 x_7 + \beta_8 x_8,$$

estimated in Exercise 17.4.

(a) Calculate $\mathbf{b}$, $\mathbf{e}'\mathbf{e}$, $\hat{\sigma}^2$, and $\hat{V}(\mathbf{b})$.

(b) Report a 95% confidence interval for the "effect of education," namely β_2.

(c) Report a 95% *joint* confidence region for the "effects of experience," β_3 and β_4.

(d) Let $\theta = \beta_3 + 2\beta_4\bar{x}_3$, where $\bar{x}_3$ is the sample mean experience (treated as nonrandom). Give an interpretation of, and a 90% confidence interval for, θ.

(e) Test, at the 5% significance level, the null hypothesis that "race does not affect earnings," that is, $\beta_5 = 0$.

(f) Test, at the 5% significance level, the joint null hypothesis that "region does not affect earnings," that is, $\beta_6 = \beta_7 = \beta_8 = 0$.

GAUSS Hint:
If S is a matrix, r is a vector of integers, and c is a vector of integers, then T = submat(S,r,c) is the submatrix of S consisting of the rows and columns enumerated in r and c, respectively.

22 Issues in Hypothesis Testing

22.1. Introduction

In this chapter, we take up a variety of practical and procedural topics associated with hypothesis testing. Among the topics are: the conversion of general hypotheses into the zero-null-subvector form, the choice of significance level, testing against one-sided alternatives, the abuse of tests, and inference when the normality assumption is not adopted.

22.2. General Linear Hypothesis

In general, a linear hypothesis takes the form $\boldsymbol{\theta} = \boldsymbol{\theta}°$, where $\boldsymbol{\theta} = \mathbf{H}\boldsymbol{\beta}$, $\mathbf{H}$ is $p \times k$ nonrandom with rank p, and $\boldsymbol{\theta}°$ is numerical. In Section 21.4, we focused on the special case $\boldsymbol{\beta}_2 = \mathbf{0}$, but other cases arise in practice.

For example, suppose that we have the demand function

$$E(Y|X_2, X_3, X_4) = \beta_1 X_1 + \beta_2 X_2 + \beta_3 X_3 + \beta_4 X_4,$$

where Y = log quantity of butter, $X_1 = 1$, X_2 = log real income, X_3 = log butter price, X_4 = log margarine price. We entertain the hypothesis that only the ratio of the two prices, not their separate levels, matters to the consumer. This says that the two log-price slopes are equal in magnitude but opposite in sign, that is, $\beta_4 = -\beta_3$. This hypothesis is expressed as $\theta = \beta_3 + \beta_4$, $\theta° = 0$.

For another example, consider the Cobb-Douglas production function,

$$E(Y|K, L, N) = \beta_1 + \beta_2 K + \beta_3 L + \beta_4 N,$$

with Y = log output, K = log capital, L = log land, N = log labor. The hypothesis of constant returns to scale says that the sum of the log-input slopes (the elasticities) is unity. It takes the form $\theta = \beta_2 + \beta_3 + \beta_4$ with $\theta^\circ = 1$.

For a third example, consider a macroeconomic consumption function, with $E(Y|X) = \alpha_1 + \gamma_1 X$ in wartime, $E(Y|X) = \alpha_2 + \gamma_2 X$ in peacetime. Here Y = consumption, and X = income. Defining the dummy variable Z, which equals 1 if war and equals 0 if peace, permits us to write the two functions together as

$$\begin{aligned} E(Y|X, Z) &= \alpha_1 Z + \gamma_1 ZX + \alpha_2(1 - Z) + \gamma_2(1 - Z)X \\ &= \beta_1 x_1 + \beta_2 x_2 + \beta_3 x_3 + \beta_4 x_4, \end{aligned}$$

say. The null hypothesis that the function is the same in war and peace is a joint linear hypothesis: $\beta_1 = \beta_3$ and $\beta_2 = \beta_4$. It is expressible as $\boldsymbol{\theta} = \boldsymbol{\theta}^\circ$, with

$$\boldsymbol{\theta} = \begin{pmatrix} 1 & 0 & -1 & 0 \\ 0 & 1 & 0 & -1 \end{pmatrix} \boldsymbol{\beta}, \qquad \boldsymbol{\theta}^\circ = \begin{pmatrix} 0 \\ 0 \end{pmatrix}.$$

For the general linear hypothesis in the CNR model, we saw (Section 21.3) that the F-test statistic is

$$v^\circ = (\mathbf{t} - \boldsymbol{\theta}^\circ)'\mathbf{D}(\mathbf{t} - \boldsymbol{\theta}^\circ)/(p\hat{\sigma}^2),$$

where $\mathbf{t} = \mathbf{Hb}$ and $\mathbf{D} = (\mathbf{H}'\mathbf{Q}^{-1}\mathbf{H})^{-1}$. For the zero-null-subvector special case, we saw (Section 21.4) that the numerator of this test statistic could be written as

$$(\mathbf{t} - \boldsymbol{\theta}^\circ)'\mathbf{D}(\mathbf{t} - \boldsymbol{\theta}^\circ) = \mathbf{b}_2'\mathbf{Q}_{22}^*\mathbf{b}_2 = \mathbf{e}^{*\prime}\mathbf{e}^* - \mathbf{e}'\mathbf{e},$$

where $\mathbf{e}^{*\prime}\mathbf{e}^*$ is the sum of squared residuals from the short regression of $\mathbf{y}$ on $\mathbf{X}_1$ alone, and $\mathbf{e}'\mathbf{e}$ is the sum of squared residuals from the long regression of $\mathbf{y}$ on $\mathbf{X} = (\mathbf{X}_1, \mathbf{X}_2)$.

In fact any linear hypothesis can be converted into a zero-null-subvector hypothesis, so that the computational convenience of short and long regressions is available in general.

Start with $\boldsymbol{\theta} = \mathbf{H}\boldsymbol{\beta}$, where the matrix $\mathbf{H}$ is $p \times k$ with rank p. We may suppose without loss of generality that the last p columns of $\mathbf{H}$ are linearly independent: that is, partitioning $\mathbf{H} = (\mathbf{H}_1, \mathbf{H}_2)$, the $p \times p$ submatrix $\mathbf{H}_2$ is nonsingular. Partition $\mathbf{X}$ and $\boldsymbol{\beta}$ conformably as $\mathbf{X} = (\mathbf{X}_1, \mathbf{X}_2)$ and $\boldsymbol{\beta} = (\boldsymbol{\beta}_1', \boldsymbol{\beta}_2')'$. Define the $k \times k$ matrix

$$\mathbf{T} = \begin{pmatrix} \mathbf{I} & \mathbf{O} \\ -\mathbf{H}_2^{-1}\mathbf{H}_1 & \mathbf{I} \end{pmatrix},$$

where the partitioning is into $k - p$ and p rows and columns. Its inverse is

$$\mathbf{T}^{-1} = \begin{pmatrix} \mathbf{I} & \mathbf{O} \\ \mathbf{H}_2^{-1}\mathbf{H}_1 & \mathbf{I} \end{pmatrix}.$$

Let

$$\mathbf{Z} = \mathbf{XT} = (\mathbf{X}_1, \mathbf{X}_2)\mathbf{T} = (\mathbf{X}_1 - \mathbf{X}_2\mathbf{H}_2^{-1}\mathbf{H}_1, \mathbf{X}_2) = (\mathbf{Z}_1, \mathbf{Z}_2),$$

which is an $n \times k$ rank-k nonrandom matrix, interpretable as the matrix of observations on transformed explanatory variables. Also let

$$\boldsymbol{\alpha} = \mathbf{T}^{-1}\boldsymbol{\beta} = \mathbf{T}^{-1}\begin{pmatrix} \boldsymbol{\beta}_1 \\ \boldsymbol{\beta}_2 \end{pmatrix} = \begin{pmatrix} \boldsymbol{\beta}_1 \\ \mathbf{H}_2^{-1}\mathbf{H}_1\boldsymbol{\beta}_1 + \boldsymbol{\beta}_2 \end{pmatrix} = \begin{pmatrix} \boldsymbol{\alpha}_1 \\ \boldsymbol{\alpha}_2 \end{pmatrix},$$

which is a $k \times 1$ vector of transformed parameters.

Now

$$\mathbf{X}\boldsymbol{\beta} = \mathbf{X}(\mathbf{TT}^{-1})\boldsymbol{\beta} = (\mathbf{XT})(\mathbf{T}^{-1}\boldsymbol{\beta}) = \mathbf{Z}\boldsymbol{\alpha} = \mathbf{Z}_1\boldsymbol{\alpha}_1 + \mathbf{Z}_2\boldsymbol{\alpha}_2,$$

so $E(\mathbf{y}) = \mathbf{X}\boldsymbol{\beta}$ is equivalent to

$$E(\mathbf{y}) = \mathbf{Z}_1\boldsymbol{\alpha}_1 + \mathbf{Z}_2\boldsymbol{\alpha}_2. \tag{22.1}$$

Further,

$$\boldsymbol{\theta} = \mathbf{H}\boldsymbol{\beta} = \mathbf{H}(\mathbf{T}\boldsymbol{\alpha}) = (\mathbf{H}_1, \mathbf{H}_2)\mathbf{T}\boldsymbol{\alpha} = \mathbf{H}_2\boldsymbol{\alpha}_2,$$

so $\boldsymbol{\theta} = \boldsymbol{\theta}^\circ$ is equivalent to $\mathbf{H}_2\boldsymbol{\alpha}_2 = \boldsymbol{\theta}^\circ$, that is, to $\boldsymbol{\alpha}_2 = \mathbf{H}_2^{-1}\boldsymbol{\theta}^\circ = \boldsymbol{\alpha}_2^\circ$, say. And $\boldsymbol{\alpha}_2 = \boldsymbol{\alpha}_2^\circ$ is equivalent to saying that Eq. (22.1) can be written as

$$E(\mathbf{y}) = \mathbf{Z}_1\boldsymbol{\alpha}_1 + \mathbf{Z}_2\boldsymbol{\alpha}_2^\circ. \tag{22.2}$$

Let $\mathbf{y}^\circ = \mathbf{y} - \mathbf{Z}_2\boldsymbol{\alpha}_2^\circ$, which may be interpreted as a vector of observations on a transformed dependent variable. Then Eq. (22.1) is equivalent to

$$E(\mathbf{y}^\circ) = \mathbf{Z}_1\boldsymbol{\alpha}_1 + \mathbf{Z}_2\boldsymbol{\alpha}_2^*,$$

where $\boldsymbol{\alpha}_2^* = \boldsymbol{\alpha}_2 - \boldsymbol{\alpha}_2^\circ$, while Eq. (22.2) is equivalent to

$$E(\mathbf{y}^\circ) = \mathbf{Z}_1\boldsymbol{\alpha}_1.$$

We have translated a general linear hypothesis into a zero-null-subvector hypothesis. Because $\mathbf{Z} = \mathbf{XT}$ with $\mathbf{T}$ nonsingular, regressing $\mathbf{y}$ on

$\mathbf{Z}$, or $\mathbf{y}^\circ$ on $\mathbf{Z}$, gives the same sum of squared residuals as regressing $\mathbf{y}$ on $\mathbf{X}$: see Exercise 17.2. So the kernel of the test statistic for $\mathbf{H}\boldsymbol{\beta} = \mathbf{0}$ can be calculated as $\mathbf{e}^{*\prime}\mathbf{e}^* - \mathbf{e}'\mathbf{e}$, where $\mathbf{e}'\mathbf{e}$ is obtained from the long regression of $\mathbf{y}$ on $\mathbf{Z}$ (or equivalently of $\mathbf{y}^\circ$ on $\mathbf{Z}$, or of $\mathbf{y}$ on $\mathbf{X}$), and $\mathbf{e}^{*\prime}\mathbf{e}^*$ is obtained from the short regression of $\mathbf{y}^\circ$ on $\mathbf{Z}_1$.

All this is easier done than said. For example, for the butter demand equation, the restriction $\beta_4 = -\beta_3$ says that

$$\begin{aligned} E(y) &= \mathbf{x}_1\beta_1 + \mathbf{x}_2\beta_2 + \mathbf{x}_3\beta_3 + \mathbf{x}_4\beta_4 \\ &= \mathbf{x}_1\beta_1 + \mathbf{x}_2\beta_2 + (\mathbf{x}_3 - \mathbf{x}_4)\beta_3 \\ &= \mathbf{z}_1\beta_1 + \mathbf{z}_2\beta_2 + \mathbf{z}_3\beta_3, \end{aligned}$$

with $\mathbf{z}_1 = \mathbf{x}_1$, $\mathbf{z}_2 = \mathbf{x}_2$, $\mathbf{z}_3 = \mathbf{x}_3 - \mathbf{x}_4$. The restricted regression is implemented by running $\mathbf{y}$ on $\mathbf{Z} = (\mathbf{z}_1, \mathbf{z}_2, \mathbf{z}_3)$.

For the Cobb-Douglas production model, the constant-returns-to-scale restriction may be implemented by regressing $y^\circ = Y - N =$ log(output/labor) on $z_1 = 1$, $z_2 = K - N =$ log(capital/labor), and $z_3 = L - N =$ log(land/labor).

For the war and peace consumption functions, the equal-slope restriction $\beta_2 = \beta_4$ says that

$$\begin{aligned} E(\mathbf{y}) &= \mathbf{x}_1\beta_1 + \mathbf{x}_2\beta_2 + \mathbf{x}_3\beta_3 + \mathbf{x}_4\beta_4 \\ &= \mathbf{x}_1\beta_1 + (\mathbf{x}_2 + \mathbf{x}_4)\beta_2 + \mathbf{x}_3\beta_3 \\ &= \mathbf{z}_1\beta_1 + \mathbf{z}_2\beta_2 + \mathbf{z}_3\beta_3, \end{aligned}$$

with $\mathbf{z}_1 = \mathbf{x}_1$, $\mathbf{z}_2 = \mathbf{x}_2 + \mathbf{x}_4$, $\mathbf{z}_3 = \mathbf{x}_3$. This restricted regression is implemented by running $\mathbf{y}$ on $\mathbf{Z} = (\mathbf{z}_1, \mathbf{z}_2, \mathbf{z}_3)$. If the equal-intercept restriction $\beta_1 = \beta_3$ is also imposed, write

$$E(\mathbf{y}) = \mathbf{w}_1\beta_1 + \mathbf{w}_2\beta_2,$$

with $\mathbf{w}_1 = \mathbf{z}_1 + \mathbf{z}_3 = \mathbf{x}_1 + \mathbf{x}_3$, $\mathbf{w}_2 = \mathbf{z}_2 = \mathbf{x}_2 + \mathbf{x}_4$. Then run $\mathbf{y}$ on $\mathbf{W} = (\mathbf{w}_1, \mathbf{w}_2)$ to impose the second restriction as well.

The approach we have been using amounts to solving out the hypothesized coefficient restrictions to get a shorter regression that can be fitted by unrestricted least squares. Since the approach is feasible for any general linear hypothesis, in practice there is no need to calculate $(\mathbf{t} - \boldsymbol{\theta}^\circ)'\mathbf{D}(\mathbf{t} - \boldsymbol{\theta}^\circ)$ directly to get the test statistic v°. Just use the residual sums of squares from the long, and an appropriate short, regression.

Here is a final example to show how the solving-out-the-restrictions approach works. Suppose that you have a pair of data sets to which

$$E(\mathbf{y}_1) = \mathbf{X}_1\boldsymbol{\beta}_1, \qquad E(\mathbf{y}_2) = \mathbf{X}_2\boldsymbol{\beta}_2$$

apply, where $\mathbf{y}_1$ is $n_1 \times 1$, $\mathbf{X}_1$ is $n_1 \times k$, $\mathbf{y}_2$ is $n_2 \times 1$, and $\mathbf{X}_2$ is $n_2 \times k$. You want to test the null hypothesis $\boldsymbol{\beta}_1 = \boldsymbol{\beta}_2$. Assemble the data together as

$$E(\mathbf{y}) = \begin{pmatrix} E(\mathbf{y}_1) \\ E(\mathbf{y}_2) \end{pmatrix} = \begin{pmatrix} \mathbf{X}_1\boldsymbol{\beta}_1 \\ \mathbf{X}_2\boldsymbol{\beta}_2 \end{pmatrix} = \begin{pmatrix} \mathbf{X}_1 & \mathbf{O} \\ \mathbf{O} & \mathbf{X}_2 \end{pmatrix} \begin{pmatrix} \boldsymbol{\beta}_1 \\ \boldsymbol{\beta}_2 \end{pmatrix} = \mathbf{X}\boldsymbol{\beta},$$

say. If the null $\boldsymbol{\beta}_1 = \boldsymbol{\beta}_2$ ($= \boldsymbol{\beta}^\circ$, say) is true, then

$$E(\mathbf{y}) = \begin{pmatrix} \mathbf{X}_1 \\ \mathbf{X}_2 \end{pmatrix} \boldsymbol{\beta}^\circ = \mathbf{X}^\circ\boldsymbol{\beta}^\circ,$$

say. The relevant sums of squared residuals are obtainable from a long regression ($\mathbf{y}$ on the $2k$ columns of $\mathbf{X}$) and a short regression ($\mathbf{y}$ on the k columns of $\mathbf{X}^\circ$). Provided that the CNR model applies to each sample, with the same σ^2, while the two samples are independent, the difference between those sums of squared residuals is the kernel of the appropriate F-statistic. This special case of a standard F-test is sometimes referred to as a "test for structural change," or as a "Chow test." Incidentally, the long-regression sum of squared residuals can be calculated by adding together the sums of squared residuals obtained in separate LS regressions of $\mathbf{y}_1$ on $\mathbf{X}_1$, and $\mathbf{y}_2$ on $\mathbf{X}_2$.

22.3. One-Sided Alternatives

To test the single-parameter null hypothesis $\beta_j = \beta_j^\circ$ against the alternative that $\beta_j \neq \beta_j^\circ$, we have learned to use the t-statistic given by $u_j^\circ = (b_j - \beta_j^\circ)/\hat{\sigma}_{b_j}$, rejecting the null iff $|u_j^\circ| > c$, where $G(c) = 0.975$ with $G(\cdot)$ being the cdf of the $t(n - k)$ distribution.

Now suppose, as occurs in some economic contexts, that the known alternative to $\beta_j = \beta_j^\circ$ is *one-sided*, say $\beta_j > \beta_j^\circ$. A *one-tailed* version of the t-test can be used: reject the null $\beta_j = \beta_j^\circ$ iff $u_j^\circ > c^*$, where $G(c^*) = 0.95$. This variant is sensible. Heuristically, it would be foolish to reject $\beta_j = \beta_j^\circ$ *in favor of* $\beta_j > \beta_j^\circ$ when the sample has $b_j < \beta_j^\circ$. More formally, the one-tailed test has more power than the two-tailed test, for all $\beta_j > \beta_j^\circ$

—which is the only region where power is wanted in the present situation: see Exercise 22.4.

We have learned, as an equivalent to the t-test, the F-test that uses the statistic $v_j^\circ = (b_j - \beta_j^\circ)^2/\hat{\sigma}_{b_j}^2$, rejecting the null if $v_j^\circ > d$ where $G_1(d) = 0.95$, with $G_1(\cdot)$ being the cdf of the $F(1, n - k)$ distribution. The two approaches are equivalent because $v_j^\circ = (u_j^\circ)^2$ and $d = c^2$. But the F-statistic $v_j^\circ = (u_j^\circ)^2$ disregards the sign of $b_j - \beta_j^\circ$, so it is not attractive for use when the alternative is one-sided.

For a joint hypothesis with one-sided alternatives, no t-test is available. The F-statistic

$$v^\circ = (\mathbf{t} - \boldsymbol{\theta}^\circ)'\mathbf{D}(\mathbf{t} - \boldsymbol{\theta}^\circ)/(p\hat{\sigma}^2),$$

treats positive and negative misses symmetrically, so it is not attractive for tests against one-sided alternatives. For a discussion of appropriate procedures, see Gouriéroux et al. (1982) and Wolak (1987).

22.4. Choice of Significance Level

Suppose that you are asked to test the null hypothesis $\beta_j = 0$ against the alternative $\beta_j \neq 0$, in a sample with $n - k = 120$. You obtain the test statistic $u_j^\circ = 1.82$. Critical values from the $\mathcal{N}(0, 1)$ table are $c = 1.96$ at the 5% level and $c = 1.64$ at the 10% level. With $1.64 < 1.82 < 1.96$, the null would be accepted at the 5% level, but rejected at the 10% level. The same piece of evidence that will accept $\beta_j = 0$ at the 5% level will reject it at the 10% level. The interval between 1.64 and 1.96 is a "zone of opportunity." Indeed, whatever numerical value the sample delivers, a diligent researcher can force acceptance by setting the significance level low enough (e.g., 1% or 0.5%) or can force rejection by setting the significance level high enough (e.g., 10% or 20%).

How *should* a researcher choose the significance level? Econometrics texts offer little, if any, guidance. In statistics texts, the discussion focuses on the power of the test—the probability of rejecting the null hypothesis as a function of the true parameter value.

Generally power declines as the significance level declines: see Exercise 22.4. Moving from the 5% to the 1% significance level not only reduces the probability of rejecting a true null, but also reduces the probability of rejecting a false null. The first reduction is desirable, the second is undesirable.

There is a trade-off. To resolve the trade-off, statistics texts recommend a cost-benefit calculation: if the net cost of accepting a false null is less than the net cost of rejecting a true null, then choose a low significance level. Although this cost-benefit approach should be congenial to economists, the 5% level is almost always used in the empirical economics literature. It is hardly plausible that distinct cost-benefit calculations underlie that ubiquitous level. Occasionally, the 10% and 1% levels are used. Reading closely, you may well be able to spot the occasions on which those levels replace 5%. If an author really wants to accept the null, she may switch to the 1% level; if an author really wants to reject the null, he may switch to the 10% level. When such switches do not suffice, you may see such language as "barely significant at the 1% level" (a hint that the author really wants to accept) or "almost significant at the 10% level" (a hint that the author really wants to reject).

This state of affairs may seem very unsatisfactory, but the textbook recommendation of a cost-benefit calculation is not appealing either. For academic research reports, neither the costs nor the benefits of the test decision are clear. It is rare for an economic agent to undertake real-world action upon reading a test outcome reported in a journal article. At most what may happen is that readers' beliefs shift in the light of the evidence. So, in almost all applied economic contexts, the significance level is necessarily a matter of convention rather than of calculation.

It follows that readers should not take an author's announcement of significance or nonsignificance as authoritative. Regardless of the author's choice of significance level and announcement of a decision, sensible readers will have to decide for themselves whether the evidence is weighty or fragile. Regardless of how the author phrases the test decision, the burden remains on readers to assess whether the sample evidence against the null (the magnitude of the test statistic) is strong enough to induce a change in their beliefs.

A couple of lessons for writers emerge:

• It is usually bad practice to say "significant [or nonsignificant] at the 5% level," without reporting the magnitude of the test statistic. (It is even worse practice to announce "significance" or "nonsignificance" without specifying a null hypothesis. In particular, the zero null may not be the interesting null.)

• A useful alternative to the test statistic is a report of its "*P*-value," or "marginal significance level," which is the level at which the observed test statistic would be just significant. For example, suppose that a $\chi^2(p)$ test is conducted, the cdf being $G_p(\cdot)$. If w° is the observed test statistic, then its P-value is $\alpha^\circ = 1 - G_p(w^\circ)$. The null would be rejected at all significance levels higher than α°, and accepted at all significance levels lower than α°. So the P-value gives readers more information than is contained in the binary report "accept" or "reject."

22.5. Statistical versus Economic Significance

A strong case can be made that hypothesis testing is widely abused in empirical economics: see McCloskey (1985). In many research reports, the author's conclusions emphasize the statistical significance, rather than the economic significance, of the coefficient estimates. Yet, a coefficient estimate may be "very significantly different from unity" (by the t-test), while that difference is economically trivial. Or the difference may be "not significantly different from unity" but have an economically substantial magnitude.

It is certainly desirable to know how reliable a coefficient estimate is, that is, to know its standard error. But that desirability does not suffice to justify a hypothesis test, which involves measuring the estimate relative to its standard error. Rather, the confidence interval for β_j, constructed from the point estimate b_j and its standard error $\hat{\sigma}_{b_j}$, will be the proper target in most research.

When a null, say, $\beta_j = 1$, is specified, the likely intent is that β_j is *close* to 1, so close that for practical purposes it may be treated *as if it were* 1. But whether 1.1 is "practically the same as" 1.0 is a matter of economics, not of statistics. One cannot resolve the matter by relying on a hypothesis test, because the test statistic $(b_j - 1)/\hat{\sigma}_{b_j}$ measures the estimated coefficient in standard error units, which are not the meaningful units in which to measure the economic parameter $\beta_j - 1$. It may be a good idea to reserve the term "significance" for the statistical concept, adopting "substantial" for the economic concept.

There is a further objection to the common practice of indiscriminately reporting all the "t-statistics" for a regression: it encourages rank-ordering of the explanatory variables with respect to their "importance." What does it mean to say that in a multiple regression one explanatory variable is "more important" than another?

A simple example may help to address this question. Suppose that this estimated regression is reported:

$$\hat{y} = 50 + 2x_2 - 1x_3.$$

A naive reader might conclude that x_2 is "more important" than x_3 because its coefficient is larger in magnitude. A more sophisticated reader would recognize that the magnitude of the coefficients can be changed arbitrarily by changing the units in which the variables are measured. So he might ask for the standard errors. Being told that the standard errors for b_2 and b_3 are both 0.5, so their "t-statistics" are 4 and -2, he might conclude that x_2 is "more important" than x_3 because its "t-statistic" is larger in magnitude. But that conclusion is not sensible if in fact the variables are y = weight (in pounds), x_2 = height (in inches), x_3 = exercise (in hours per week), and the regression is to be used by a physician to advise an overweight patient. Would either the physician or the patient be edified to learn that height is "more important" than exercise in explaining variation in weight?

The moral of this example is that statistical measures of "importance" are a diversion from the proper target of the research—estimation of relevant parameters—to the task of "explaining variation" in the dependent variable.

22.6. Using Asymptotics

In the CNR model, provided that $n - k$ is large, there is no need to refer to the t- and F-tables when σ^2 is unknown. Recall the two asymptotic results shown in Section 18.3:

(1) If $u \sim t(n)$, then $u \xrightarrow{D} \mathcal{N}(0, 1)$.
(2) If $v \sim F(m, n)$, then $mv \xrightarrow{D} \chi^2(m)$.

Applied to the CNR model, (1) implies that there is no objection, when $n - k$ is large, to treating

$$\hat{z}_j = u_j = (b_j - \beta_j)/\hat{\sigma}_{b_j}$$

as if it were

$$z_j = (b_j - \beta_j)/\sigma_{b_j}.$$

In the same manner, (2) implies that there is no objection, when $n - k$ is large, to treating

$$\hat{w} = pv = (\mathbf{t} - \boldsymbol{\theta})'\mathbf{D}(\mathbf{t} - \boldsymbol{\theta})/\hat{\sigma}^2$$

as if it were

$$w = (\mathbf{t} - \boldsymbol{\theta})'\mathbf{D}(\mathbf{t} - \boldsymbol{\theta})/\sigma^2.$$

For example, with $n - k = 200$ and $p = 2$, the exact 5% critical value $d_p = 3.04$ from the F-table gives $pd_p = 6.08$ as the critical value for $\hat{w} = pv$, while the approximation will use $c_p = 5.99$ from the chi-square table.

This simplification applies to hypothesis tests as well as to confidence region construction.

22.7. Inference without Normality Assumption

From Chapter 19 on, the theory has relied on normality of $\mathbf{y}$. In practice, researchers routinely use the t- and F-procedures without making an explicit normality assumption. A better practice might be to use the normal and chi-square approximations of Section 22.6, for an asymptotic theory appropriate to the CR model implies that $\mathbf{b}$ is asymptotically normal, that $\hat{z}_j$ is asymptotically $\mathcal{N}(0, 1)$, and that $\hat{w}$ is asymptotically $\chi^2(p)$. Without normality, there is no presumption that the t- and F-tables offer better approximations to the exact distributions of those statistics even when the sample size is small.

To develop an asymptotic distribution theory that is appropriate to the CR model without normality, additional specification is needed. How does the $\mathbf{X}$ matrix develop as n increases? That is, how are the additional rows of $\mathbf{X}$ generated? In random sampling from a multivariate population, further specification is unnecessary, because random sampling extends itself automatically. But with our stratified sampling scheme, some additional assumptions are required. The most natural assumption is

$$\lim(\mathbf{Q}/n) = \boldsymbol{\Phi},$$

where $\boldsymbol{\Phi}$ is positive definite. (Here lim is shorthand for limit as $n \to \infty$.) To see the implications of this assumption, first rewrite $V(\mathbf{b})$ as

$$V(\mathbf{b}) = \sigma^2\mathbf{Q}^{-1} = (\sigma^2/n)(\mathbf{Q}/n)^{-1}.$$

If $\lim(\mathbf{Q}/n) = \mathbf{\Phi}$, with $\mathbf{\Phi}$ positive definite (hence invertible), then $\lim(\mathbf{Q}/n)^{-1} = \mathbf{\Phi}^{-1}$. Since $\lim(\sigma^2/n) = 0$, that would imply $\lim V(\mathbf{b}) = \mathbf{O}$. Since $E(\mathbf{b}) = \boldsymbol{\beta}$ for every n, it would follow (by the multivariate version of convergence in mean square) that $\mathbf{b} \xrightarrow{P} \boldsymbol{\beta}$, and $\mathbf{b}$ would be a consistent estimator of $\boldsymbol{\beta}$.

Suppose further that the $\epsilon_i = y_i - \mathbf{x}_i'\boldsymbol{\beta}$ are independent and identically distributed—which is stronger than the uncorrelated and identical expectation and variance assumptions of the original CR model. Then it can be shown (by a multivariate extension of the Central Limit Theorem) that

$$\mathbf{b} \overset{A}{\sim} \mathcal{N}[\boldsymbol{\beta}, (\sigma^2/n)\mathbf{\Phi}^{-1}].$$

Similarly it can be shown that $\hat{\sigma}^2 \xrightarrow{P} \sigma^2$. The net result is that the asymptotic approximations of Section 22.6 will apply even without assuming normality for $\mathbf{y}$. See Amemiya (1985, pp. 95–101), Judge et al. (1988, pp. 264–270), or Greene (1990, pp. 312–318).

Henceforth when we report asymptotic properties in models with nonstochastic $\mathbf{X}$, we shall be presuming that additional assumptions of the type introduced here are met.

Exercises

22.1 Suppose that the CNR model applies to $E(y) = x_1\beta_1 + x_2\beta_2 + x_3\beta_3 + x_4\beta_4$. Let $z = x_3 + x_4$. For a sample of 124 observations, regressing y on (x_1, x_2, x_3, x_4) gives 60 as the sum of squared residuals, while regressing y on (x_1, x_2, z) gives 64 as the sum of squared residuals. Test at the 5% significance level the null hypothesis $\beta_3 = \beta_4$ against the two-sided alternative $\beta_3 \neq \beta_4$.

22.2 Suppose that the CNR model applies to $E(y) = x_1\beta_1 + x_2\beta_2 + x_3\beta_3 + x_4\beta_4$, where y = log output, $x_1 = 1$, x_2 = log capital, x_3 = log land, and x_4 = log labor. Let $w = y - x_4$, $z_1 = x_1$, $z_2 = x_2 - x_4$, $z_3 = x_3 - x_4$. For a sample of 104 firms, regressing y on (x_1, x_2, x_3, x_4) gives 70 as the sum of squared residuals, while regressing w on (z_1, z_2, z_3) gives 80 as the sum of squared residuals.

(a) Test at the 5% significance level the null hypothesis $\beta_2 + \beta_3 + \beta_4 = 1$ against the two-sided alternative $\beta_2 + \beta_3 + \beta_4 \neq 1$.

(b) Let $v = y - x_2$, $t_1 = x_1$, $t_2 = x_3 - x_2$, $t_3 = x_4 - x_2$. If v is regressed on (t_1, t_2, t_3), what sum of squared residuals will be obtained?

22.3 The CNR model applies to $E(y) = x_1\beta_1 + x_2\beta_2 + x_3\beta_3 + x_4\beta_4 + x_5\beta_5$. A researcher regresses y on $(x_1, x_2, x_3, x_4, x_5)$, and also regresses w on (z_1, z_2), where $w = y - x_4$, $z_1 = x_1$, $z_2 = x_2 + x_3$.

(a) State the joint null hypothesis that is testable by a comparison of the sum of squared residuals from those two regressions.
(b) What is the "numerator degrees of freedom" parameter for that test?

22.4 The regression slope b in a CNR model is distributed $\mathcal{N}(\beta, 1)$. The null hypothesis $\beta = 0$ will be tested by using the statistic $z° = b/\sigma_b$.

(a) For a conventional (two-sided alternative) situation, consider running the test at the 10% and 5% levels. Write and run a program that tabulates the power of the two tests at these nine values of the true parameter β:

$$-2 \quad -1.5 \quad -1 \quad -0.5 \quad 0 \quad 0.5 \quad 1 \quad 1.5 \quad 2.$$

(b) What does your table say about the effect of significance level on power?
(c) Now consider a one-sided-alternative situation, the alternative being $\beta > 0$. Specify an appropriate one-tailed procedure that uses $z°$ and operates at the 5% level. Tabulate the power of the test at the nine values of β given above.
(d) Comparing your results in (c) with those in (a), what do you conclude about the relative merits of one-tailed and two-tailed tests at the same significance level?

22.5 For the earnings function estimated in Exercise 21.6, consider b_5, the coefficient on race. Is it significantly different from zero? Is it large, that is, substantial in economic terms?

23 Multicollinearity

23.1. Introduction

Multicollinearity, or simply *collinearity*, refers to correlation among the explanatory variables in multiple regression. As in Section 19.4, let us focus on the slope coefficients β_j ($j = 2, \ldots, k$) in a CR model that includes a constant as x_1. The estimated slopes are the b_j, whose variances are

$$\sigma_{b_j}^2 = \sigma^2/q_{jj}^* = \sigma^2/\mathbf{x}_j^{*\prime}\mathbf{x}_j^* = \sigma^2 \Big/ \left[(1 - R_j^2) \sum_i (x_{ij} - \bar{x}_j)^2\right],$$

where R_j^2 is the coefficient of determination in the auxiliary regression of x_j on all the other x's. The condition that $\mathbf{X}$ have full column rank rules out *exact collinearity*: because rank($\mathbf{X}$) $= k$, no x_j can be an exact linear function of the other x's, so no R_j^2 will equal 1. But the rank condition does not rule out high collinearity—one or more R_j^2's that are close to 1. Indeed, many economic data sets do show high auxiliary R_j^2's, and virtually none show zero R_j^2's. From the variance formula, we see that ceteris paribus, a high auxiliary R_j^2 makes for a large $\sigma_{b_j}^2$. As Judge et al. (1988, p. 882) write:

> Multicollinearity is defined as the existence of one or more near-exact linear relations among the columns of the regressor matrix $\mathbf{X}$. The consequences of multicollinearity are that the sampling distributions of the coefficient estimators may have such large variances that the coefficient estimates are unstable from sample to sample. Thus they may be too unreliable to be useful.

When its variance is large, the estimator will be imprecise, the sample value may well be far away from the true value, the confidence interval for β_j will be wide, very diverse hypotheses about β_j will all be acceptable, hypothesis tests on β_j will have little power, and b_j will not be significantly different from "anything." In short, our best estimate of β_j will not be very good, and the sample will have told us little about the true value of β_j.

All these unpleasant things are fully reflected in the standard error of b_j, just as they would be if R_j^2 were zero while the variation of x_j, namely $\Sigma_i(x_{ij} - \bar{x}_j)^2$, were small and/or the (conditional) variance of the dependent variable, namely σ^2, were large. The LS estimate b_j is still the MVLUE, its standard error is still correct, and the conventional confidence interval and hypothesis tests are still valid.

Nevertheless, in empirical research papers one comes across complaints such as "the standard errors are inflated because of collinearity," or "this variable is really significant but multicollinearity makes it look insignificant."

To evaluate such complaints, consider a simpler situation: estimating a univariate population mean when the sample size is small. Suppose that a random variable y has expectation μ and variance σ^2. In random sampling, sample size n, the MVLUE of μ is the sample mean $\bar{y}$, whose variance is $V(\bar{y}) = \sigma^2/n$. If n is small, then ceteris paribus, $V(\bar{y})$ is large. If $V(\bar{y})$ is large, then our estimator of μ is imprecise, the estimate $\bar{y}$ may well be far away from the true μ, the confidence interval for μ will be wide, very diverse hypotheses about μ will all be acceptable, hypothesis tests on μ will have little power, and $\bar{y}$ will not be significantly different from "anything." In short, our best estimate of μ will not be very good, and the sample will have told us little about the true value of μ.

So the problem of multicollinearity when estimating a conditional expectation function in a multivariate population is quite parallel to the problem of small sample size when estimating the expectation of a univariate population. But researchers faced with the latter problem do not usually dramatize the situation, as some appear to do when faced with multicollinearity.

23.2. Textbook Discussions

It may be that econometrics textbooks contribute to the dramatization of multicollinearity by giving elaborate attention to the subject. Johnston (1984) writes:

> The prevalent case in so much econometric work, especially with time series data, is one of high but not exact multicollinearity. This raises three questions: 1. What effects to expect from multicollinearity. 2. How to detect the degree of multicollinearity. 3. What remedial action to take. (p. 245)

Among the effects to expect:

> A common result is to find regressions possibly with a very high overall R^2, but with some (or many) individual coefficients apparently insignificant. The high R^2 arises when the y vector is close to the hyperplane generated by the $\mathbf{x}_j$ vectors and the apparently insignificant coefficients arise because the $\mathbf{x}_j$ vectors are nearly linearly dependent. (pp. 248–249)

However:

> It is also possible to find a high R^2 and highly significant t values on individual coefficients, even though multicollinearity is serious. This can arise if individual coefficients happen to be numerically well in excess of the true value, so that the effect still shows up in spite of the inflated standard error and/or because the true value itself is so large that even an estimate on the downside still shows up as significant. (p. 249)

Among the detection devices is $|\mathbf{X}'\mathbf{X}|$. This determinant

> declines in value with increasing collinearity, tending to zero as collinearity becomes exact. While a useful warning signal, we have no calibration scale for assessing what is serious and what is very serious. (p. 249)

As for remedies:

> More data is no help in multicollinearity if it is simply "more of the same." What matters is the structure of the $\mathbf{X}'\mathbf{X}$ matrix, and this will only be improved by adding data which are less collinear than before. However, there is often no easy way for an econometrician to get better data. The data are produced by the functioning of the economic system, and the collinearities reflect the nature of that system. (p. 250)

Turning to another text, we find that Judge et al. (1988, chap. 21) devote over twenty-five pages to multicollinearity. They point out that coefficients may appear to be nonsignificantly different from zero, and hence variables may be dropped from the regression, not because the variables have no effect, but rather because the sample is inadequate to estimate the effects precisely. This can happen even though the multiple R^2 is high enough to indicate that the full regression has significant explanatory power.

They argue that methods are required to detect the presence, severity, and form or nature of multicollinearity. They review some methods used to decide that the multicollinearity is severe: the simple correlation between a pair of explanatory variables exceeds 0.8 or 0.9, or the simple correlation exceeds the R^2 of the main regression. Such cutoff points are, they warn, arbitrary, and "pairwise correlations can give no insight into more complex interrelationships" when more than two explanatory variables are involved (p. 869).

Other methods are discussed: the determinant of $\mathbf{X}'\mathbf{X}$, variance inflation factors, auxiliary regressions, Theil's multicollinearity effect, and matrix decompositions. In the decomposition approach, relatively small characteristic roots of $\mathbf{X}'\mathbf{X}$ indicate near-linear dependencies among the explanatory variables, and the associated characteristic vectors identify the dependencies themselves. They remark (p. 870) that "analysis of the characteristic roots and vectors of the $\mathbf{X}'\mathbf{X}$ matrix can reveal much about the presence and nature of multicollinearity." They view the decomposition approach as the best of the available devices, but caution that it does not provide a complete solution: fixing a cutoff point for relative smallness is just a rule of thumb, and the method may fail to isolate multiple linear dependencies from one another.

Judge et al. go on to discuss several strategies for mitigating the effects of severe multicollinearity, while emphasizing that none of those strategies is completely safe.

23.3. Micronumerosity

Econometrics texts devote many pages to the problem of multicollinearity in multiple regression, but they say little about the closely analogous problem of small sample size in estimating a univariate mean. Perhaps that imbalance is attributable to the lack of an exotic polysyllabic name for "small sample size." If so, we can remove that impediment by introducing the term *micronumerosity*.

Suppose an econometrician set out to write a chapter about small sample size in sampling from a univariate population. Judging from what is now written about multicollinearity, the chapter might look like this:

Micronumerosity

The extreme case, "exact micronumerosity," arises when $n = 0$, in which case the sample estimate of μ is not unique. (Technically, there is a violation of the rank condition $n > 0$: the matrix 0 is singular.) The extreme case is easy enough to recognize. "Near micronumerosity" is more subtle, and yet very serious. It arises when the rank condition $n > 0$ is barely satisfied. Near micronumerosity is very prevalent in empirical economics.

Consequences of micronumerosity

The consequences of micronumerosity are serious. Precision of estimation is reduced. There are two aspects of this reduction: estimates of μ may have large errors, and not only that, but $V(\bar{y})$ will be large.

Investigators will sometimes be led to accept the hypothesis $\mu = 0$ because $u^\circ = \bar{y}/\hat{\sigma}_{\bar{y}}$ is small, even though the true situation may be not that $\mu = 0$ but simply that the sample data have not enabled us to pick μ up.

The estimate of μ will be very sensitive to sample data, and the addition of a few more observations can sometimes produce drastic shifts in the sample mean.

The true μ may be sufficiently large for the null hypothesis $\mu = 0$ to be rejected, even though $V(\bar{y}) = \sigma^2/n$ is large because of micronumerosity. But if the true μ is small (although nonzero) the hypothesis $\mu = 0$ may mistakenly be accepted.

Testing for micronumerosity

Tests for the presence of micronumerosity require the judicious use of various fingers. Some researchers prefer a single finger, others use their toes, still others let their thumbs rule.

A generally reliable guide may be obtained by counting the number of observations. Most of the time in econometric analysis, when n is close to zero, it is also far from infinity.

Several test procedures develop critical values n^*, such that micronumerosity is a problem only if n is smaller than n^*. But those procedures are questionable.

Remedies for micronumerosity
If micronumerosity proves serious in the sense that the estimate of μ has an unsatisfactorily low degree of precision, we are in the statistical position of not being able to make bricks without straw. The remedy lies essentially in the acquisition, if possible, of larger samples from the same population.

But more data are no remedy for micronumerosity if the additional data are simply "more of the same." So obtaining lots of small samples from the same population will not help.

If we return from this fantasy to reality, several lessons may be drawn.

• Multicollinearity is no more (or less) serious than micronumerosity. Exact multicollinearity ($R_j^2 = 1$) is a close analogue of exact micronumerosity ($n = 0$). When a research article complains about multicollinearity, readers ought to see whether the complaints would be convincing if "micronumerosity" were substituted for "multicollinearity."

• For example, if a test for exact multicollinearity is reported, the null hypothesis being $R_j^2 = 1$, readers ought to consider whether they would test the null hypothesis $n = 0$. Or if a test for orthogonality is reported, the null hypothesis being $R_j^2 = 0$, readers ought to consider whether they would test the null hypothesis that n is large. It is quite sensible to measure n, but would one want to undertake a statistical test on the true value of n?

• For another example, if a rule is proposed to decide whether the collinearity is severe (how large R_j^2 has to be before one says that there is a multicollinearity problem), readers ought to consider whether it is plausible to develop a rule that decides how small n has to be before one says that there is a small-sample-size problem.

23.4. When Multicollinearity Is Desirable

Multicollinearity may make the estimates of individual β_j's imprecise, while facilitating the precise estimation of particular combinations of the elements of $\boldsymbol{\beta}$. Suppose that we have estimated

$$E(y) = \beta_1 + x_2\beta_2 + x_3\beta_3,$$

by

$$\hat{y} = b_1 + x_2 b_2 + x_3 b_3.$$

Let $\theta = \beta_2 + \beta_3$, which is estimated by $t = b_2 + b_3$. The variances of the estimates are

$$\sigma_{b_2}^2 = \sigma^2 q^{22}, \qquad \sigma_{b_3}^2 = \sigma^2 q^{33}, \qquad \sigma_t^2 = \sigma^2(q^{22} + q^{33} + 2q^{23}),$$

where the q^{hj} are elements of $\mathbf{Q}^{-1}$. Take the special case where $\sigma^2 = 1$, and

$$\mathbf{Q}_{22}^* = \begin{pmatrix} 1 & r \\ r & 1 \end{pmatrix}, \qquad \mathbf{Q}^{22} = (\mathbf{Q}_{22}^*)^{-1} = (1 - r^2)^{-1}\begin{pmatrix} 1 & -r \\ -r & 1 \end{pmatrix}.$$

Here r is the sample correlation between x_2 and x_3. We have

$$\sigma_{b_2}^2 = 1/(1 - r^2) = \sigma_{b_3}^2, \qquad \sigma_t^2 = 2/(1 + r).$$

If $r = 0$, there is no collinearity, and

$$\sigma_{b_2}^2 = \sigma_{b_3}^2 = 1, \qquad \sigma_t^2 = 2.$$

But if $r = 0.9$, there is strong collinearity, and

$$\sigma_{b_2}^2 = \sigma_{b_3}^2 = 1/0.19 = 5.3, \qquad \sigma_t^2 = 2/1.9 = 1.05.$$

In this example, collinearity hinders precise inference about β_2 and β_3 separately, but facilitates precise inference about their sum $\theta = \beta_2 + \beta_3$. So if we happen to be interested in that particular θ, then the high positive collinearity is desirable. For further discussion, see Conlisk (1971).

23.5. Remarks

• In the CR model, all the consequences of multicollinearity are reflected in $V(\mathbf{b}) = \sigma^2\mathbf{Q}^{-1}$, or in its unbiased estimator $\hat{V}(\mathbf{b}) = \hat{\sigma}^2\mathbf{Q}^{-1}$. Researchers should not be concerned with whether or not "there really is collinearity." They may well be concerned with whether the variances of the coefficient estimates are too large—for whatever reason—to provide useful estimates of the regression coefficients.

• Multicollinearity is just one of the possible sources of high $\sigma^2_{b_j}$. For estimation of β_j, what is desirable per se is not low collinearity (small R_j^2) but rather low coefficient variance (small $\sigma^2_{b_j}$).

• A sensible researcher may well want to calculate the auxiliary R_j^2's, but it is unlikely that she will want to test hypotheses about their true magnitudes.

• To say that "standard errors are inflated by multicollinearity" is to suggest that they are artificially, or spuriously, large. But in fact they are appropriately large: the coefficient estimates actually would vary a lot from sample to sample. This may be regrettable but it is not spurious.

• To say that "the coefficient is really significant but multicollinearity makes it look insignificant" is to confuse statistical significance with economic significance: see Section 22.5.

Exercises

23.1 These results were found for LS regression of y = executive salaries on x_1 = sales and x_2 = profits, across a sample of 102 firms:

$$\hat{y} = \underset{(0.83)}{0.50x_1} + \underset{(0.83)}{0.40x_2}, \qquad \mathbf{e}'\mathbf{e} = 250, \qquad \mathbf{X}'\mathbf{X} = \begin{pmatrix} 10 & 8 \\ 8 & 10 \end{pmatrix}.$$

(All variables had been expressed as deviations about means for convenience.) Assume that the CNR model applies to the salary function $E(y) = \beta_1 x_1 + \beta_2 x_2$. Evidently, the high collinearity between sales and profits has prevented precise estimation of the parameters of the salary function. To eliminate this problem, it has been proposed that we proceed as follows. First, regress profits on sales, and obtain the residuals x_2^*. Second, regress y on x_1 and x_2^* to estimate the parameters of the salary function. Denote the results of the second step by $\hat{y}^* = c_1 x_1 + c_2 x_2^*$.

(a) Calculate c_1 and c_2, and calculate their standard errors.
(b) Evaluate the proposal as a device for eliminating collinearity.
(c) Evaluate the proposal as a device for obtaining more precise parameter estimates.

23.2 The CR model applies and the **X** matrix shows high collinearity. The sample size is doubled by getting two observations on y, rather than one, at each of the rows of the original **X**.

(a) What happens to the degree of collinearity?
(b) What happens to the variance of the LS coefficients?
(c) Comment on the claim that more data is no remedy for the multicollinearity problem if the data are simply "more of the same."

23.3 Suppose that $\boldsymbol{\mu} = E(\mathbf{y}) = \mathbf{X}\boldsymbol{\beta}$, where $\boldsymbol{\mu}$ and $\mathbf{X}$ are known and $\boldsymbol{\beta}$ is unknown.

(a) Under what condition on the rank of $\mathbf{X}$ is $\boldsymbol{\beta}$ uniquely determined?
(b) Comment on the relevance of this result to the multicollinearity problem.

24 Regression Strategies

24.1. Introduction

In empirical research, it is common practice to run several versions of a regression. We will explore some reasons for this practice and consider how the resulting estimates may be interpreted.

24.2. Shortening a Regression

Suppose that as a result of high collinearity (or for some other reason), the LS coefficient estimates are not precise enough to be useful. What should be done? The appropriate response will depend upon the objective of the research. If the objective were to "explain the variation in y," that is, to get a good fit, that is, to get a high R^2, then there would be no good reason to be concerned with the individual b_j's. And so there would be no good reason to be bothered by large standard errors and "nonsignificance" of the individual coefficients.

But suppose that the primary research objective is to learn about $\boldsymbol{\beta}_1$ in the model

$$E(\mathbf{y}) = \mathbf{X}_1\boldsymbol{\beta}_1 + \mathbf{X}_2\boldsymbol{\beta}_2.$$

For example, consider the household demand for butter, where y = expenditures on butter; the k_1 variables in $\mathbf{X}_1$ include the constant, income, butter price, and margarine price, whose coefficients are of interest; and the $k_2 \times 1$ variables in $\mathbf{X}_2$ include family size, occupation, and location, which are included as "control" variables.

We run the long regression $\hat{\mathbf{y}} = \mathbf{X}_1\mathbf{b}_1 + \mathbf{X}_2\mathbf{b}_2$, intending to use $\mathbf{b}_1$ as the estimator of $\boldsymbol{\beta}_1$. If the CR model holds, then

$$E(\mathbf{b}_1) = \boldsymbol{\beta}_1, \qquad V(\mathbf{b}_1) = \sigma^2\mathbf{Q}^{11} = \sigma^2(\mathbf{Q}^*_{11})^{-1} = \sigma^2(\mathbf{X}_1'\mathbf{M}_2\mathbf{X}_1)^{-1}.$$

The estimated variance matrix of $\mathbf{b}_1$ is $\hat{V}(\mathbf{b}_1) = \hat{\sigma}^2\mathbf{Q}^{11}$. Suppose that the diagonal elements of $\hat{V}(\mathbf{b}_1)$ are so large that $\mathbf{b}_1$ is not adequately informative about the parameter vector $\boldsymbol{\beta}_1$.

A natural response is to shorten the regression, that is, to run $\mathbf{y}$ on $\mathbf{X}_1$ alone, reporting $\mathbf{b}_1^*$ instead of $\mathbf{b}_1$ as the estimate of $\boldsymbol{\beta}_1$. To motivate that response, recall from Eqs. (17.7)–(17.8) that

$$E(\mathbf{b}_1^*) = \boldsymbol{\beta}_1 + \mathbf{F}\boldsymbol{\beta}_2,$$

$$V(\mathbf{b}_1^*) = \sigma^2(\mathbf{X}_1'\mathbf{X}_1)^{-1},$$

with $\mathbf{F} = \mathbf{A}_1\mathbf{X}_2$. As in Eq. (17.9), the variance comparison is clear-cut:

$$V(\mathbf{b}_1) = V(\mathbf{b}_1^*) + \mathbf{F}V(\mathbf{b}_2)\mathbf{F}',$$

where $\mathbf{F}V(\mathbf{b}_2)\mathbf{F}'$ is nonnegative definite. So $V(\mathbf{b}_1) \geq V(\mathbf{b}_1^*)$, regardless of the value of $\boldsymbol{\beta}_2$. The bias of $\mathbf{b}_1^*$ as an estimator of $\boldsymbol{\beta}_1$, namely $\mathbf{F}\boldsymbol{\beta}_2$, vanishes if $\boldsymbol{\beta}_2 = \mathbf{0}$, that is, if the omitted explanatory variables are irrelevant.

From this perspective, one can identify at least three distinct rationales for reporting $\mathbf{b}_1^*$ rather than $\mathbf{b}_1$, that is, for using the short, rather than the long, regression:

(1) We believe that $\boldsymbol{\beta}_2 = \mathbf{0}$. Excluding $\mathbf{X}_2$, as the short regression does, introduces no bias, and does reduce the variance of the estimator of $\boldsymbol{\beta}_1$. Indeed if $\boldsymbol{\beta}_2 = \mathbf{0}$, then $\mathbf{b}_1^*$ is the MVLUE of $\boldsymbol{\beta}_1$, because the CR model will apply to $E(\mathbf{y}) = \mathbf{X}_1\boldsymbol{\beta}_1$.

(2) We do not believe that $\boldsymbol{\beta}_2 = \mathbf{0}$, but we have lowered our aspiration level. Rather than insisting on estimating $\boldsymbol{\beta}_1$ and $\boldsymbol{\beta}_2$ separately, we will be content with an estimate of $\boldsymbol{\beta}_1^* = \boldsymbol{\beta}_1 + \mathbf{F}\boldsymbol{\beta}_2$. Indeed $\mathbf{b}_1^*$ is the MVLUE of that parameter combination.

(3) We do not believe that $\boldsymbol{\beta}_2 = \mathbf{0}$, nor will we be content with estimating $\boldsymbol{\beta}_1^*$, but we have lowered our aspiration level in a different way. Rather than insisting on an unbiased estimator of $\boldsymbol{\beta}_1$, we will be content with a biased estimator, provided that its bias is sufficiently offset by reduced variance.

24.3. Mean Squared Error

We focus on rationale (3) from the list above. The idea is that it is plausible to prefer a biased estimator to an unbiased one provided that the former's variance is sufficiently small.

To assess the available trade-off between bias and variance, we generalize the mean squared error criterion introduced in Section 11.3. If a random vector $\mathbf{t}$ has expectation vector $E(\mathbf{t})$ and variance matrix $V(\mathbf{t})$, then, as an estimator of the parameter vector $\boldsymbol{\theta}$, its *mean squared error matrix* is

$$S(\mathbf{t}; \boldsymbol{\theta}) \equiv E[(\mathbf{t} - \boldsymbol{\theta})(\mathbf{t} - \boldsymbol{\theta})'] = V(\mathbf{t}) + [E(\mathbf{t} - \boldsymbol{\theta})][E(\mathbf{t} - \boldsymbol{\theta})]'.$$

The minimum mean squared error (MSE) criterion for choosing an estimator of $\boldsymbol{\theta}$ says that we should prefer $\mathbf{t}_2$ to $\mathbf{t}_1$ if $S(\mathbf{t}_1; \boldsymbol{\theta}) \geq S(\mathbf{t}_2; \boldsymbol{\theta})$ in the matrix sense, that is, if $S(\mathbf{t}_1; \boldsymbol{\theta}) - S(\mathbf{t}_2; \boldsymbol{\theta})$ is nonnegative definite.

For the short- and long-regression estimators of $\boldsymbol{\beta}_1$, we have

$$\mathbf{S} \equiv S(\mathbf{b}_1; \boldsymbol{\beta}_1) = V(\mathbf{b}_1),$$

$$\mathbf{S}^* \equiv S(\mathbf{b}_1^*; \boldsymbol{\beta}_1) = V(\mathbf{b}_1^*) + \mathbf{F}\boldsymbol{\beta}_2\boldsymbol{\beta}_2'\mathbf{F}'.$$

Subtracting gives

$$\mathbf{D} = \mathbf{S} - \mathbf{S}^* = \mathbf{F}V(\mathbf{b}_2)\mathbf{F}' - \mathbf{F}\boldsymbol{\beta}_2\boldsymbol{\beta}_2'\mathbf{F}' = \mathbf{F}[V(\mathbf{b}_2) - \boldsymbol{\beta}_2\boldsymbol{\beta}_2']\mathbf{F}'.$$

By the MSE criterion, $\mathbf{b}_1^*$ is preferable to $\mathbf{b}_1$ if the matrix $\mathbf{D}$ is nonnegative definite, a sufficient condition for which is that the matrix $[V(\mathbf{b}_2) - \boldsymbol{\beta}_2\boldsymbol{\beta}_2']$ is nonnegative definite. Heuristically, this condition says that the magnitude of $\boldsymbol{\beta}_2$ is small relative to the variance matrix of its estimator $\mathbf{b}_2$.

Take the special case where $k_2 = 1$. Now $\mathbf{b}_2$ and $\boldsymbol{\beta}_2$ are scalars, and $V(\mathbf{b}_2) - \boldsymbol{\beta}_2\boldsymbol{\beta}_2' = \sigma_{b_2}^2 - \beta_2^2$, so $\mathbf{D}$ is nonnegative definite if $\tau_2^2 \equiv (\beta_2/\sigma_{b_2})^2 \leq 1$. For this scalar case we have a clean conclusion: on the MSE criterion, prefer $\mathbf{b}_1^*$ to $\mathbf{b}_1$ iff $\tau_2^2 = (\beta_2/\sigma_{b_2})^2 \leq 1$. This gives a particular precise meaning to the notion that the bias is small enough to be offset by reduced variance. Specialize further to the case where $k_1 = k_2 = 1$. Here $\mathbf{b}_1$ is also a scalar, and

$$S = \sigma_{b_1}^2 = \sigma^2/q_{11}^*$$

$$S^* = \sigma_{b_1*}^2 + F^2\beta_2^2 = \sigma^2/q_{11} + (q_{12}/q_{11})^2\beta_2^2.$$

As functions of β_2, S is constant while S^* is linear in β_2^2. At $\beta_2 = 0$, $S^* \leq$

S because $q^*_{11} \le q_{11}$. As β_2 departs from zero in either direction, S^* increases, equaling S at $\beta_2 = \pm\sigma_{b_2}$ (i.e., at $\tau_2 = \pm 1$), and thereafter exceeding S. The short-regression estimator is preferable provided that β_2 is sufficiently close to zero.

Example. Take $\sigma^2 = 1$, and $q_{11} = q_{22} = 1$, so $r = q_{12}$ lies between -1 and 1. Then $q^*_{11} = (1 - r^2)$, whence

$$S = 1/(1 - r^2), \qquad S^* = 1 + r^2\beta_2^2.$$

Figure 24.1 takes $r = 0.5$ and plots S and S^* against β_2^2. The curve marked S^{**} will be explained later.

These special cases illustrate a tension that is almost inevitable in those empirical research situations in which the primary objective is to learn about a subset of the regression coefficients. The tension is between shortening and lengthening the regression, between "under-

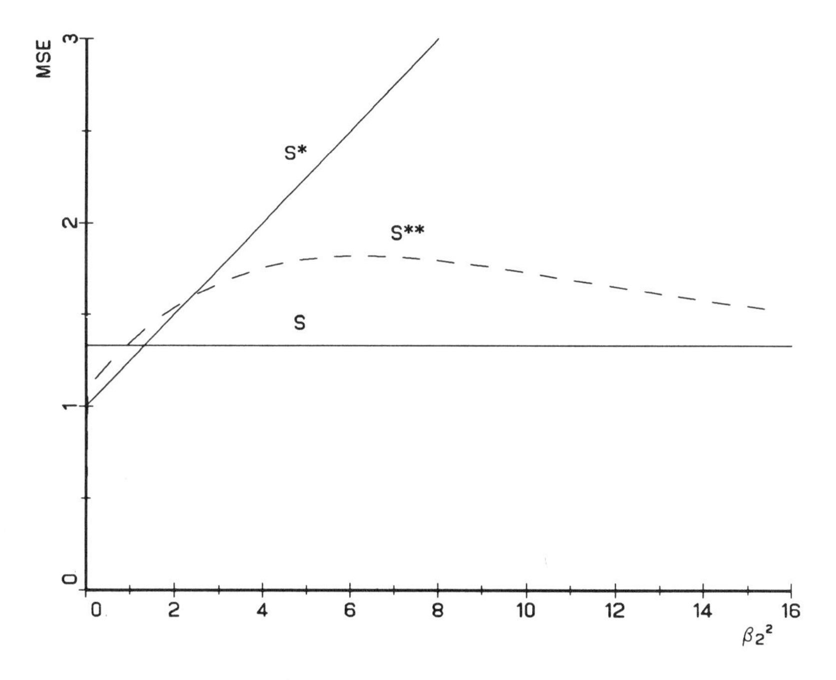

Figure 24.1 Pretest estimation mean squared errors: S, S^*, S^{**} = MSE's for b_1, b_1^*, b_1^{**}.

specifying" and "overspecifying" a regression function, between bias and variance. There is an incentive to exclude control variables to reduce variance, but doing so may introduce bias. There is an incentive to include control variables to avoid bias, but doing so may increase variance. The MSE criterion offers a particular evaluation of the available trade-off.

24.4. Pretest Estimation

Continue with the $k_2 = 1$ case. If τ_2^2 were known, the choice between $\mathbf{b}_1^*$ and $\mathbf{b}_1$ would, on the MSE criterion, be clear-cut. But with τ_2^2 unknown in practice, how shall we implement the MSE criterion? It is natural to use the sample to learn about the value of τ_2^2, and the natural estimator of

$$\tau_2^2 = (\beta_2/\sigma_{b_2})^2$$

is

$$t_2^2 = (b_2/\hat{\sigma}_{b_2})^2.$$

But this is precisely v°, the F-statistic (squared t-statistic) that we would use to test the null hypothesis $\beta_2 = 0$ against the alternative $\beta_2 \neq 0$.

If we were testing the null $\beta_2 = 0$, large values of v° would lead to its rejection, small values to its acceptance. In the present context, we do not wish to test $\beta_2 = 0$ (which is equivalent to $\tau_2^2 = 0$), but we will use the same statistic v°, to serve as an indicator of whether $\tau_2^2 \leq 1$. Evidently, small values of the statistic will favor small values of τ_2^2.

We have arrived at a particular *regression strategy* for estimation of β_1, namely *pretest estimation*. Generalizing to the case where $k_2 > 1$, we may spell out the procedure as follows.

(i) Choose some cutoff value d.
(ii) Run the long regression of $\mathbf{y}$ on $(\mathbf{X}_1, \mathbf{X}_2)$, obtaining $\mathbf{b}_1$, $\mathbf{b}_2$, and $\hat{\sigma}^2$.
(iii) Calculate $v^\circ = \mathbf{b}_2'\mathbf{Q}_{22}^*\mathbf{b}_2/(k_2\hat{\sigma}^2)$.
(iv) If $v^\circ > d$, then report $\mathbf{b}_1$ as the estimate of $\boldsymbol{\beta}_1$.
(v) If $v^\circ \leq d$, then run the short regression of $\mathbf{y}$ on $\mathbf{X}_1$, and report its $\mathbf{b}_1^*$ as the estimate of $\boldsymbol{\beta}_1$.

The cutoff value d may be the critical value associated with some significance level in the F-distribution, although we are not testing the null $\boldsymbol{\beta}_2 = \mathbf{0}$.

In any sample, either $\mathbf{b}_1$ or $\mathbf{b}_1^*$ is *selected* as the estimate. Formally, the pretest estimator, say $\mathbf{b}_1^{**}$, may be written as a weighted average of the short- and long-regression estimators. Let $z = 1$ if $v^\circ \leq d$, and $z = 0$ if $v^\circ > d$. Then the pretest estimator is

$$\mathbf{b}_1^{**} = (1 - z)\mathbf{b}_1 + z\mathbf{b}_1^*.$$

As a guide to thinking about the distribution of $\mathbf{b}_1^{**}$, consider two examples, drawn from outside the regression context, that illustrate how selection affects the distribution of sample statistics.

Example. Suppose that X and Y are independent Bernoulli variables, each having parameter p. Let $Z = \max(X, Y)$. Then

$$\begin{aligned}\Pr(Z = 1) &= \Pr(X = 1, Y = 1) + \Pr(X = 1, Y = 0) + \Pr(X = 0, Y = 1) \\ &= \qquad p^2 \qquad + \qquad p(1 - p) \qquad + \qquad (1 - p)p \\ &= p + p(1 - p).\end{aligned}$$

So Z is a Bernoulli variable with parameter $p^* = p + p(1 - p)$. Then $E(Z) = p^* > p = E(X) = E(Y)$. For example, if $p = 0.05$, then $E(Z) = 0.0975$.

Example. Suppose that X and Y are independent standard normal variables. Let $Z = \max(X, Y)$. Let $f(\cdot)$ and $F(\cdot)$ denote the standard normal pdf and cdf. Then the cdf of Z is

$$G(z) = \Pr(Z \leq z) = \Pr(X \leq z \cap Y \leq z) = \Pr(X \leq z)\Pr(Y \leq z) = F^2(z).$$

Clearly, the probability that Z exceeds some value c, namely $1 - G(c) = 1 - F^2(c)$, is greater than the probability $1 - F(c)$ that X (or Y) exceeds that value. The pdf of Z, namely

$$g(z) = \partial G(z)/\partial z = 2F(z)f(z),$$

is plotted in Figure 24.2. Observe how the selection shifts the distribution (and hence the expectation) to the right.

Returning to the regression context, we recognize that, because of the selection, the distribution theory for the pretest estimator is more complicated than that for either $\mathbf{b}_1$ or $\mathbf{b}_1^*$. For an introduction to the

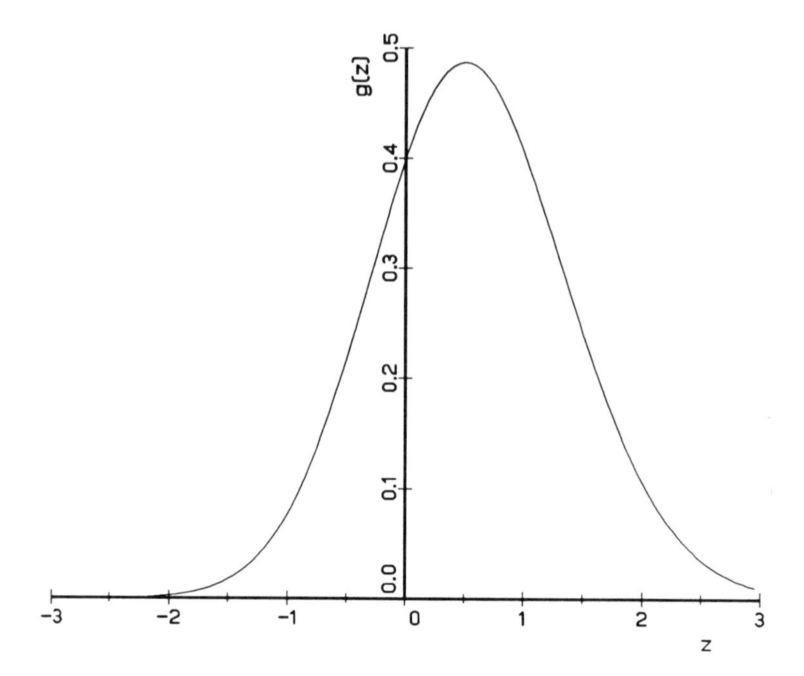

Figure 24.2 Pdf of maximum of two independent $\mathcal{N}(0, 1)$ variables.

theory, see Wallace and Ashar (1972) and Judge et al. (1988, pp. 832–835).

Example. Resume the example of Section 24.3, with $\sigma^2 = 1$, $q_{11} = q_{22} = 1$, $q_{12} = r = 0.5$. Suppose that the CNR model applies, and for convenience suppose that σ^2 is known, so that the chi-square statistic $w^\circ = (b_2/\sigma_{b_2})^2$ may be used instead of the F-statistic v°. Then S^{**}, the MSE for this ideal version of the pretest estimator b_1^{**}, can be calculated fairly readily from the properties of the bivariate normal distribution: see Exercise 24.1. In Figure 24.1, the curve S^{**} is drawn for $d = 3.84$, which corresponds to a nominal 5% significance level test of $\beta_2 = 0$. As this example indicates, there is no range of β_2^2 over which the pretest estimator dominates the other two estimators. So we get no clear-cut guidance about the attractiveness of pretest estimation.

Explicitly or implicitly, the pretest strategy is followed by many empirical researchers. In journal articles, you will often find that the author has "experimented" with some alternative specifications before arriving at a final, preferred, regression. The experimentation is often of the type considered here, that is, shortening a regression when some coefficients in the long regression are "not significantly different from zero." Other restrictions on a regression are sometimes employed, for example, constant returns to scale. The motivation is the same: restricted estimates may be biased, but have smaller variance. Indeed, as shown in Section 22.2, any set of linear restrictions can be translated into a zero subvector form, so the analysis applies directly.

The usual computer output (conventional standard errors calculated for the selected regression) will not do justice to the pretest strategy. In any sample, the strategy will select either $\mathbf{b}_1$ or $\mathbf{b}_1^*$, and the conventional standard errors (from the long or short regression respectively) simply do not take account of the stochastic nature of that selection. Readers should at least be aware of the exploratory process that led the author to the final, selected, regression.

As a writer, it is a good idea to put yourself in the position of a prospective reader: provide the information that you would want to have if you were the reader. For some suggestions, see Leamer (1983).

24.5. Regression Fishing

There is a popular style of regression analysis that, though reminiscent of pretest estimation, is quite distinct in character. Here again the data consist of $\mathbf{y}$, $\mathbf{X}_1$, $\mathbf{X}_2$, but now the researcher has no particular interest in either $\boldsymbol{\beta}_1$ or $\boldsymbol{\beta}_2$. Instead he wants to "explain the variation in the dependent variable" using only a few explanatory variables. In starkest form, the procedure is:

(i) Run $\mathbf{y}$ on $\mathbf{X}_1$, obtain $\mathbf{b}_1^*$, $\hat{V}(\mathbf{b}_1^*)$, and $R^2(1)$.
(ii) Run $\mathbf{y}$ on $\mathbf{X}_2$, obtain $\mathbf{b}_2^*$, $\hat{V}(\mathbf{b}_2^*)$, and $R^2(2)$.
(iii) If $R^2(1) > R^2(2)$, then report the first short regression.
If $R^2(1) \leq R^2(2)$, then report the second short regression.

(For convenience, we have allowed the summer vector to appear in both $\mathbf{X}_1$ and $\mathbf{X}_2$.) Alternatively described: run the long regression, test $\boldsymbol{\beta}_1 = \mathbf{0}$, also test $\boldsymbol{\beta}_2 = \mathbf{0}$, then report the "more significant" of the two short regressions.

Naturally, the coefficients in the reported regression will tend to be statistically significant when assessed by conventional standards. But those standards are clearly inappropriate. To report and interpret a selected model as if it were an unselected model is incorrect, as the examples in Section 24.4 illustrate. More convincing, perhaps, is an example drawn from Lovell (1983). Suppose that you have a set of k null hypotheses, each of which is tested at the significance level α. Suppose that all the null hypotheses are true. What is the probability of getting at least one rejection? That is, what are the chances of getting at least one nominally significant result? If the test statistics are independent, then

$$\Pr(\text{at least one rejection}) = 1 - \Pr(\text{all accepted}) = 1 - (1 - \alpha)^k.$$

For example, with $k = 10$ and $\alpha = 0.10$, the probability is $1 - (0.90)^{10} = 0.65$, which means that the actual significance level is 65% rather than the nominal 10%. Unless you want to test at the 65% level, you should not consider such an outcome to be statistically significant. It is hardly surprising, and perhaps not even interesting, to obtain a nominally significant outcome by fishing.

Exercises

24.1 Some of the flavor of the distribution theory for the pretest estimator can be captured in a simpler context. Suppose that the random variable $y \sim \mathcal{N}(\mu, 1)$, and that we suspect that μ is near zero. A single observation will be drawn. These three estimators of μ:

$$m = y, \qquad m^* = 0, \qquad m^{**} = \begin{cases} m^* & \text{if } y \leq 1 \\ m & \text{if } y > 1 \end{cases},$$

play the roles of b_1, b_1^*, and b_1^{**} respectively.

(a) Show that the expectation of m^{**} is $E(m^{**}) = \pi_0\mu + \pi_1$, where

$$\pi_0 = F(-\theta_1) + F(-\theta_2), \qquad \pi_1 = f(\theta_2) - f(\theta_1),$$

$$\theta_1 = 1 + \mu, \qquad \theta_2 = 1 - \mu,$$

and $f(\cdot)$ and $F(\cdot)$ denote the $\mathcal{N}(0, 1)$ pdf and cdf. Hints: (i) If $y \sim \mathcal{N}(\mu, 1)$, then $t = y - \mu \sim \mathcal{N}(0, 1)$; (ii) for the $\mathcal{N}(0, 1)$

pdf, $f(t) = (2\pi)^{-1/2} \exp(-t^2/2)$, the first derivative is $f'(t) = \partial f(t)/\partial t = -tf(t)$.

(b) Show that the variance of m^{**} is

$$V(m^{**}) = \pi_0(1 - \pi_0)\mu^2 + (\pi_2 - \pi_1^2) + 2\mu\pi_1(1 - \pi_0),$$

where

$$\pi_2 = \pi_0 - \mu\pi_1 + f(\theta_1) + f(\theta_2).$$

Hint: For the $\mathcal{N}(0, 1)$ pdf, the second derivative is $f''(t) = \partial f'(t)/\partial t = -[tf'(t) + f(t)] = (t^2 - 1)f(t)$.

(c) Tabulate the MSE's of m, m^*, and m^{**} as functions of μ.

(d) Comment on the results.

25 *Regression with* **X** *Random*

25.1. Introduction

We now drop the assumption that the explanatory variables are nonstochastic, and provide models that may be relevant in random sampling from a multivariate population. For least squares linear regression we report exact results when the population conditional expectation function is linear, and then asymptotic results when the linearity assumption is absent. The analysis is a direct generalization of the analysis in Sections 13.1 and 13.2, which referred to random sampling from a bivariate population.

25.2. Neoclassical Regression Model

Once again the data consist of an $n \times 1$ vector $\mathbf{y}$ and an $n \times k$ matrix $\mathbf{X} = (\mathbf{x}_1, \ldots, \mathbf{x}_k)$. The *neoclassical regression*, or NeoCR, *model* consists of these assumptions:

(25.1) $E(\mathbf{y}|\mathbf{X}) = \mathbf{X}\boldsymbol{\beta}$,

(25.2) $V(\mathbf{y}|\mathbf{X}) = \sigma^2\mathbf{I}$,

(25.3) $\mathbf{X}$ stochastic,

(25.4) $\text{rank}(\mathbf{X}) = k$.

Here "$|\mathbf{X}$" means conditional on the matrix $\mathbf{X}$. The most direct interpretation of these assumptions is that a CR model holds conditional on every value of $\mathbf{X}$.

To provide a framework for the NeoCR model, return to the population specification of Section 16.1. Suppose that there is a multivariate probability distribution for the random vector $(y, x_2, \ldots, x_k)'$, with pdf or pmf $f(y, x_2, \ldots, x_k)$. Expectations, variances, and covariances are defined in the usual manner:

$$E(y) = \mu_y, \qquad V(y) = \sigma_y^2, \qquad C(x_h, x_j) = \sigma_{hj}, \qquad C(x_j, y) = \sigma_{jy},$$

and so forth. Suppose further that the conditional expectation function of y given the x's is linear:

$$E(y|x_2, \ldots, x_k) = \beta_1 + \beta_2 x_2 + \cdots + \beta_k x_k,$$

and that the conditional variance function of y given the x's is constant:

$$V(y|x_2, \ldots, x_k) = \sigma^2.$$

We write these compactly as

$$E(y|\mathbf{x}) = \mathbf{x}'\boldsymbol{\beta}, \qquad V(y|\mathbf{x}) = \sigma^2,$$

where $\mathbf{x} = (x_1, \ldots, x_k)'$ with $x_1 = 1$, and $\boldsymbol{\beta} = (\beta_1, \ldots, \beta_k)'$.

Now we sample randomly from the multivariate population. That is, n independent drawings, $(y_1, \mathbf{x}_1'), \ldots, (y_n, \mathbf{x}_n')$, are made, giving the observed sample data $(\mathbf{y}, \mathbf{X})$. The rows of the observed data matrix, namely the $(y_i, \mathbf{x}_i')$, are independent and identically distributed. (Caution: $\mathbf{x}_i'$ denotes the ith row of $\mathbf{X}$, not the transpose of the ith column of $\mathbf{X}$.) In contrast to the CR model, in the NeoCR model the $\mathbf{X}$ matrix is random.

Because the y_i's are identically distributed, we have $E(y_i) = \mu_y$ and $V(y_i) = \sigma_y^2$ for all i. So for the $n \times 1$ random vector $\mathbf{y}$, we have $E(\mathbf{y}) = \mathbf{s}\mu_y$, where $\mathbf{s}$ is the $n \times 1$ summer vector. This too contrasts with the CR model, where $E(\mathbf{y}) = \mathbf{X}\boldsymbol{\beta}$ and the expectations of the y_i's differ from one another. The conditional pdf $g(y|\mathbf{x})$ is the same at all observations, so

$$E(y_i|\mathbf{x}_i) = \mathbf{x}_i'\boldsymbol{\beta}, \qquad V(y_i|\mathbf{x}_i) = \sigma^2 \qquad (i = 1, \ldots, n).$$

Those are consequences of the fact that the $(y_i, \mathbf{x}_i')$ are identically distributed. Now consider the consequences of the fact that they are independent. For specificity, consider the pdf of the first observation on y conditional on the first two observations on the $k \times 1$ vector $\mathbf{x}$:

$$g^*(y_1|\mathbf{x}_1, \mathbf{x}_2) = f^*(y_1, \mathbf{x}_1, \mathbf{x}_2)/h^*(\mathbf{x}_1, \mathbf{x}_2).$$

Because $(y_1, \mathbf{x}_1')$ is independent of $\mathbf{x}_2$, we have

$$f^*(y_1, \mathbf{x}_1, \mathbf{x}_2) = f(y_1, \mathbf{x}_1)h(\mathbf{x}_2),$$

and because $\mathbf{x}_1$ and $\mathbf{x}_2$ are independent and identically distributed, we have

$$h^*(\mathbf{x}_1, \mathbf{x}_2) = h(\mathbf{x}_1)h(\mathbf{x}_2). \tag{25.5}$$

So

$$g^*(y_1|\mathbf{x}_1, \mathbf{x}_2) = f(y_1, \mathbf{x}_1)/h(\mathbf{x}_1) = g(y_1|\mathbf{x}_1), \tag{25.6}$$

which says that the distribution of y_1 conditional on $(\mathbf{x}_1, \mathbf{x}_2)$ is identical to the distribution of y_1 conditional on only $\mathbf{x}_1$. By the same logic,

$$g^*(y_2|\mathbf{x}_1, \mathbf{x}_2) = f(y_2, \mathbf{x}_2)/h(\mathbf{x}_2) = g(y_2|\mathbf{x}_2). \tag{25.7}$$

Proceed to the joint pdf of y_1, y_2 conditional on $\mathbf{x}_1$ and $\mathbf{x}_2$:

$$g^{**}(y_1, y_2|\mathbf{x}_1, \mathbf{x}_2) = f^{**}(y_1, \mathbf{x}_1, y_2, \mathbf{x}_2)/h^*(\mathbf{x}_1, \mathbf{x}_2).$$

Because $(y_1, \mathbf{x}_1')$ and $(y_2, \mathbf{x}_2')$ are independent and identically distributed, we have

$$f^{**}(y_1, \mathbf{x}_1, y_2, \mathbf{x}_2) = f(y_1, \mathbf{x}_1)f(y_2, \mathbf{x}_2),$$

and we also have Eq. (25.5). Consequently,

$$\begin{aligned} g^{**}(y_1, y_2|\mathbf{x}_1, \mathbf{x}_2) &= g(y_1|\mathbf{x}_1)g(y_2|\mathbf{x}_2) \\ &= g^*(y_1|\mathbf{x}_1, \mathbf{x}_2)g^*(y_2|\mathbf{x}_1, \mathbf{x}_2), \end{aligned} \tag{25.8}$$

using Eqs. (25.6)–(25.7). This says that, conditional on $\mathbf{x}_1$ and $\mathbf{x}_2$, the joint pdf of y_1 and y_2 equals the product of their marginal pdf's. In other words, the variables y_1 and y_2 are independent conditional on $\mathbf{x}_1$ and $\mathbf{x}_2$, as well as unconditionally.

When two distributions are the same, their expectations and variances are the same, so from Eq. (25.6) we conclude that

$$E(y_1|\mathbf{x}_1, \mathbf{x}_2) = E(y_1|\mathbf{x}_1) = \mathbf{x}_1'\boldsymbol{\beta},$$

$$V(y_1|\mathbf{x}_1, \mathbf{x}_2) = V(y_1|\mathbf{x}_1) = \sigma^2.$$

When two variables are independent, they are uncorrelated, so from Eq. (25.8) we conclude that

$$C(y_1, y_2|\mathbf{x}_1, \mathbf{x}_2) = 0.$$

The same conclusions follow when we condition on $\mathbf{x}_3, \ldots, \mathbf{x}_n$ along with $\mathbf{x}_1, \mathbf{x}_2$. And conditioning on all n rows $\mathbf{x}_1', \ldots, \mathbf{x}_n'$ is equivalent to conditioning on the matrix $\mathbf{X}$. So

$$E(y_1|\mathbf{X}) \equiv E(y_1|\mathbf{x}_1, \mathbf{x}_2, \ldots, \mathbf{x}_n) = E(y_1|\mathbf{x}_1) = \mathbf{x}_1'\boldsymbol{\beta},$$

$$V(y_1|\mathbf{X}) \equiv V(y_1|\mathbf{x}_1, \mathbf{x}_2, \ldots, \mathbf{x}_n) = V(y_1|\mathbf{x}_1) = \sigma^2,$$

$$C(y_1, y_2|\mathbf{X}) = 0.$$

There is nothing special about the first two observations in this regard, not even their adjacency. So for $i = 1, \ldots, n$:

$$E(y_i|\mathbf{X}) \equiv E(y_i|\mathbf{x}_1, \mathbf{x}_2, \ldots, \mathbf{x}_n) = E(y_i|\mathbf{x}_i) = \mathbf{x}_i'\boldsymbol{\beta},$$

$$V(y_i|\mathbf{X}) \equiv V(y_i|\mathbf{x}_1, \mathbf{x}_2, \ldots, \mathbf{x}_n) = V(y_i|\mathbf{x}_i) = \sigma^2,$$

$$C(y_i, y_h|\mathbf{X}) = 0, \qquad i \neq h.$$

Assembling these results for $y_1, \ldots, y_n$ into results for the vector $\mathbf{y}$, we have

$$E(\mathbf{y}|\mathbf{X}) = \mathbf{X}\boldsymbol{\beta}, \qquad V(\mathbf{y}|\mathbf{X}) = \sigma^2\mathbf{I},$$

which are precisely the assumptions in Eqs. (25.1) and (25.2) of the NeoCR model. As for the rank condition, Eq. (25.4), there is a technical qualification. In random sampling there is always the possibility of drawing an $\mathbf{X}$ matrix that does not have full column rank: for example, all n of the $\mathbf{x}_i'$'s might turn out to be identical. Even if the population variance matrix of the x's is nonsingular, obtaining a short-ranked $\mathbf{X}$ has positive probability when the x's are all discrete. To dispose of that complication, adopt the convention that any sample with rank($\mathbf{X}$) $< k$ is discarded. Then Eq. (25.4) applies.

With that understanding, random sampling from the multivariate population specified here supports the NeoCR model.

It is not the only scheme that would do so. Inspection of the argument above will show that to arrive at Eqs. (25.1)–(25.2), there is no need for the successive observations on the explanatory variables to be independently or even identically distributed: see Section 13.5 for discussion in the bivariate case. What *is* ruled out in the NeoCR model is the presence of the lagged dependent variable among the explanatory variables. For in that case, y_1 will be an element of $\mathbf{x}_2$, so $E(y_1|\mathbf{x}_1, \mathbf{x}_2) = y_1 \neq E(y_1|\mathbf{x}_1)$, whence $E(\mathbf{y}|\mathbf{X}) \neq \mathbf{X\beta}$: see Section 26.3. It is best to think of $E(y_i|\mathbf{x}_i) = \mathbf{x}_i'\boldsymbol{\beta}$ as a necessary, but not sufficient, condition for the NeoCR model to hold.

25.3. Properties of Least Squares Estimation

It is easy to assess the properties of the LS statistics $\mathbf{b} = \mathbf{Ay}$, $\mathbf{e} = \mathbf{My}$, $\hat{\sigma}^2 = \mathbf{e}'\mathbf{e}/(n-k)$, and $\hat{V}(\mathbf{b}) = \hat{\sigma}^2\mathbf{Q}^{-1}$, in the NeoCR model. The matrices $\mathbf{Q}$, $\mathbf{A}$, and $\mathbf{M}$, being functions of $\mathbf{X}$, are now random, but conditional on $\mathbf{X}$, they are constant. We calculate:

$$E(\mathbf{b}|\mathbf{X}) = E(\mathbf{Ay}|\mathbf{X}) = \mathbf{A}E(\mathbf{y}|\mathbf{X}) = \mathbf{AX\beta} = \boldsymbol{\beta},$$

$$V(\mathbf{b}|\mathbf{X}) = V(\mathbf{Ay}|\mathbf{X}) = \mathbf{A}V(\mathbf{y}|\mathbf{X})\mathbf{A}' = \sigma^2\mathbf{AA}' = \sigma^2\mathbf{Q}^{-1},$$

$$E(\mathbf{e}|\mathbf{X}) = E(\mathbf{My}|\mathbf{X}) = \mathbf{M}E(\mathbf{y}|\mathbf{X}) = \mathbf{MX\beta} = \mathbf{0},$$

$$V(\mathbf{e}|\mathbf{X}) = V(\mathbf{My}|\mathbf{X}) = \mathbf{M}V(\mathbf{y}|\mathbf{X})\mathbf{M}' = \sigma^2\mathbf{MM}' = \sigma^2\mathbf{M}.$$

From these it follows that

$$E(\mathbf{e}'\mathbf{e}|\mathbf{X}) = \sigma^2 \operatorname{tr}(\mathbf{M}) = \sigma^2(n-k),$$

$$E(\hat{\sigma}^2|\mathbf{X}) = \sigma^2,$$

$$E[\hat{V}(\mathbf{b})|\mathbf{X}] = E(\hat{\sigma}^2\mathbf{Q}^{-1}|\mathbf{X}) = E(\hat{\sigma}^2|\mathbf{X})\mathbf{Q}^{-1} = \sigma^2\mathbf{Q}^{-1}.$$

So, conditional on any value of the matrix $\mathbf{X}$, the LS statistics $\mathbf{b}$, $\hat{\sigma}^2$, and $\hat{V}(\mathbf{b})$ remain unbiased. This is a direct consequence of the fact that the NeoCR model effectively assumes that a CR model holds conditional on every value of $\mathbf{X}$.

Proceeding to unconditional moments, use the Law of Iterated Expectations (T8, Section 5.2) to calculate:

$$E(\mathbf{b}) = E_{\mathbf{X}}[E(\mathbf{b}|\mathbf{X})] = E_{\mathbf{X}}(\boldsymbol{\beta}) = \boldsymbol{\beta},$$

$$V(\mathbf{b}) = E_{\mathbf{X}}[V(\mathbf{b}|\mathbf{X})] + V_{\mathbf{X}}[E(\mathbf{b}|\mathbf{X})] = E_{\mathbf{X}}(\sigma^2\mathbf{Q}^{-1}) + \mathbf{O}$$
$$= \sigma^2E(\mathbf{Q}^{-1}),$$

$$E(\hat{\sigma}^2) = E_{\mathbf{X}}[E(\hat{\sigma}^2|\mathbf{X})] = E_{\mathbf{X}}(\sigma^2) = \sigma^2,$$

$$E[\hat{V}(\mathbf{b})] = E_{\mathbf{X}}\{E[\hat{V}(\mathbf{b})|\mathbf{X}]\} = E_{\mathbf{X}}(\sigma^2\mathbf{Q}^{-1}) = \sigma^2E(\mathbf{Q}^{-1}).$$

We see that the LS statistics $\mathbf{b}$, $\hat{\sigma}^2$, and $\hat{V}(\mathbf{b})$ are unbiased unconditionally as well. So the LS coefficients and their accompanying standard errors are still appropriate. As for optimality, a version of the Gauss-Markov Theorem applies in the NeoCR model: in the class of estimators that, conditional on every $\mathbf{X}$, are linear and unbiased, the LS estimator has minimum variance.

We see that LS estimation retains its attractiveness in the NeoCR model. Some results do differ. For example, in an analysis of the short regression, the matrices $\mathbf{F} = \mathbf{A}_1\mathbf{X}_2$ and $\mathbf{Q}_{22}^* = \mathbf{X}_2^{*\prime}\mathbf{X}_2^*$ are now random, so that

$$E(\mathbf{b}_1^*) = \boldsymbol{\beta}_1 + E(\mathbf{F})\boldsymbol{\beta}_2, \qquad E(\mathbf{e}^{*\prime}\mathbf{e}^*) = \sigma^2(n - k_1) + \boldsymbol{\beta}_2' E(\mathbf{Q}_{22}^*)\boldsymbol{\beta}_2.$$

Nevertheless, the main conclusion of the analysis is that the key properties of LS estimators carry over when $\mathbf{X}$ is allowed to be random. Nothing in the randomness of the explanatory variables per se creates an objection to LS estimation.

25.4. Neoclassical Normal Regression Model

If we strengthen the NeoCR model by assuming that, conditional on $\mathbf{X}$, the random vector $\mathbf{y}$ is multivariate normal, we obtain the *neoclassical normal regression*, or NeoCNR, *model*:

$$\mathbf{y}|\mathbf{X} \sim \mathcal{N}(\mathbf{X}\boldsymbol{\beta}, \sigma^2\mathbf{I}), \qquad \mathbf{X} \text{ stochastic}, \qquad \text{rank}(\mathbf{X}) = k.$$

The framework for this is random sampling from a multivariate population in which the conditional distribution of y given the x's is $y|\mathbf{x} \sim \mathcal{N}(\mathbf{x}'\boldsymbol{\beta}, \sigma^2)$.

All the distribution results in the CNR model now hold, conditional on $\mathbf{X}$. For example, let b_j be an element of $\mathbf{b}$. Then

$$b_j|\mathbf{X} \sim \mathcal{N}(\beta_j, \sigma^2 q^{jj}).$$

Observe that the conditional distribution of b_j does depend on $\mathbf{X}$ via q^{jj}: the conditional distributions are all normal with the same expectation but with different variances.

The marginal distribution of b_j, being a mixture of those different conditional normals, will not be normal. Nevertheless, the confidence interval and region constructions, and the hypothesis test procedures, developed in Chapters 19–22, remain valid. To see why, let z_j =

$(b_j - \beta_j)/(\sigma\sqrt{q^{jj}})$. Then $z_j|\mathbf{X} \sim \mathcal{N}(0, 1)$ for all $\mathbf{X}$, so the marginal distribution of z_j is also $\mathcal{N}(0, 1)$. Because it is the z_j variable that is used to develop the confidence interval and hypothesis tests, those procedures remain valid. For example, let $z_j^{\circ} = (b_j - \beta_j^{\circ})/(\sigma\sqrt{q^{jj}})$. Then, if the null hypothesis $\beta_j = \beta_j^{\circ}$ is true,

$$\Pr[(z_j^{\circ} > c)|\mathbf{X}] = \Pr[(z_j > c)|\mathbf{X}] = 1 - F(c),$$

where $F(\cdot)$ is the standard normal cdf. This probability does not vary with $\mathbf{X}$, so if the null is true, then $\Pr(z_j^{\circ} > c) = 1 - F(c)$. The same logic applies to the statistic $u_j = (b_j - \beta_j)/(\hat{\sigma}\sqrt{q^{jj}})$. For, $u_j|\mathbf{X} \sim t(n - k)$ for all $\mathbf{X}$, which implies that $u_j \sim t(n - k)$ unconditionally. And the same logic applies to the chi-square and F statistics. Confidence levels and significance levels are exactly as they were in the CNR model.

In summary, we have not been misled by concentrating attention heretofore on the stratified-on-x sampling scheme. Rather, we have merely avoided writing "$|\mathbf{X}$" throughout.

25.5. Asymptotic Properties of Least Squares Estimation

In Chapter 13, for random sampling from a bivariate population, we reviewed asymptotic results for the sample linear projection slope that did not rely on normality, or on linearity of the CEF, or on homoskedasticity. Those results generalize to cover LS estimation in random sampling from a multivariate population.

It is convenient to revise our notation, isolating the constant from the other explanatory variables. In the population, the $k \times 1$ random vector $\mathbf{z} = (\mathbf{x}', y)'$ has expectation vector and variance matrix:

$$E(\mathbf{z}) = \boldsymbol{\mu} = \begin{pmatrix} \boldsymbol{\mu}_x \\ \mu_y \end{pmatrix} = \begin{pmatrix} E(\mathbf{x}) \\ E(y) \end{pmatrix},$$

$$V(\mathbf{z}) = \boldsymbol{\Sigma} = \begin{pmatrix} \boldsymbol{\Sigma}_{xx} & \boldsymbol{\sigma}_{xy} \\ \boldsymbol{\sigma}'_{xy} & \sigma_{yy} \end{pmatrix} = \begin{pmatrix} V(\mathbf{x}) & C(\mathbf{x}, y) \\ C(y, \mathbf{x}) & V(y) \end{pmatrix}.$$

Consider the best linear predictor of y given $\mathbf{x}$ in the population, $E^*(y|\mathbf{x}) = \alpha + \mathbf{x}'\boldsymbol{\beta}$. The equations determining its slope vector (see Section 14.1) can be assembled into $C(\mathbf{x}, u) = \mathbf{0}$, where $u = y - (\alpha + \mathbf{x}'\boldsymbol{\beta})$. That is,

$$C(\mathbf{x}, y) = C(\mathbf{x}, \mathbf{x}'\boldsymbol{\beta}) = C(\mathbf{x}, \mathbf{x})\boldsymbol{\beta} = V(\mathbf{x})\boldsymbol{\beta},$$

or $\boldsymbol{\Sigma}_{xx}\boldsymbol{\beta} = \boldsymbol{\sigma}_{xy}$. Provided that $\boldsymbol{\Sigma}_{xx} = V(\mathbf{x})$ is nonsingular, the population BLP slope vector is

$$\boldsymbol{\beta} = (\boldsymbol{\Sigma}_{xx})^{-1}\boldsymbol{\sigma}_{xy}.$$

Now turn to the sample. Let $\mathbf{X}_2$ be the $n \times (k - 1)$ matrix of observations on the nonconstant explanatory variables, $\mathbf{x}_1$ be the $n \times 1$ summer vector, and $\mathbf{y}$ be the $n \times 1$ vector of observations on the dependent variable. From the discussion of deviations from means in residual regression theory (Section 17.4), the sample LS slope vector can be written as

$$\mathbf{b} = (\mathbf{X}_2^{*\prime}\mathbf{X}_2^*)^{-1}\mathbf{X}_2^{*\prime}\mathbf{y}^*,$$

where $\mathbf{X}_2^* = \mathbf{M}_1\mathbf{X}_2$, $\mathbf{y}^* = \mathbf{M}_1\mathbf{y}$, $\mathbf{M}_1 = \mathbf{I} - (1/n)\,\mathbf{x}_1\mathbf{x}_1'$, or for that matter as

$$\mathbf{b} = (\mathbf{X}_2^{*\prime}\mathbf{X}_2^*/n)^{-1}(\mathbf{X}_2^{*\prime}\mathbf{y}^*/n).$$

Now recognize that $\mathbf{X}_2^{*\prime}\mathbf{X}_2^*/n = \mathbf{S}_{xx}$, the $(k - 1) \times (k - 1)$ *sample variance matrix* of the x's, while $\mathbf{X}_2^{*\prime}\mathbf{y}^*/n = \mathbf{s}_{xy}$, the $(k - 1) \times 1$ *sample covariance vector* of the x's with y. Thus the sample LS slope vector is

$$\mathbf{b} = (\mathbf{S}_{xx})^{-1}\mathbf{s}_{xy},$$

which is the obvious analog estimator of the population BLP slope vector

$$\boldsymbol{\beta} = (\boldsymbol{\Sigma}_{xx})^{-1}\boldsymbol{\sigma}_{xy}.$$

In random sampling, sample moments converge in probability to the corresponding population moments. So $\mathbf{S}_{xx} \xrightarrow{P} \boldsymbol{\Sigma}_{xx}$ and $\mathbf{s}_{xy} \xrightarrow{P} \boldsymbol{\sigma}_{xy}$. By a multivariate version of S2 (Section 9.5) it follows that

$$\mathbf{b} \xrightarrow{P} \boldsymbol{\beta},$$

so the LS slope vector $\mathbf{b}$ is a consistent estimator of the population slope vector $\boldsymbol{\beta}$. Similarly, the LS intercept $a = \bar{y} - \bar{\mathbf{x}}'\mathbf{b}$ is a consistent estimator of the population BLP intercept $\alpha = E(y) - [E(\mathbf{x})]'\boldsymbol{\beta} = \mu_y - \boldsymbol{\mu}_x'\boldsymbol{\beta}$.

Proceeding, by using multivariate versions of the CLT and the Delta method it can be shown that

$$\mathbf{b} \overset{A}{\sim} \mathcal{N}(\boldsymbol{\beta}, \boldsymbol{\Phi}/n),$$

with

$$\mathbf{\Phi} = (\mathbf{\Sigma}_{xx})^{-1}E(\mathbf{x}^*\mathbf{x}^{*\prime}u^2)(\mathbf{\Sigma}_{xx})^{-1},$$

$$\mathbf{x}^* = \mathbf{x} - \boldsymbol{\mu}_x, \qquad u = y - (\alpha + \mathbf{x}'\boldsymbol{\beta}).$$

This generalizes the result (Section 13.1) for the bivariate case, namely $b \overset{A}{\sim} \mathcal{N}(\beta, \phi^2/n)$ with $\phi^2 = E(x^{*2}u^2)/V^2(x)$.

This asymptotic theory for the sample LS coefficients holds with no assumption on the form of the CEF. If the population CEF *is* linear, then the LS estimators are unbiased as well as consistent. If also the population conditional variance function is constant, then $E(u^2|\mathbf{x}) = V(y|\mathbf{x}) = \sigma^2$, and

$$E(\mathbf{x}^*\mathbf{x}^{*\prime}u^2) = E_{\mathbf{x}}[E(\mathbf{x}^*\mathbf{x}^{*\prime}u^2|\mathbf{x})] = E_{\mathbf{x}}[\mathbf{x}^*\mathbf{x}^{*\prime}E(u^2|\mathbf{x})]$$
$$= E(\mathbf{x}^*\mathbf{x}^{*\prime}\sigma^2) = \sigma^2E(\mathbf{x}^*\mathbf{x}^{*\prime}) = \sigma^2V(\mathbf{x}) = \sigma^2\mathbf{\Sigma}_{xx}.$$

Then $\mathbf{\Phi}$ will reduce to $\sigma^2(\mathbf{\Sigma}_{xx})^{-1}$, and the asymptotic distribution will simplify to

$$\mathbf{b} \overset{A}{\sim} \mathcal{N}[\boldsymbol{\beta}, (\sigma^2/n)(\mathbf{\Sigma}_{xx})^{-1}].$$

The asymptotic results serve to justify, as approximations when random sampling from any multivariate population, the conventional normal-theory confidence regions, confidence intervals, and hypothesis tests.

In practice, the elements of $\mathbf{\Phi}$ will have to be estimated. We know how to do this for the linear-homoskedastic case. For the general case, $\mathbf{S}_{xx}$ provides a consistent estimator of $\mathbf{\Sigma}_{xx}$. Further, let $\mathbf{x}_i^{**\prime}$ denote the ith row of $\mathbf{X}_2^*$, and e_i denote the ith element of the LS residual vector $\mathbf{e}$. Then

$$(1/n)\sum_{i=1}^{n}(\mathbf{x}_i^{**}\mathbf{x}_i^{**\prime}e_i^2)$$

provides a consistent estimator of $E(\mathbf{x}^*\mathbf{x}^{*\prime}u^2)$. The square roots of the diagonal elements of the resulting estimated $\mathbf{\Phi}$ matrix serve as standard errors of the LS coefficients when estimating a population BLP, in the absence of assumptions of linearity of CEF and homoskedasticity. They may be referred to as "general-heteroskedasticity-corrected" standard errors.

Exercises

25.1 Suppose that x and y are bivariate–normally distributed with $E(y|x) = \alpha + \beta x$, $V(y|x) = \sigma^2$, and $V(x) = \sigma_x^2$. In random sampling, sample size n from this population, let b be the sample slope and let s_x^2 be the sample variance of x. Let

$$z = \sqrt{n}(b - \beta)/(\sigma/s_x), \qquad w = ns_x^2/\sigma_x^2, \qquad u = \sqrt{(n-1)}(b - \beta)/(\sigma/\sigma_x).$$

(a) Show that $z \sim \mathcal{N}(0, 1)$, that $w \sim \chi^2(n - 1)$, and that z and w are independent.

(b) Show that $u \sim t(n - 1)$.

(c) Explain how the result in (b) completely specifies the marginal distribution of the sample slope in terms of parameters and sample size.

26 Time Series

26.1. Departures from Random Sampling

We digress from regression analysis in order to introduce some basic ideas on *time series*. We deal with a single variable y, on which we have a set of n observations y_t for $t = 1, \ldots, n$. Here t indexes time, measured discretely.

Figures 26.1, 26.2, and 26.3 display three sets of 100 observations on a variable y, with t measured on the horizontal axis, and y_t on the vertical axis.

Figure 26.1 was produced as follows. For $t = 0, 1, \ldots, 100$, observations u_t were independently drawn from the $\mathcal{N}(0, 1)$ distribution. Then, for $t = 1, \ldots, 100$, we set $y_t = u_t$. So the y_t series is a size-100 random sample from the $\mathcal{N}(0, 1)$ distribution. The joint distribution of any adjacent pair of y's is SBVN(0), so $E(y_t|y_{t-1}) = E(y_t) = 0$ regardless of the value of y_{t-1}: the conditional expectation of y_t given y_{t-1} does not vary with y_{t-1}. This lack of predictability is manifest in the jagged and irregular time path of Figure 26.1.

Now, the typical economic time series does not look at all like that figure, but may well (perhaps after removing a linear time trend) display a relatively smooth and wavelike course such as that in Figures 26.2 and 26.3. If so, it must be inappropriate to view such a series as a random sample from a univariate population. (The $\mathcal{N}(0, 1)$ population is being used only for convenience.) If we plan to work with economic time series data, we need an observational scheme that departs from random sampling. There are two distinct types of departure from random sampling: the observations may not be independent, or they may not be identically distributed.

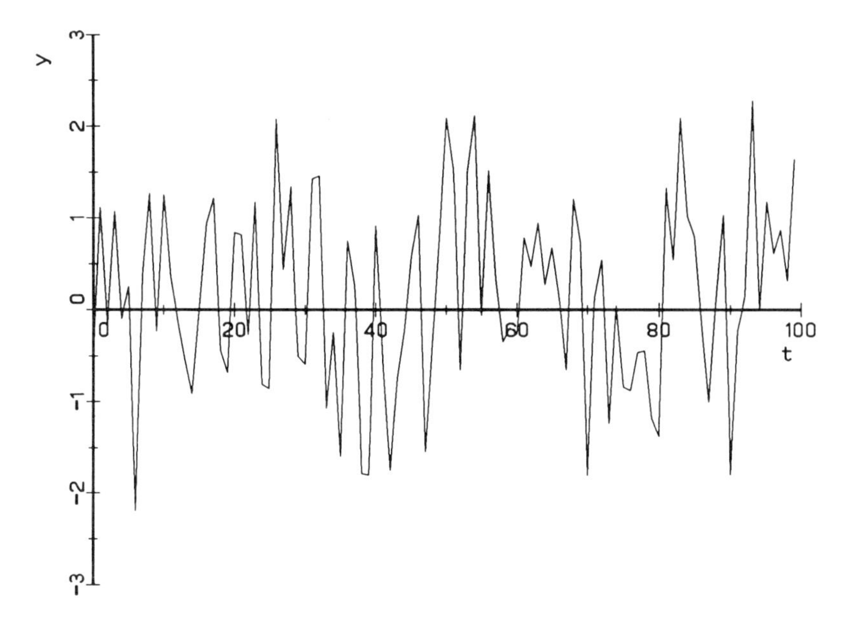

Figure 26.1 Time series 1.

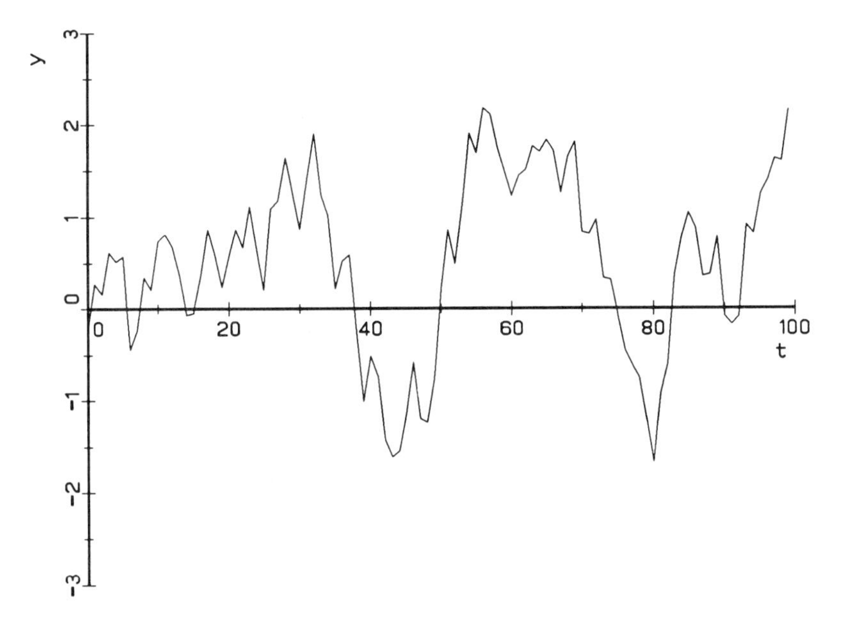

Figure 26.2 Time series 2.

Figure 26.2 was produced as follows. For $t = 0, 1, \ldots, 100$, observations u_t were independently drawn from a $\mathcal{N}(0, 1)$ distribution. (In fact, the same numerical u_t values were used as for Figure 26.1.) Then we set $y_0 = u_0$ and generated the remaining y_t's recursively as

$$y_t = \rho y_{t-1} + \sigma u_t \qquad (t = 1, \ldots, 100),$$

with $\rho = 0.9$ and $\sigma = \sqrt{(1 - \rho^2)}$.

Because $y_1 = \rho y_0 + \sigma u_1$ is linear in the two independent $\mathcal{N}(0, 1)$ variables y_0 and u_1, with $\rho^2 + \sigma^2 = 1$, it follows that $y_1 \sim \mathcal{N}(0, 1)$. Then because $y_2 = \rho y_1 + \sigma u_2$ is linear in the two independent $\mathcal{N}(0, 1)$ variables y_1 and u_2, we see that $y_2 \sim \mathcal{N}(0, 1)$. Proceeding in this manner we see that each y_t is distributed $\mathcal{N}(0, 1)$, so they are identically distributed. But they are not independent. For example, the covariance between y_1 and y_2 is

$$C(y_1, y_2) = C(y_1, \rho y_1 + \sigma u_2) = \rho V(y_1) + \sigma C(y_1, u_2) = \rho,$$

and the covariance between y_1 and y_3 is

$$C(y_1, y_3) = C(y_1, \rho y_2 + \sigma u_3) = \rho C(y_1, y_2) + \sigma C(y_1, u_3) = \rho^2.$$

What we have is a set of random variables that are identically, but not independently, distributed.

Focusing on an adjacent pair (y_t, y_{t-1}), we find that their joint distribution is SBVN(ρ), because

$$\begin{pmatrix} y_{t-1} \\ y_t \end{pmatrix} = \begin{pmatrix} 1 & 0 \\ \rho & \sigma \end{pmatrix} \begin{pmatrix} y_{t-1} \\ u_t \end{pmatrix},$$

where y_{t-1} and u_t are independent $\mathcal{N}(0, 1)$ variables. So, by bivariate normal theory, $E(y_t|y_{t-1}) = \rho y_{t-1}$: the conditional expectation of y_t given y_{t-1} does vary with y_{t-1}. This predictability of an observation from its predecessor, manifest as a relatively smooth wave in Figure 26.2, suffices to distinguish the series from a random sample.

Turning to Figure 26.3, it may come as a surprise to learn that the observations plotted there were independently drawn. The figure was produced as follows. For $t = 0, \ldots, 100$, the u_t were independently drawn from a $\mathcal{N}(0, 1)$ distribution. (In fact, the same u_t values were used as for Figures 26.1 and 26.2.) Then we set

$$y_t = \mu_t + \sigma u_t \qquad (t = 1, \ldots, 100),$$

where $\sigma = 1/3$ and $\mu_t = (4/3)\sin(7.2t)$, with the angle measured in degrees. The μ_t's are nonstochastic, so $y_t \sim \mathcal{N}(\mu_t, \sigma^2)$. The y_t's are independent because the u_t's are independent. But their expectations μ_t differ, so they are not identically distributed. What we have is a set of random variables that are independently, but not identically, distributed.

Focusing on an adjacent pair (y_t, y_{t-1}), we see that their joint distribution is $\text{BVN}(\mu_t, \mu_{t-1}, \sigma^2, \sigma^2, 0)$ with $\sigma^2 = 1/9$. So, by bivariate normal theory, $E(y_t|y_{t-1}) = \mu_t$: the conditional expectation of y_t does not vary with y_{t-1}. In that sense an observation is not predictable from its predecessor. The regularity in Figure 26.3 is attributable to the sine wave pattern in the deterministic μ_t series, not to any dependence among the random variables y_t.

A useful message emerges from the rough similarity of Figures 26.2 and 26.3: for a real-world economic time series, it may not be self-evident which type of departure from random sampling is relevant. Perhaps the observations are dependent, being *autocorrelated*. Perhaps they are independent, with a *changing expectation* μ_t. In the latter case,

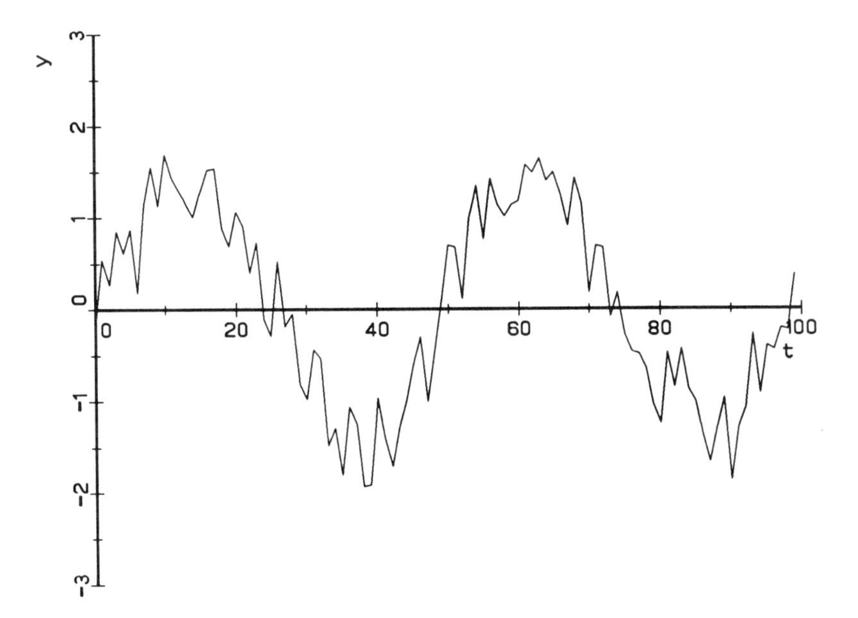

Figure 26.3 Time series 3.

if the expectation is expressible as a linear function of observable explanatory variables, then a CR model might apply. Of course, both departures may occur simultaneously.

26.2. Stationary Population Model

Let $\mathbf{y} = (y_1, \ldots, y_t, \ldots, y_n)'$ be an observed vector that displays the characteristic regularity of economic time series. One setup that produces such $\mathbf{y}$'s is a CR model, with $\boldsymbol{\mu} = E(\mathbf{y}) = \mathbf{X}\boldsymbol{\beta}$, $\boldsymbol{\Sigma} = V(\mathbf{y}) = \sigma^2\mathbf{I}$, and the rows of $\mathbf{X}$ showing a regular development over time.

Let us focus on the other departure from random sampling. We will suppose that the y_t's have identical expectations and variances, but some nonzero covariances. Then the elements of $\boldsymbol{\mu} = E(\mathbf{y})$ are all the same, and the diagonal elements of $\boldsymbol{\Sigma} = V(\mathbf{y})$ are all the same, but at least some off-diagonal elements of $\boldsymbol{\Sigma}$ are nonzero. In Section 28.3 we will combine the two departures, allowing $\boldsymbol{\mu} = \mathbf{X}\boldsymbol{\beta}$ as well as a nondiagonal $\boldsymbol{\Sigma}$ matrix.

Here we confine attention to an important special case. Assume that $E(\mathbf{y}) = \boldsymbol{\mu} = \mathbf{x}_1\mu$, where $\mathbf{x}_1$ is the $n \times 1$ summer vector and μ is a scalar, and that $V(\mathbf{y}) = \boldsymbol{\Sigma}$, with

$$\boldsymbol{\Sigma} = \begin{pmatrix} \gamma_0 & \gamma_1 & \gamma_2 & \cdots & \gamma_{n-1} \\ \gamma_1 & \gamma_0 & \gamma_1 & \cdots & \gamma_{n-2} \\ \gamma_2 & \gamma_1 & \gamma_0 & \cdots & \gamma_{n-3} \\ \cdot & \cdot & \cdot & & \cdot \\ \cdot & \cdot & \cdot & & \cdot \\ \cdot & \cdot & \cdot & & \cdot \\ \gamma_{n-1} & \gamma_{n-2} & \gamma_{n-3} & \cdots & \gamma_0 \end{pmatrix}$$

We have introduced the notation

$$\gamma_j = C(y_t, y_{t-j}) \qquad (j = 0, \pm 1, \pm 2, \ldots, \pm(n-1)).$$

Here γ_j is the jth *autocovariance*, with $\gamma_0 = C(y_t, y_t) = V(y_t)$ being the variance. Further,

$$\rho_j = C(y_t, y_{t-j})/[\sqrt{V(y_t)}\sqrt{V(y_{t-j})}] = \gamma_j/\gamma_0$$

is the jth *autocorrelation*, with $\rho_0 = 1$. Observe that $\gamma_{-j} = \gamma_j$ because $C(y_t, y_{t+j}) = C(y_{s-j}, y_s) = C(y_s, y_{s-j})$, with $s = t + j$. By the same token, $\rho_{-j} = \rho_j$.

A distinctive feature of the $\boldsymbol{\Sigma}$ matrix above is that the covariance between any two elements of $\mathbf{y}$ depends only on the absolute difference between their subscripts, that is on the time distance between them. We refer to this specification for $\boldsymbol{\mu}$ and $\boldsymbol{\Sigma}$ as the *stationary population*, or SP, *model*. The term "stationary," as used here, refers to constancy across t. We have the expectations $E(y_t)$, the variances $V(y_t)$, and the autocovariances $C(y_t, y_{t-j})$ being stationary, that is, constant over t. So the autocorrelations are also stationary. Of course, the autocovariances and autocorrelations need not be constant across j.

A stricter form of stationarity holds if the entire joint probability distribution of any subset of the variables depends only upon the differences in their time subscripts. Then, for example, the joint distribution of (y_1, y_5, y_7) is the same as that of (y_{22}, y_{26}, y_{28}), and of course the marginal distribution of each variable is the same. The terminology varies in the literature: "stationary" may be reserved for the stricter form, with our weaker form being referred to as weakly, or covariance, or second-order, or wide-sense, stationary.

26.3. Conditional Expectation Functions

At this point, we are prepared to interpret an $n \times 1$ vector $\mathbf{y}$, the observed time series, as a single drawing from an n-variate population with $E(\mathbf{y}) = \boldsymbol{\mu}$ and $V(\mathbf{y}) = \boldsymbol{\Sigma}$ as above. It is natural to inquire about CEF's in the population, in particular, about the conditional expectation of y at time t given one or more past values of y. For the sake of convenience, we will suppose that $\mathbf{y}$ is multinormally distributed so that all CEF's are linear. Normality is not crucial to the analysis; the gist of the results will apply to BLP's if the CEF's are not linear.

As usual, the coefficients of a linear CEF are expressible in terms of population expectations, variances, and covariances. The general formulas of Section 14.1 and Section 25.5 specialize here because of stationarity.

For example, consider the CEF of y_t given y_{t-1}, namely $E(y_t|y_{t-1}) = \alpha + \beta y_{t-1}$. Here $\beta = \gamma_1/\gamma_0 = \rho_1$ and $\alpha = (1 - \rho_1)\mu$. For a richer example, consider the CEF of y_t given y_{t-1} and y_{t-2}:

$$E(y_t|y_{t-1}, y_{t-2}) = \beta_0 + \beta_1 y_{t-1} + \beta_2 y_{t-2}.$$

Because of stationarity, we get $\beta_0 = (1 - \beta_1 - \beta_2)\mu$, and

$$\begin{pmatrix} \gamma_0 & \gamma_1 \\ \gamma_1 & \gamma_0 \end{pmatrix} \begin{pmatrix} \beta_1 \\ \beta_2 \end{pmatrix} = \begin{pmatrix} \gamma_1 \\ \gamma_2 \end{pmatrix}.$$

Dividing through by γ_0 gives

$$\begin{pmatrix} 1 & \rho_1 \\ \rho_1 & 1 \end{pmatrix} \begin{pmatrix} \beta_1 \\ \beta_2 \end{pmatrix} = \begin{pmatrix} \rho_1 \\ \rho_2 \end{pmatrix},$$

the solution to which is

$$\beta_1 = \rho_1(1 - \rho_2)/(1 - \rho_1^2), \qquad \beta_2 = (\rho_2 - \rho_1^2)/(1 - \rho_1^2).$$

Observe that the autocorrelations suffice to determine the slopes and, along with the expectation, to determine the intercept. Observe also that the CEF's are themselves stationary: they do not change with t.

Special cases of the SP model arise when specific assumptions are made about the pattern of the autocovariances across j. Such assumptions have implications for the pattern of slopes in the CEF's.

Example. Suppose that $\gamma_2 = 0$, so $\rho_2 = 0$. Then in $E(y_t|y_{t-1})$, the slope is $\beta = \rho_1$, while in $E(y_t|y_{t-1}, y_{t-2})$, the slopes are

$$\beta_1 = \rho_1/(1 - \rho_1^2), \qquad \beta_2 = -\rho_1^2/(1 - \rho_1^2) = -\rho_1\beta_1.$$

Example. Suppose that $\gamma_2/\gamma_1 = \gamma_1/\gamma_0$, so $\rho_2 = \rho_1^2$. Then in $E(y_t|y_{t-1})$, the slope is $\beta = \rho_1$, while in $E(y_t|y_{t-1}, y_{t-2})$, the slopes are

$$\beta_1 = \rho_1, \qquad \beta_2 = 0.$$

In the second example, we see a coincidence of short and long regression slopes that does not appear in the first example. This suggests that one may be able to discriminate between alternative special cases of $\mathbf{\Sigma}$ by examination of several CEF's.

In practice, the CEF's will be unknown, and it will be of interest to estimate them. Given a sample from the SP model, that is, an observed $\mathbf{y}$-vector, we may consider estimating the parameters of a population CEF. The natural procedure is sample LS linear regression of y_t on the corresponding set of past values.

For the sake of concreteness, suppose that we wish to estimate $E(y_t|y_{t-1}) = \alpha + \beta y_{t-1}$. With n observations in hand, the usable data consist of the $(n - 1) \times 1$ vectors $\mathbf{y} = (y_2, \ldots, y_n)'$, $\mathbf{x}_1 = (1, \ldots, 1)'$, and $\mathbf{x}_2 = (y_1, \ldots, y_{n-1})'$. Let $\mathbf{X} = (\mathbf{x}_1, \mathbf{x}_2)$ and $\boldsymbol{\beta} = (\alpha, \beta)'$. The LS coefficient vector is $\mathbf{b} = (\mathbf{X}'\mathbf{X})^{-1}\mathbf{X}'\mathbf{y}$. To assess its sampling distribution, observe first that the $\mathbf{X}$ matrix contains elements of the random vector $\mathbf{y}$. So $\mathbf{X}$ cannot be constant, which means that a CR model is not applicable. Will a NeoCR model be applicable? With $\mathbf{x}_t = (1, x_{t,2})' = (1, y_{t-1})'$, it is true that $E(y_t|\mathbf{x}_t) = \alpha + \beta y_{t-1} = \mathbf{x}_t'\boldsymbol{\beta}$. However, $\mathbf{x}_{t+1} = (1, x_{t+1,2}) = (1, y_t)'$ contains the actual value of y_t, so that

$$E(y_t|\mathbf{x}_t, \mathbf{x}_{t+1}) = E(y_t|y_{t-1}, y_t) = y_t \neq E(y_t|\mathbf{x}_t) = \mathbf{x}_t'\boldsymbol{\beta}.$$

Consequently, $E(y_t|\mathbf{X}) \neq \mathbf{X}\boldsymbol{\beta}$, whence $E(\mathbf{y}|\mathbf{X}) \neq \mathbf{X}\boldsymbol{\beta}$.

Despite the linearity of the CEF, a strict requirement of the NeoCR model is violated. Consequently, $E(\mathbf{b}|\mathbf{X}) = \mathbf{A}E(\mathbf{y}|\mathbf{X})$ does not reduce to $\mathbf{AX}\boldsymbol{\beta} = \boldsymbol{\beta}$, and hence there is no presumption that $E(\mathbf{b}) = E_{\mathbf{X}}[E(\mathbf{b}|\mathbf{X})] = \boldsymbol{\beta}$. Indeed $\mathbf{b}$ is a biased estimator of $\boldsymbol{\beta}$, and no unbiased estimator of $\boldsymbol{\beta}$ is available.

Nevertheless, under general conditions, $\mathbf{b}$ is a consistent estimator of $\boldsymbol{\beta}$, and in fact the asymptotic theory for LS estimation (Section 25.5) applies. The conditions include a specification of how additional observations are produced, that is how the $\boldsymbol{\mu}$ vector and $\boldsymbol{\Sigma}$ matrix develop as n increases.

26.4. Stationary Processes

We seek an underlying framework that will support the SP model for an observed $n \times 1$ vector $\mathbf{y}$, and that will readily extend as n increases. Consider, then, an infinite sequence of random variables ordered in time: y_t for $t = \ldots, -2, -1, 0, 1, 2, \ldots$. The index t denotes time, measured discretely. We refer to such an infinite sequence as a *stochastic process*. Each of the variables in the sequence has an expectation and variance, and each pair of them has a covariance. Now suppose that all the variables have the same expectation and variance, and further that the covariance between any pair of them depends only on the absolute difference between their subscripts (that is, on the length of time between them). We refer to such an infinite sequence as a *stationary stochastic process*. (Here again, the stricter concept of stationarity may arise, and the terminology varies.) Evidently, the SP model will apply to any n successive variables in such a sequence.

How can such an infinite sequence of random variables be generated? Here are two leading examples of mechanisms that produce stationary stochastic processes.

First-Order Moving-Average Process

Suppose that for $t = \ldots, -2, -1, 0, 1, 2, \ldots$, the values of y_t are determined by

$$(26.1) \quad y_t = \phi_0 + \phi u_{t-1} + u_t,$$

where the u_t's are independent drawings on a random variable u with $E(u) = 0$ and $V(u) = \sigma^2$. Because the u's are independent and identically distributed, it follows that the y's are identically distributed, and indeed that the joint distribution of any pair of y's depends only upon the difference in their time subscripts. Similarly for any triplet, and so forth. This case is called the *first-order moving-average*, or MA(1), *process*.

The first and second moments of an MA(1) process are readily derived. From Eq. (26.1) and the assumptions on the u_t, it follows for every t that:

$$(26.2a) \quad \mu = \phi_0 + \phi E(u_{t-1}) + E(u_t) = \phi_0,$$

$$(26.2b) \quad \gamma_0 = V(\phi u_{t-1} + u_t) = \phi^2\sigma^2 + \sigma^2 = (1 + \phi^2)\sigma^2,$$

$$(26.2c) \quad \gamma_1 = C(\phi u_{t-1} + u_t, \phi u_{t-2} + u_{t-1}) = \phi\sigma^2,$$

while

$$\gamma_2 = C(\phi u_{t-1} + u_t, \phi u_{t-3} + u_{t-2}) = 0,$$

and similarly $\gamma_3 = \gamma_4 = \cdots = 0$. So the autocorrelations of the MA(1) process are $\rho_1 = \gamma_1/\gamma_0 = \phi/(1 + \phi^2)$, and $\rho_j = 0$ for $j = 2, 3, \ldots$. For any $\mathbf{y}$ containing n successive values of y, the SP model applies: the $\mathbf{\Sigma}$ matrix has all elements zero except along the main diagonal and along the strips just above and below that main diagonal.

First-Order Autoregressive Process

Suppose that for $t = \ldots, -2, -1, 0, 1, 2, \ldots$, the values of y_t are determined by

$$(26.3) \quad y_t = \theta_0 + \theta y_{t-1} + u_t,$$

where the u_t's are independent drawings on a random variable u with $E(u) = 0$ and $V(u) = \sigma^2$, and $|\theta| < 1$. By repeated back-substitution, we find

$$y_t = \theta_0 \sum_{s=0}^{\infty} \theta^s + \sum_{s=0}^{\infty} \theta^s u_{t-s} = \theta_0/(1 - \theta) + \sum_{s=0}^{\infty} \theta^s u_{t-s}.$$

Observe how the condition $|\theta| < 1$ was used to ensure convergence of the infinite series. Because the u's are independent and identically distributed, it follows that the y's are identically distributed, and indeed that the joint distribution of any subset of y's will depend only on the differences in their time subscripts. This case is called the *first-order autoregressive*, or AR(1), *process*.

Taking that stationarity for granted, the first and second moments of the AR(1) process are readily derived. Observe that u_t is independent of $y_{t-1}, y_{t-2}, \ldots$. From Eq. (26.3) and the assumptions on the u_t, it follows for every t that:

(26.4a) $\mu = \theta_0 + \theta E(y_{t-1}) + E(u_t)$

$$\Rightarrow \quad \mu = \theta_0 + \theta\mu \quad \Rightarrow \quad \mu = \theta_0/(1 - \theta),$$

(26.4b) $\gamma_0 = V(\theta y_{t-1} + u_t) = \theta^2\gamma_0 + \sigma^2 \quad \Rightarrow \quad \gamma_0 = \sigma^2/(1 - \theta^2),$

(26.4c) $\gamma_1 = C(\theta y_{t-1} + u_t, y_{t-1}) = \theta V(y_{t-1}) = \theta\gamma_0,$

while

$$\gamma_2 = C(\theta y_{t-1} + u_t, y_{t-2}) = \theta C(y_{t-1}, y_{t-2}) = \theta\gamma_1 = \theta^2\gamma_0,$$

and similarly $\gamma_3 = \theta^3\gamma_0$, $\gamma_4 = \theta^4\gamma_0$, and so forth. So the autocorrelations of the AR(1) process are $\rho_1 = \theta$, $\rho_2 = \theta^2, \ldots, \rho_j = \theta^j, \ldots$. For any $\mathbf{y}$ containing n successive values of y, the SP model applies: the $\mathbf{\Sigma}$ matrix has all elements nonzero, and declining in a particular way as we go away from the main diagonal.

In both of these leading examples, the y's are identically, but not independently, distributed. In the MA(1) process, the dependence is confined to adjacent y's, while in the AR(1) process it extends indefinitely, although with $|\theta| < 1$, the correlations do taper off in magnitude as the time distance between the variables increases.

More elaborate cases can be constructed, starting again with an infinite sequence of independent and identically distributed u's with $E(u) = 0$. Thus, there is the *second-order moving-average*, or MA(2), *process*:

$$y_t = \phi_0 + \phi_1 u_{t-1} + \phi_2 u_{t-2} + u_t,$$

and the *second-order autoregressive*, or AR(2), *process*:

$$y_t = \theta_0 + \theta_1 y_{t-1} + \theta_2 y_{t-2} + u_t,$$

with $\theta_1 + \theta_2 < 1$, $\theta_2 - \theta_1 < 1$, and $\theta_2 > -1$ to ensure stationarity. Mixed cases may also be constructed. For example, the ARMA(1, 2) process has

$$y_t = \theta_0 + \theta y_{t-1} + \phi_1 u_{t-1} + \phi_2 u_{t-2} + u_t,$$

with $|\theta| < 1$. These elaborations allow for more complex patterns of autocorrelation, and hence more flexible patterns of CEF slopes, than appeared in our leading examples.

A couple of remarks:

• In our examples, the u's were taken to be independent and identically distributed, so the stricter form of stationarity prevailed. In fact, uncorrelated u's with constant expectation and variance would suffice for most purposes.

• In many contexts, the assumption of an infinitely long past history is unattractive. Stationarity for $t = 1, 2, \ldots,$ can still be ensured by appropriate choice of initial conditions. Thus an MA(1) process can be started up with u_0, and an AR(1) process can be started up with $y_0 = \theta_0/(1 - \theta) + u_0/\sqrt{(1 - \theta^2)}$.

26.5. Sampling and Estimation

A stationary stochastic process may be characterized in terms of the *process parameters* which consist of the ϕ's and/or θ's and σ^2. It may also be characterized in terms of the *population moments*, which consist of μ, $\gamma_0, \gamma_1, \gamma_2, \ldots$ (or equivalently $\mu, \gamma_0, \rho_1, \rho_2, \ldots$). The CEF coefficients, for any set of conditioning variables, can be derived from those. Either set of parameters may be viewed as interesting features of the population. A representation in terms of the process parameters will be more parsimonious, especially when a low-order AR, MA, or ARMA specification applies.

Suppose that the parameters are unknown, but n consecutive observations $y_1, \ldots, y_n$ generated by the process are available. We view $\mathbf{y} = (y_1, \ldots, y_n)'$ as a single drawing from a multivariate population with $E(\mathbf{y}) = \boldsymbol{\mu}$ and $V(\mathbf{y}) = \boldsymbol{\Sigma}$ as in the SP model. The analogy principle suggests that we estimate the population moments by the corresponding sample moments, and then convert those into estimates of the process parameters. Define the sample mean, sample variance, and first sample autocovariance by

(26.5a) $m = \sum_{t=1}^{n} y_t/n,$

(26.5b) $c_0 = \sum_{t=1}^{n} (y_t - m)^2/n,$

(26.5c) $c_1 = \sum_{t=2}^{n} (y_t - m)(y_{t-1} - m)/(n - 1).$

Similarly, the second sample autovariance is

$$c_2 = \sum_{t=3}^{n} (y_t - m)(y_{t-2} - m)/(n - 2),$$

and so forth. The sample autocorrelations are $r_j = c_j/c_0$. These sample moments can serve as estimators of the population moments μ, γ_0, γ_j, and ρ_j.

How are they converted into estimators of the process parameters? There is an extensive literature on inferring the type of process by inspection of the sample autocorrelations: see Judge et al. (1988, pp. 684–705) for an introduction. But here we suppose that the type is known, and to illustrate, confine attention to the first-order cases.

For the MA(1) case, the natural estimators of the process parameters are the solutions to the sample counterparts of Eqs. (26.2a–c):

(26.6a) $m = \hat{\phi}_0,$

(26.6b) $c_0 = (1 + \hat{\phi}^2)\hat{\sigma}^2,$

(26.6c) $c_1 = \hat{\phi}\hat{\sigma}^2.$

More explicitly,

$$\hat{\phi}_0 = m, \qquad \hat{\phi} = [1 - \sqrt{(1 - 4r_1^2)}]/(2r_1), \qquad \hat{\sigma}^2 = c_1/\hat{\phi}.$$

(See Exercise 26.2 for an explanation of the choice of root in the quadratic equation $r_1 = \hat{\phi}/(1 + \hat{\phi}^2)$.)

For the AR(1) case, the natural estimators of the process parameters are the solutions to the sample counterparts of Eqs. (26.4a–c):

(26.7a) $m = \hat{\theta}_0/(1 - \hat{\theta})$,

(26.7b) $c_0 = \hat{\sigma}^2/(1 - \hat{\theta}^2)$,

(26.7c) $c_1 = \hat{\theta}c_0$.

More explicitly,

$$\hat{\theta} = r_1, \qquad \hat{\theta}_0 = (1 - r_1)m, \qquad \hat{\sigma}^2 = (1 - r_1^2)c_0.$$

In both cases, the resulting estimators of process parameters will be consistent (by S2, Section 9.5), if the sample moments m, c_0, and c_1 converge in probability to the corresponding population moments μ, γ_0, and γ_1. This convergence will occur in general even though the observations are not obtained by random sampling from the univariate distribution of y.

To illustrate the convergence argument, we assess the sample mean when the data are generated by the MA(1) process. We have $\mathbf{y} = (y_1, \ldots, y_n)'$ as the sample observation vector, with $E(\mathbf{y}) = \mathbf{x}_1\mu$ and $V(\mathbf{y}) = \mathbf{\Sigma}$, where $\mathbf{x}_1$ is the $n \times 1$ summer vector, and

$$\mathbf{\Sigma} = \begin{pmatrix} \gamma_0 & \gamma_1 & 0 & 0 & \cdots & 0 \\ \gamma_1 & \gamma_0 & \gamma_1 & 0 & \cdots & 0 \\ 0 & \gamma_1 & \gamma_0 & \gamma_1 & \cdots & 0 \\ 0 & 0 & \gamma_1 & \gamma_0 & \cdots & 0 \\ \cdot & \cdot & \cdot & \cdot & & \cdot \\ \cdot & \cdot & \cdot & \cdot & & \cdot \\ \cdot & \cdot & \cdot & \cdot & & \cdot \\ 0 & 0 & 0 & 0 & \cdots & \gamma_0 \end{pmatrix}.$$

The sample mean is $m = \bar{y} = (1/n)\,\mathbf{x}_1'\mathbf{y}$, so

$$E(\bar{y}) = (1/n)\,\mathbf{x}_1'E(\mathbf{y}) = (1/n)\,\mathbf{x}_1'\mathbf{x}_1\mu = \mu,$$

$$V(\bar{y}) = (1/n^2)\mathbf{x}_1'V(\mathbf{y})\mathbf{x}_1 = (1/n^2)\mathbf{x}_1'\mathbf{\Sigma}\mathbf{x}_1.$$

Now $\mathbf{x}_1'\mathbf{\Sigma}\mathbf{x}_1$ is the sum of all the elements in the $\mathbf{\Sigma}$ matrix, which by inspection of the display above is $\mathbf{x}_1'\mathbf{\Sigma}\mathbf{x}_1 = n\gamma_0 + 2(n - 1)\gamma_1$. So

$$V(\bar{y}) = (\gamma_0/n)(1 + 2\rho_1 - 2\rho_1/n).$$

This differs from the random sampling result $V(\bar{y}) = \gamma_0/n$, but still converges to zero as n goes to infinity. So $\bar{y}$ converges in mean square, hence in probability, to μ. For further calculations in this style, see Goldberger (1964, pp. 142–153).

Under general conditions, the convergence argument extends to other sample moments, to the AR(1) process, and indeed to higher-order processes. In that event, the analog estimators of the process parameters will be consistent.

At this point, we can recognize why LS linear regression provides consistent estimation of the CEF parameters. The normal equations that determine the LS coefficients in terms of observed sums of squares and sums of cross-products are essentially the sample counterparts of the equations that determine the CEF parameters in terms of population moments. For example, suppose that with n observations in hand, we run the LS linear regression of y_t on $(1, y_{t-1})$. There are $n - 1$ usable observations. Let

$$m^* = \sum_{t=2}^{n} y_t/(n - 1), \qquad m^{**} = \sum_{t=2}^{n} y_{t-1}/(n - 1).$$

As long as n is at least moderately large, these will differ only trivially from each other, and from the m defined in Eq. (26.5a). The LS slope will be $b = s_{xy}/s_x^2$, where

$$s_x^2 = \sum_{t=2}^{n} (y_{t-1} - m^{**})^2/(n - 1),$$

$$s_{xy} = \sum_{t=2}^{n} (y_{t-1} - m^{**})(y_t - m^*)/(n - 1).$$

As long as n is at least moderately large, these will differ only trivially from the c_0 and c_1 defined in Eqs. (26.5b–c). Being practically the same as c_1/c_0, the LS slope will, under general conditions, converge in probability to $\gamma_1/\gamma_0 = \beta$. The argument extends to CEF's with several lagged values of y as conditioning variables.

26.6. Remarks

• For stationary stochastic processes, convergence of sample moments to the corresponding population moments is not inevitable. Consider the *equicorrelated* process:

$$y_t = \theta_0 + u_t + v,$$

where the u's are independent drawings from a distribution with $E(u) = 0$ and $V(u) = \sigma^2$, while the random variable v is independent of the u's with $E(v) = 0$ and $V(v) = \tau^2$. Then $\mu = E(y_t) = \theta_0$, $V(y_t) = \sigma^2 + \tau^2$ for all t, and further $C(y_t, y_{t-j}) = \tau^2$ for all t *and* for all $j > 0$. All the off-diagonal elements of the $\mathbf{\Sigma}$ matrix are equal to τ^2. The population autocorrelations are $\rho_j = \tau^2/(\sigma^2 + \tau^2)$ for all $j > 0$. For the sample mean m we calculate $E(m) = (1/n)\mathbf{x}_1'E(\mathbf{y}) = \mu$, and

$$V(m) = (1/n^2)\mathbf{x}_1'\mathbf{\Sigma}\mathbf{x}_1 = \sigma^2/n + (n - 1)\tau^2/n,$$

which does not vanish as n goes to infinity. While unbiased, the sample mean is not consistent. Variants of this model are used in the analysis of panel data: see Greene (1990, chap. 16).

• Nonstationary stochastic processes also arise in economic analysis. The simplest example is the *random walk*:

$$y_t = y_{t-1} + u_t,$$

where the u's are independent drawings from a distribution with $E(u) = 0$ and $V(u) = \sigma^2$. It is clear that $E(y_t|y_{t-1}) = y_{t-1}$, although the variance of the process must increase with t. In this case, differencing the series will produce stationarity.

Exercises

26.1 Show that for any MA(1) process, $|\rho_1| \leq 1/2$. Is that true also for an AR(1) process?

26.2 Consider these two models for an MA(1) process:

$$y_t = \phi_0 + \phi u_{t-1} + u_t,$$

where the u's are independent drawings on a variable u with $E(u) = 0$ and $V(u) = \sigma^2$, and

$$y_t = \phi_0 + \phi^* u^*_{t-1} + u^*_t,$$

where the u^*'s are independent drawings on a variable u^* with $E(u^*) = 0$ and $V(u^*) = \sigma^{*2}$.

(a) Suppose that $\phi^* = 1/\phi$ and $\sigma^{*2} = \phi^2\sigma^2$. Show that the two models produce the same population moments $\mu, \gamma_0, \gamma_1, \gamma_2, \ldots$.
(b) To rule out the ambiguity found in (a), it is convenient to require that $|\phi| \leq 1$. With that in mind, explain the choice of the estimator proposed for ϕ in Section 26.5.

26.3 Suppose that $y_t = \alpha + \beta y_{t-1} + u_t$, where the u_t are independent $\mathcal{N}(0, \sigma^2)$ variables. You know that $\alpha = 10$, $\beta = 3/5$, and $\sigma^2 = 2$. You are told that $y_2 = 50$. Find the best prediction of y_4.

26.4 Suppose that $y_t = 1 + 0.8u_{t-1} + 0.6u_{t-2} + u_t$, where the u's are independent $\mathcal{N}(0, 1)$ variables. For this MA(2) process, find

$$E(y_t|y_{t-1}), \qquad E(y_t|y_{t-1}, y_{t-2}), \qquad E(y_t|y_{t-1}, y_{t-2}, y_{t-3}).$$

26.5 Suppose that $y_t = 1 + 0.4y_{t-1} + 0.3y_{t-2} + u_t$, where the u's are independent $\mathcal{N}(0, 1)$ variables. For this AR(2) process, find

$$E(y_t|y_{t-1}), \qquad E(y_t|y_{t-1}, y_{t-2}), \qquad E(y_t|y_{t-1}, y_{t-2}, y_{t-3}).$$

26.6 Suppose that u_t $(t = 0, \ldots, 100)$ are independent drawings from the $\mathcal{N}(0, 1)$ distribution. Consider these three models for an observed time series y_t $(t = 1, \ldots, 100)$:

Model 1: $y_t = u_t$.
Model 2: $y_t = py_{t-1} + su_t$, where $y_0 = u_0$, $p = 0.9$, $s = \sqrt{(1 - p^2)}$.
Model 3: $y_t = (4/3)\sin(7.2t) + (1/3)u_t$ (angle measured in degrees).

(a) Write a program that generates 101 independent drawings from the $\mathcal{N}(0, 1)$ distribution, and then produces for each model in turn a y_t series.
(b) Complete the program by regressing y_t on $(1, y_{t-1})$ for each of your three y_t series. For each regression, report $\mathbf{X}'\mathbf{X}$, $\mathbf{X}'\mathbf{y}$, $\mathbf{b}$, and the conventional standard errors for $\mathbf{b}$.
(c) Is any of the three slopes surprising? Comment briefly. If you suspect that it is just a coincidence, run your program again to see if it recurs.

GAUSS Hints:

(1) u = rndn(n,1) gives an n × 1 vector of random drawings from the standard normal distribution.

(2) t = seqa(a,d,n) gives an n × 1 vector whose elements are a, a + d, a + 2d, . . . , a + (n − 1) d.

(3) If d is an n × 1 vector of angles measured in degrees, then x = d*pi/180 is the n × 1 vector of the angles measured in radians.

(4) If x is an n × 1 vector of angles measured in radians, then v = sin(x) is the n × 1 vector whose elements are the sines of the elements of x.

(5) If v is an n × 1 vector, while c and d are scalars, then y = recserar(v,c,d) is the n × 1 vector whose elements are $y_1 = c$, $y_2 = d\ y_1 + v_2$, $y_3 = d\ y_2 + v_3$, $y_4 = d\ y_3 + v_4$, etc.

26.7 Table A.5 contains a data set of annual observations for the United States, 1956–1980, as taken from Mirer (1988, pp. 24–25). The variables are:

V1 = Identification number (1, . . . , 25)
V2 = Year − 1900
V3 = GNP price index (100 in 1972)
V4 = Real GNP
V5 = Real gross private domestic investment
V6 = Real personal consumption
V7 = Real disposable personal income
V8 = Change in GNP price index
V9 = Change in consumer price index
V10 = Unemployment rate
V11 = Money stock (M1)
V12 = Treasury bill rate
V13 = Corporate bond rate (Moody's Aaa).

Note: In this data set, V4, V5, V6, V7 are in billions of 1972 dollars; V11 is in billions of current dollars; V8, V9, V12, V13 are in percent per year; V10 is in percent. This data set is presumed to be available as an ASCII file labeled TIM.

Run three linear regressions:

(a) Money autoregression: y_t on 1, y_{t-1}.

(b) Money demand function: y_t on 1, z_t.
(c) Residual autoregression: e_t on 1, e_{t-1}.

Here y = log of real money stock = ln[(V11/V3)100], z = log of real GNP = ln(V4), and e = residual from (b). There will be 25 observations for (b), and 24 observations for (a) and (c). For each regression, report $\mathbf{X}'\mathbf{X}$, $\mathbf{X}'\mathbf{y}$, $\mathbf{b}$, and the conventional standard errors for $\mathbf{b}$.

26.8 In Exercise 26.7 the slope in the residual autoregression (c) turned out to be substantially less than that in the money autoregression (a). Comment on this result in the light of your results in Exercise 26.6.

27 *Generalized Classical Regression*

27.1. Generalized Classical Regression Model

We now generalize the classical regression model to allow the n observations on y to have different variances and to be correlated. Again the data consist of the $n \times 1$ random vector $\mathbf{y}$ and the $n \times k$ nonstochastic matrix $\mathbf{X}$. The *generalized classical regression*, or GCR, *model* consists of these four assumptions:

(27.1) $E(\mathbf{y}) = \mathbf{X}\boldsymbol{\beta}$,

(27.2) $V(\mathbf{y}) = \boldsymbol{\Sigma}$, with $\boldsymbol{\Sigma}$ positive definite,

(27.3) $\mathbf{X}$ nonstochastic,

(27.4) $\text{rank}(\mathbf{X}) = k$.

The only change from the CR model is in the second assumption: the elements of $\mathbf{y}$ now may have different variances and nonzero covariances. (Positive definiteness simply rules out situations where one of the y_i's is an exact linear function of the others.) Special cases of the GCR model arise when specific assumptions are made on those variances and covariances.

27.2. Least Squares Estimation

We begin with LS estimation, which produces coefficients $\mathbf{b} = \mathbf{A}\mathbf{y}$ and residuals $\mathbf{e} = \mathbf{M}\mathbf{y}$, with $\mathbf{A} = \mathbf{Q}^{-1}\mathbf{X}'$, $\mathbf{Q} = \mathbf{X}'\mathbf{X}$, and $\mathbf{M} = \mathbf{I} - \mathbf{X}\mathbf{A}$. By linear function rules,

$$E(\mathbf{b}) = \mathbf{A}E(\mathbf{y}) = \mathbf{AX\beta} = \boldsymbol{\beta},$$

$$V(\mathbf{b}) = \mathbf{A}V(\mathbf{y})\mathbf{A}' = \mathbf{A\Sigma A}' = \mathbf{Q}^{-1}\mathbf{RQ}^{-1},$$

where

$$\mathbf{R} = \mathbf{X}'\boldsymbol{\Sigma}\mathbf{X}.$$

The LS coefficient vector **b** remains unbiased, but its variance matrix is no longer a scalar multiple of $\mathbf{Q}^{-1}$.

Using linear function rules again, we have

$$E(\mathbf{e}) = \mathbf{MX\beta} = \mathbf{0}, \qquad V(\mathbf{e}) = \mathbf{M\Sigma M}'.$$

So the expected sum of squared residuals is

$$E(\mathbf{e}'\mathbf{e}) = \text{tr}[V(\mathbf{e})] = \text{tr}(\mathbf{M\Sigma M}') = \text{tr}(\mathbf{M}'\mathbf{M\Sigma}) = \text{tr}(\mathbf{M\Sigma}),$$

whence the adjusted mean squared residual $\hat{\sigma}^2 = \mathbf{e}'\mathbf{e}/(n - k)$ has expectation

$$E(\hat{\sigma}^2) = \text{tr}(\mathbf{M\Sigma})/(n - k).$$

This expectation involves a mixture of the elements of $\boldsymbol{\Sigma}$, rather than the single parameter σ^2 as in the CR model. (Indeed, in the GCR model there may not be a natural parameter called σ^2.) Observe that

$$\mathbf{M\Sigma} = (\mathbf{I} - \mathbf{XA})\boldsymbol{\Sigma} = \boldsymbol{\Sigma} - \mathbf{XQ}^{-1}\mathbf{X}'\boldsymbol{\Sigma},$$

$$\text{tr}(\mathbf{XQ}^{-1}\mathbf{X}'\boldsymbol{\Sigma}) = \text{tr}(\mathbf{Q}^{-1}\mathbf{X}'\boldsymbol{\Sigma}\mathbf{X}) = \text{tr}(\mathbf{Q}^{-1}\mathbf{R}).$$

So

$$\text{tr}(\mathbf{M\Sigma}) = \text{tr}(\boldsymbol{\Sigma}) - \text{tr}(\mathbf{Q}^{-1}\mathbf{R}),$$

an expression which is convenient for computational purposes.

Proceeding to the usual estimator of $V(\mathbf{b})$, namely $\hat{V}(\mathbf{b}) = \hat{\sigma}^2\mathbf{Q}^{-1}$, we have

$$E[\hat{V}(\mathbf{b})] = [\text{tr}(\mathbf{M\Sigma})/(n - k)]\mathbf{Q}^{-1},$$

which clearly differs from $V(\mathbf{b}) = \mathbf{Q}^{-1}\mathbf{RQ}^{-1}$. The familiar estimator of the variance matrix of **b** is biased, so the conventional standard errors are not correct measures of imprecision, and consequently the confidence region and hypothesis test procedures of Chapters 19–22 will not be valid. For some examples of the bias see Exercises 28.1 and 28.4.

There are several ways to proceed at this point. One might retain the LS coefficient vector $\mathbf{b}$ as the estimator of $\boldsymbol{\beta}$, and seek a correct estimator for its variance matrix, in order to permit valid inferences. Or—and this is the line taken here—one might seek a better estimator of $\boldsymbol{\beta}$. The possibility of a better estimator is open, because the Gauss-Markov Theorem (Section 15.4), which established the MVLUE property of LS, relied on the assumption that $V(\mathbf{y}) = \sigma^2\mathbf{I}$.

27.3. Generalized Least Squares Estimation

For estimation of $\boldsymbol{\beta}$, the key result is

AITKEN'S THEOREM. In the GCR model, with $\boldsymbol{\Sigma}$ known, the MVLUE of $\boldsymbol{\beta}$ is the *generalized least squares* (or GLS) *estimator*, $\mathbf{b}^* = \mathbf{A}^*\mathbf{y}$, where $\mathbf{A}^* = (\mathbf{X}'\boldsymbol{\Sigma}^{-1}\mathbf{X})^{-1}\mathbf{X}'\boldsymbol{\Sigma}^{-1}$.

Proof. Observe that the $k \times n$ matrix $\mathbf{A}^*$ is nonstochastic, has rank k, and satisfies

$$\mathbf{A}^*\mathbf{X} = \mathbf{I}, \qquad \mathbf{A}^*\boldsymbol{\Sigma}\mathbf{A}^{*\prime} = (\mathbf{X}'\boldsymbol{\Sigma}^{-1}\mathbf{X})^{-1}.$$

It follows by linear function rules that

$$E(\mathbf{b}^*) = \boldsymbol{\beta}, \qquad V(\mathbf{b}^*) = (\mathbf{X}'\boldsymbol{\Sigma}^{-1}\mathbf{X})^{-1},$$

so $\mathbf{b}^*$ is a linear unbiased estimator of $\boldsymbol{\beta}$. To show that $\mathbf{b}^*$ has minimum variance in the class of linear unbiased estimators in the GCR model, we first show that the GCR model is equivalent to a CR model in transformed data, and then that $\mathbf{b}^*$ is the LS estimator in that CR model.

Recall a construction used for another purpose in Section 18.4. Because $\boldsymbol{\Sigma}$ is positive definite, we can write $\boldsymbol{\Sigma} = \mathbf{C}\boldsymbol{\Lambda}\mathbf{C}'$, where $\mathbf{C}$ is orthonormal and $\boldsymbol{\Lambda}$ is diagonal with all diagonal elements positive. Let $\boldsymbol{\Lambda}^*$ be the diagonal matrix with the reciprocal square roots of the diagonal elements of $\boldsymbol{\Lambda}$ on its diagonal, and let $\mathbf{H} = \mathbf{C}\boldsymbol{\Lambda}^*\mathbf{C}'$, so $\mathbf{H}'\mathbf{H} = \boldsymbol{\Sigma}^{-1}$ and $\mathbf{H}\boldsymbol{\Sigma}\mathbf{H}' = \mathbf{I}$. Let $\mathbf{y}^* = \mathbf{H}\mathbf{y}$ and $\mathbf{X}^* = \mathbf{H}\mathbf{X}$; these should be viewed as observations on transformed variables. The $n \times n$ matrix $\mathbf{H}$ is nonstochastic and nonsingular, so from Eqs. (27.1)–(27.4) we deduce that:

(27.1*) $E(\mathbf{y}^*) = \mathbf{H}E(\mathbf{y}) = \mathbf{H}\mathbf{X}\boldsymbol{\beta} = \mathbf{X}^*\boldsymbol{\beta}$,

(27.2*) $V(\mathbf{y}^*) = \mathbf{H}V(\mathbf{y})\mathbf{H}' = \mathbf{H}\boldsymbol{\Sigma}\mathbf{H}' = \mathbf{I}$,

(27.3*) $\mathbf{X}^* = \mathbf{HX}$ is nonstochastic,

(27.4*) $\text{rank}(\mathbf{X}^*) = \text{rank}(\mathbf{X}) = k$.

Taken together, these say that a CR model (with $\sigma^2 = 1$) applies to the transformed data $(\mathbf{y}^*, \mathbf{X}^*)$. Because $\mathbf{H}$ is nonsingular, the argument reverses, so the GCR model for the original data $(\mathbf{y}, \mathbf{X})$ is *equivalent* to a CR model for the transformed data $(\mathbf{y}^*, \mathbf{X}^*)$.

The parameter vector $\boldsymbol{\beta}$ is unaffected by the transformation. By the Gauss-Markov Theorem itself, among all linear functions of $\mathbf{y}^*$ that are unbiased for $\boldsymbol{\beta}$, the one with minimum variance is the coefficient vector in LS linear regression of $\mathbf{y}^*$ on $\mathbf{X}^*$, namely

$$\mathbf{c}^* = (\mathbf{X}^{*\prime}\mathbf{X}^*)^{-1}\mathbf{X}^{*\prime}\mathbf{y}^*.$$

But $\mathbf{X}^{*\prime}\mathbf{X}^* = (\mathbf{X}'\mathbf{H}')(\mathbf{HX}) = \mathbf{X}'\boldsymbol{\Sigma}^{-1}\mathbf{X}$, and $\mathbf{X}^{*\prime}\mathbf{y}^* = (\mathbf{X}'\mathbf{H}')(\mathbf{Hy}) = \mathbf{X}'\boldsymbol{\Sigma}^{-1}\mathbf{y}$, so $\mathbf{c}^* = \mathbf{b}^*$. Finally, because $\mathbf{H}$ is nonsingular, the class of linear functions of $\mathbf{y}^*$ is the same as the class of linear functions of $\mathbf{y}$. We conclude that in the GCR model, $\mathbf{b}^*$ is the MVLUE of $\boldsymbol{\beta}$. ■

27.4. Remarks on GLS Estimation

The following remarks may aid the interpretation and implementation of GLS estimation.

• Why is $\mathbf{b}^*$ referred to as the "generalized least squares" estimator? Being the LS coefficient vector for the transformed data, $\mathbf{b}^*$ is the value for $\mathbf{c}$ that minimizes the criterion $\phi^*(\mathbf{c}) = \mathbf{u}^{*\prime}\mathbf{u}^*$, where

$$\mathbf{u}^* = \mathbf{y}^* - \mathbf{X}^*\mathbf{c} = \mathbf{H}(\mathbf{y} - \mathbf{Xc}) = \mathbf{Hu},$$

with $\mathbf{u} = \mathbf{y} - \mathbf{Xc}$. But $\mathbf{H}'\mathbf{H} = \boldsymbol{\Sigma}^{-1}$, so $\phi^*(\mathbf{c}) = \mathbf{u}'\boldsymbol{\Sigma}^{-1}\mathbf{u}$. This criterion is a positive definite quadratic form in $\mathbf{u}$, which is indeed a generalization of the sum of squares, $\mathbf{u}'\mathbf{u}$.

• The device of transforming the data, used to establish the optimality of $\mathbf{b}^*$, also provides a computational routine. With $\boldsymbol{\Sigma}$ known, to do GLS estimation, transform the data and run LS.

• The transformation is not unique. For example, instead of using $\mathbf{H} = \mathbf{C}\boldsymbol{\Lambda}^*\mathbf{C}'$, we can use $\mathbf{H}^* = \boldsymbol{\Lambda}^*\mathbf{C}'$, because

$$\mathbf{H}^{*\prime}\mathbf{H}^* = \mathbf{C}\boldsymbol{\Lambda}^{*\prime}\boldsymbol{\Lambda}^*\mathbf{C}' = \mathbf{C}\boldsymbol{\Lambda}^{-1}\mathbf{C}' = (\mathbf{C}\boldsymbol{\Lambda}\mathbf{C}')^{-1} = \boldsymbol{\Sigma}^{-1},$$

$$\mathbf{H}^*\boldsymbol{\Sigma}\mathbf{H}^{*\prime} = \boldsymbol{\Lambda}^*\mathbf{C}'(\mathbf{C}\boldsymbol{\Lambda}\mathbf{C}')\mathbf{C}\boldsymbol{\Lambda}^* = \boldsymbol{\Lambda}^*\boldsymbol{\Lambda}\boldsymbol{\Lambda}^* = \mathbf{I}.$$

So a CR model (with $\sigma^2 = 1$) will apply to the data $\mathbf{y}^{**} = \mathbf{H}^*\mathbf{y}$, $\mathbf{X}^{**} = \mathbf{H}^*\mathbf{X}$. Despite the nonuniqueness of the transformation, the estimator $\mathbf{b}^*$ will be unique, as is readily confirmed. For any GCR model, there are in fact many nonsingular matrices that will transform it into a CR model. All produce the same $\mathbf{b}^*$, so it is strictly a matter of convenience as to which version of $\mathbf{H}$ should be used in practice.

• To obtain the GLS estimator, it suffices to know $\mathbf{\Sigma}$ up to a scalar multiple. Suppose that $\mathbf{\Sigma} = \sigma^2\mathbf{\Omega}$, where $\mathbf{\Omega}$ is known but the scalar σ^2 is unknown. Since $\mathbf{\Omega}$ is positive definite, we can find an $\mathbf{H}^\circ$ such that $\mathbf{H}^{\circ\prime}\mathbf{H}^\circ = \mathbf{\Omega}^{-1}$ and $\mathbf{H}^\circ\mathbf{\Omega}\mathbf{H}^{\circ\prime} = \mathbf{I}$. Let $\mathbf{y}^\circ = \mathbf{H}^\circ\mathbf{y}$ and $\mathbf{X}^\circ = \mathbf{H}^\circ\mathbf{X}$. Then

$$E(\mathbf{y}^\circ) = \mathbf{H}^\circ E(\mathbf{y}) = \mathbf{H}^\circ\mathbf{X}\boldsymbol{\beta} = \mathbf{X}^\circ\boldsymbol{\beta},$$

$$V(\mathbf{y}^\circ) = \mathbf{H}^\circ V(\mathbf{y})\mathbf{H}^{\circ\prime} = \mathbf{H}^\circ\mathbf{\Sigma}\mathbf{H}^{\circ\prime} = \mathbf{H}^\circ(\sigma^2\mathbf{\Omega})\mathbf{H}^{\circ\prime} = \sigma^2\mathbf{I},$$

and $\mathbf{X}^\circ$ is nonstochastic with full column rank. So a CR model (with σ^2 unknown) applies to $(\mathbf{y}^\circ, \mathbf{X}^\circ)$. The LS estimator on the transformed data is

$$\mathbf{b}^\circ = (\mathbf{X}^{\circ\prime}\mathbf{X}^\circ)^{-1}\mathbf{X}^{\circ\prime}\mathbf{y}^\circ.$$

But $\mathbf{X}^{\circ\prime}\mathbf{X}^\circ = \mathbf{X}'\mathbf{H}^{\circ\prime}\mathbf{H}^\circ\mathbf{X} = \mathbf{X}'\mathbf{\Omega}^{-1}\mathbf{X} = \mathbf{X}'(\mathbf{\Sigma}/\sigma^2)^{-1}\mathbf{X} = \sigma^2\mathbf{X}'\mathbf{\Sigma}^{-1}\mathbf{X}$, and similarly $\mathbf{X}^{\circ\prime}\mathbf{y}^\circ = \sigma^2\mathbf{X}'\mathbf{\Sigma}^{-1}\mathbf{y}$, so $\mathbf{b}^\circ = \mathbf{b}^*$. When such an $\mathbf{H}^\circ$ is used, it will remain to estimate σ^2 and $V(\mathbf{b}^*)$, but for that task all of CR theory applies. So knowledge of $\mathbf{\Sigma}$ up to proportionality suffices to calculate the GLS estimator $\mathbf{b}^*$ and to estimate its variance matrix unbiasedly.

• With respect to goodness of fit: let

$$\phi(\mathbf{c}) = \mathbf{u}'\mathbf{u}, \qquad \phi^*(\mathbf{c}) = \mathbf{u}^{*\prime}\mathbf{u}^* = \mathbf{u}'\mathbf{\Sigma}^{-1}\mathbf{u},$$

where $\mathbf{u} = \mathbf{y} - \mathbf{X}\mathbf{c}$ and $\mathbf{u}^* = \mathbf{y}^* - \mathbf{X}^*\mathbf{c}$. Clearly $\phi(\mathbf{b}) \leq \phi(\mathbf{b}^*)$ (with equality iff $\mathbf{b} = \mathbf{b}^*$) so LS gives a better fit than GLS to the original data. By the same token, $\phi^*(\mathbf{b}^*) \leq \phi^*(\mathbf{b})$ (with equality iff $\mathbf{b} = \mathbf{b}^*$), so GLS gives a better fit to the transformed data.

• It makes no sense to compare $\phi(\mathbf{b})$ with $\phi^*(\mathbf{b}^*)$, as is occasionally done; the two criteria may not even be in the same units. A fortiori, it makes no sense to compare an R^{2*} calculated from the LS regression of $\mathbf{y}^*$ on $\mathbf{X}^*$, with the R^2 from the LS regression of $\mathbf{y}$ on $\mathbf{X}$; the dependent variables $\mathbf{y}$ and $\mathbf{y}^*$ are different, and furthermore the $\mathbf{X}^*$ matrix may not contain a summer vector even when $\mathbf{X}$ does.

27.5. Feasible Generalized Least Squares Estimation

As we have seen, calculation of the GLS estimator $\mathbf{b}^*$ requires knowledge of $V(\mathbf{y}) = \boldsymbol{\Sigma}$ at least up to proportionality:

$$\mathbf{b}^* = \mathbf{A}^*\mathbf{y},$$

where

$$\mathbf{A}^* = (\mathbf{X}'\boldsymbol{\Omega}^{-1}\mathbf{X})^{-1}\mathbf{X}'\boldsymbol{\Omega}^{-1} = (\mathbf{X}'\boldsymbol{\Sigma}^{-1}\mathbf{X})^{-1}\mathbf{X}'\boldsymbol{\Sigma}^{-1},$$

and $\boldsymbol{\Omega}$, a scalar multiple of $\boldsymbol{\Sigma} = V(\mathbf{y})$, is known. In practice $\boldsymbol{\Omega}$ will be unknown, so that GLS estimation will not be feasible. A *feasible generalized least squares*, or FGLS, *estimator* of $\boldsymbol{\beta}$ is defined by

$$\hat{\mathbf{b}}^* = \hat{\mathbf{A}}^*\mathbf{y}, \qquad \text{where } \hat{\mathbf{A}}^* = (\mathbf{X}'\hat{\boldsymbol{\Sigma}}^{-1}\mathbf{X})^{-1}\mathbf{X}'\hat{\boldsymbol{\Sigma}}^{-1},$$

with $\hat{\boldsymbol{\Sigma}}$ being an estimator of $\boldsymbol{\Sigma}$.

The properties of an FGLS coefficient estimator $\hat{\mathbf{b}}^*$ depend on the properties of the variance-matrix estimator $\hat{\boldsymbol{\Sigma}}$. The key result is that if $\hat{\boldsymbol{\Sigma}}$ is a consistent estimator of $\boldsymbol{\Sigma}$, then under general conditions the FGLS estimator $\hat{\mathbf{b}}^*$ has the same asymptotic distribution as the GLS estimator $\mathbf{b}^*$. For discussion of the conditions, see Greene (1990, pp. 388–390) or Judge et al. (1988, pp. 352–356). That is to say, for large n, the distinction between the distributions of $\hat{\mathbf{b}}^*$ and $\mathbf{b}^*$ is negligible.

For some insight into this conclusion, recall two previous situations.

(1) In random sampling from a bivariate population, the sample LP slope (which uses deviations from the sample means) and the ideal sample LP slope (which uses deviations from the population means) have the same asymptotic distribution: see Section 10.5.

(2) In the CNR model, the statistics

$$u_j = (b_j - \beta_j)/\hat{\sigma}_{b_j} \quad \text{and} \quad z_j = (b_j - \beta_j)/\sigma_{b_j}$$

have the same asymptotic distribution, namely $\mathcal{N}(0, 1)$: see Section 22.6. These examples are suggestive, because they show that replacing an unknown parameter (μ_x in the first case; σ^2 in the second) by a consistent estimator may make the statistic feasible to calculate without affecting the asymptotic distribution.

To obtain a consistent estimator of $\boldsymbol{\Sigma}$ for use in FGLS estimation, it is natural to rely on the residual vector $\mathbf{e}$ from LS regression of $\mathbf{y}$ on $\mathbf{X}$. After all, $\boldsymbol{\Sigma} = V(\mathbf{y}) = V(\boldsymbol{\epsilon})$ where $\boldsymbol{\epsilon} = \mathbf{y} - \mathbf{X}\boldsymbol{\beta}$, and $\mathbf{e} = \mathbf{y} - \mathbf{X}\mathbf{b}$. How the residual vector should be used, and whether the quality of the resulting estimator of $\mathbf{\Sigma}$ is adequate to ensure that FGLS has the same asymptotic distribution as GLS, depends on the context of special cases. Those special cases are defined by the "structuring" of $\mathbf{\Sigma}$, where "structuring" means specifying that $\mathbf{\Sigma} = \mathbf{\Sigma}(\boldsymbol{\theta})$, where $\boldsymbol{\theta}$ is an unknown parameter vector with a relatively small number of elements. We will explore two leading special cases in Chapter 28. For the present, observe that unless such knowledge is available, no consistent estimator of $\mathbf{\Sigma}$ will be obtainable. After all, there are $n(n + 1)/2$ distinct elements in $\mathbf{\Sigma}$, and one can hardly be optimistic about estimating so many distinct parameters (along with the k elements of $\boldsymbol{\beta}$) when only n observations are in hand.

27.6. Extensions of the GCR Model

The GCR model may be extended into a *generalized neoclassical regression model*, by allowing $\mathbf{X}$ to be stochastic:

$$E(\mathbf{y}|\mathbf{X}) = \mathbf{X}\boldsymbol{\beta},$$

$$V(\mathbf{y}|\mathbf{X}) = \mathbf{\Sigma}, \text{ with } \Sigma \text{ positive definite},$$

$$\mathbf{X} \text{ random},$$

$$\text{rank}(\mathbf{X}) = k.$$

Again no new theory is required. Just transform into a NeoCR model via an appropriate $\mathbf{H}$ matrix, and apply the theory of Chapter 25.

The GCR may be strengthened into a *generalized classical normal regression model*, by adding a normality assumption:

$$\mathbf{y} \sim \mathcal{N}(\mathbf{X}\boldsymbol{\beta}, \mathbf{\Sigma}),$$

$$\mathbf{\Sigma} \text{ positive definite},$$

$$\mathbf{X} \text{ nonstochastic},$$

$$\text{rank}(\mathbf{X}) = k.$$

No new theory is required here. If $\mathbf{\Sigma}$ is known up to proportionality, then transform the data into a CNR model via an appropriate $\mathbf{H}$ matrix (which will also preserve normality) and apply the statistical inference theory of Chapters 19–22. Among other things, $\mathbf{b}^*$ will be the ML estimator. If $\mathbf{\Sigma}$ is unknown, but structured as $\mathbf{\Sigma}(\boldsymbol{\theta})$ in the sense defined above, then the likelihood function may be maximized with respect to $\boldsymbol{\beta}$ and $\boldsymbol{\theta}$ jointly. This will produce a different estimator of $\boldsymbol{\beta}$ than that obtained by FGLS (which first estimates $\boldsymbol{\theta}$, and then uses the implied estimate of $\boldsymbol{\Sigma} = \boldsymbol{\Sigma}(\boldsymbol{\theta})$ to estimate $\boldsymbol{\beta}$). Under general conditions, this ML estimator will have the same asymptotic distribution as the FGLS and GLS estimators. For discussion, see Amemiya (1985, pp. 190–191, 200–203).

Exercises

27.1 Suppose that the GCR model applies to $E(\mathbf{y}) = \mathbf{X}\boldsymbol{\beta}$, $V(\mathbf{y}) = \boldsymbol{\Sigma}$. Let $\mathbf{b}$, $\mathbf{e}$, and $\hat{\mathbf{y}}$ denote the coefficient, residual, and fitted-value vectors for LS regression of $\mathbf{y}$ on $\mathbf{X}$, and let $\mathbf{b}^*$ denote the GLS coefficient vector. For each of the following statements, determine whether it is true or false:

(a) The covariance matrix of $\mathbf{b}$ and $\mathbf{b}^*$ is equal to the variance matrix of $\mathbf{b}^*$.
(b) If $\mathbf{t}$ is a linear unbiased estimator of $\boldsymbol{\beta}$, then the covariance matrix of $\mathbf{t}$ and $\mathbf{b}^*$ is equal to the variance matrix of $\mathbf{b}^*$.
(c) The covariance of each element of $\hat{\mathbf{y}}$ with the corresponding element of $\mathbf{e}$ may be nonzero, but the sum of those covariances is zero.

27.2 Determine whether the following statement is true or false: Suppose that the CR model applies to $E(\mathbf{y}) = \mathbf{X}\boldsymbol{\beta}$, that $\mathbf{T}$ is a nonstochastic nonsingular matrix, and that $\mathbf{y}^* = \mathbf{T}\mathbf{y}$, $\mathbf{X}^* = \mathbf{T}\mathbf{X}$; then GLS regression of $\mathbf{y}^*$ on $\mathbf{X}^*$ gives the same coefficient estimates as LS regression of $\mathbf{y}$ on $\mathbf{X}$.

27.3 With data drawn from a GCR model, a researcher first ran LS regression using her own LS program to obtain coefficients and standard errors. Then she was given the true $\boldsymbol{\Sigma}$ (up to a scalar multiple). She transformed the data appropriately, and ran LS on the transformed data, using the same program to obtain coefficients and standard errors. For several coefficients, standard errors in the second run were larger than those in the first run. Does this contradict Aitken's Theorem? Explain.

28 Heteroskedasticity and Autocorrelation

28.1. Introduction

We sketch two leading special cases of the GCR model: pure heteroskedasticity, and a nonstationary first-order autoregressive process. For more complete treatments see Greene (1990, chaps. 13, 14, 15), Judge et al. (1988, chaps. 8, 9), or Amemiya (1985, chap. 6).

28.2. Pure Heteroskedasticity

In the *pure heteroskedasticity case*, the y_i's are uncorrelated, but have different variances: the matrix $\mathbf{\Sigma}$ is diagonal, with diagonal elements $\sigma_1^2, \ldots, \sigma_i^2, \ldots, \sigma_n^2$. This case will arise when, in the underlying multivariate population, the conditional variance function $V(y|\mathbf{x})$ is not constant across $\mathbf{x}$. An $n \times n$ matrix $\mathbf{H}$ that makes $\mathbf{H\Sigma H}' = \mathbf{I}$ is the diagonal matrix that has the $1/\sigma_i$'s on its diagonal. If the σ_i's are known, then we can transform the data by dividing all variables at the ith observation by σ_i to get

$$y_i^* = y_i/\sigma_i, \qquad x_{ij}^* = x_{ij}/\sigma_i.$$

The CR model will apply to the new data and LS regression of $\mathbf{y}^*$ on $\mathbf{X}^*$ will produce the GLS estimator $\mathbf{b}^*$.

Scalars proportional to the σ_i may be used instead. Suppose we know that $V(y|\mathbf{x}_i) = \sigma^2\omega_i^2$, where the ω_i^2 are known but σ^2 is unknown. Divide all variables at the ith observation by ω_i, and run LS on the transformed data, to obtain $\mathbf{b}^*$. An often-cited example of this arises when the variances are proportional to the square of one of the explanatory variables, say x_k (where x_k is always positive). Then division by x_{ik}, which makes the transformed variables ratios of the original variables, is the appropriate transformation.

If the σ_i^2 are not known up to proportionality, but some structuring of them is known, then FGLS may be available.

Suppose we know that the σ_i^2 have only two distinct values: ω_1^2 for observations $1, \ldots, n_1$ and ω_2^2 for observations $n_1 + 1, \ldots, n$, but we do not know the values of the two ω^2's. Then this version of FGLS is natural: Regress $\mathbf{y}$ on $\mathbf{X}$ to get the residual vector $\mathbf{e}$. Partition the residual vector $\mathbf{e}$ as $(\mathbf{e}_1', \mathbf{e}_2')'$. Let $w_1^2 = \mathbf{e}_1'\mathbf{e}_1/n_1$, and $w_2^2 = \mathbf{e}_2'\mathbf{e}_2/n_2$. Transform the data by dividing the first n_1 observations through by w_1, and dividing the remaining $n_2 = n - n_1$ observations through by w_2. Run LS on the transformed observations.

Or, suppose we know that $V(y|\mathbf{x}) = g(\mathbf{x}; \boldsymbol{\theta})$ where the function $g(\mathbf{x}; \cdot)$ is known except for the $r \times 1$ parameter vector $\boldsymbol{\theta}$, with r much less than n. One possibility here is that $V(y|\mathbf{x}) = \exp(\boldsymbol{\alpha}'\mathbf{x})$. Because $\log V(y|\mathbf{x}) = \log E(\epsilon^2|\mathbf{x}) = \boldsymbol{\alpha}'\mathbf{x}$, the following application of FGLS seems natural: Run the LS regression of y on the x's, obtaining residuals e_i. Then run the LS regression of the $\log e_i^2$ on the x's to estimate $\boldsymbol{\alpha}$, and calculate $\hat{\sigma}_i^2 = \exp(\hat{\boldsymbol{\alpha}}'\mathbf{x}_i)$. Transform the observations by dividing through by $\hat{\sigma}_i$ and run LS on the transformed data. (Note the informal flavor of this approach; after all, $V(\log y) \neq \log[V(y)]$.)

In each of these cases, the prior structuring of the σ_i^2 serves to reduce the number of unknown parameters. Only then can estimators of $\boldsymbol{\Sigma}$ be obtained that have sufficient reliablity to ensure that $\hat{\mathbf{b}}^*$ shares the asymptotic distribution of $\mathbf{b}^*$.

However, under pure heteroskedasticity, if the objective is merely to obtain valid estimates of the variances of the LS coefficients, then such structuring is not needed. Let $\hat{\sigma}_i^2 = e_i^2$, where the e_i are the residuals from LS regression. Let $\hat{\boldsymbol{\Sigma}}$ be the diagonal matrix with the $\hat{\sigma}_i^2$ on the diagonal, and let $\hat{\mathbf{R}} = \mathbf{X}'\hat{\boldsymbol{\Sigma}}\mathbf{X}$. Then $\hat{V}(\mathbf{b}) = \mathbf{Q}^{-1}\hat{\mathbf{R}}\mathbf{Q}^{-1}$ will under general conditions provide a valid estimator of $V(\mathbf{b})$. A similar procedure was introduced in discussing BLP estimation in Section 25.5.

28.3. First-Order Autoregressive Process

Suppose that $\boldsymbol{\Sigma} = \sigma^2\boldsymbol{\Omega}$, where

$$\mathbf{\Omega} = \begin{pmatrix} 1 & \rho & \rho^2 & \rho^3 & \cdots & \rho^{n-1} \\ \rho & 1 & \rho & \rho^2 & \cdots & \rho^{n-2} \\ \rho^2 & \rho & 1 & \rho & \cdots & \rho^{n-3} \\ \rho^3 & \rho^2 & \rho & 1 & \cdots & \rho^{n-4} \\ \cdot & \cdot & \cdot & \cdot & & \cdot \\ \cdot & \cdot & \cdot & \cdot & & \cdot \\ \cdot & \cdot & \cdot & \cdot & & \cdot \\ \rho^{n-1} & \rho^{n-2} & \rho^{n-3} & \rho^{n-4} & \cdots & 1 \end{pmatrix},$$

with $-1 < \rho < 1$. This says that $C(y_h, y_i) = \sigma^2\rho^{|h-i|}$. With $V(y_i) = \sigma^2\rho^{|i-i|} = \sigma^2$ for all i, the y_i's are homoskedastic. But with $\rho \neq 0$, they are correlated. We refer to this as the *first-order autoregressive*, or AR(1), *case* of the GCR model.

The successive drawings on the y_i are not independent, and indeed this AR(1) specification is intended to apply to time series data. In terms of the development in Chapter 26, we have combined a regression model for $E(\mathbf{y})$ with a time series model for $V(\mathbf{y})$. Referring to Section 26.4, we see that a mechanism that will produce the present specification is given by:

(28.1) $\quad y_i = \mu_i + \epsilon_i,$

(28.2) $\quad \epsilon_i = \rho\epsilon_{i-1} + u_i,$

where the $\mu_i = \mathbf{x}_i'\boldsymbol{\beta}$ are nonstochastic, and the u_i are independent and identically distributed with $E(u) = 0$, $V(u) = \sigma_u^2$. It follows that

$$E(y_i) = \mu_i,$$

$$V(y_i) = V(\epsilon_i) = \sigma_u^2/(1 - \rho^2) = \sigma^2,$$

$$C(y_i, y_{i-1}) = \rho\sigma^2,$$

and so forth. In terms of the discussion in Section 26.2, here a stationary population model applies to the disturbances ϵ_i: the y_i's themselves have different expectations and hence the process generating them is not stationary. Still, the parameter ρ is the population first autocorrelation coefficient of the disturbances ϵ_i, and also of the dependent variable y_i.

If ρ is known, then $\boldsymbol{\Sigma}$ is known up to proportionality, so GLS may be implemented. Define the $n \times n$ matrix

$$\mathbf{H} = \begin{pmatrix} \sqrt{(1-\rho^2)} & 0 & 0 & \cdots & 0 & 0 \\ -\rho & 1 & 0 & \cdots & 0 & 0 \\ 0 & -\rho & 1 & \cdots & 0 & 0 \\ \cdot & \cdot & \cdot & & \cdot & \cdot \\ \cdot & \cdot & \cdot & & \cdot & \cdot \\ \cdot & \cdot & \cdot & & \cdot & \cdot \\ 0 & 0 & 0 & \cdots & -\rho & 1 \end{pmatrix},$$

and observe that $\mathbf{H\Sigma H}' = \sigma^2(1-\rho^2)\mathbf{I}$, which is proportional to $\mathbf{I}$. So transformation of the data by this $\mathbf{H}$ will produce variables on which LS can be run to obtain the GLS estimator $\mathbf{b}^*$.

To clarify the transformation, let $\mathbf{y}^* = \mathbf{Hy}$, $\mathbf{X}^* = \mathbf{HX}$. Then

$$y_1^* = \sqrt{(1-\rho^2)}y_1, \qquad \mathbf{x}_1^* = \sqrt{(1-\rho^2)}\mathbf{x}_1,$$

$$y_i^* = y_i - \rho y_{i-1}, \qquad \mathbf{x}_i^* = \mathbf{x}_i - \rho\mathbf{x}_{i-1} \qquad (i = 2, \ldots, n),$$

with

$$E(y_i^*) = \mathbf{x}_i^{*\prime}\boldsymbol{\beta}, \qquad V(y_i^*) = (1-\rho^2)V(y_i) = \sigma_u^2.$$

Apart from the special treatment of the first observation, this transformation can be rationalized directly by reference to the underlying process in Eqs. (28.1)–(28.2): Lag Eq. (28.1) to get $y_{i-1} = \mu_{i-1} + \epsilon_{i-1}$, multiply that by ρ, and subtract from Eq. (28.1), to get

$$\begin{aligned} (28.3) \qquad y_i - \rho y_{i-1} &= \mu_i - \rho\mu_{i-1} + \epsilon_i - \rho\epsilon_{i-1} \\ &= \mathbf{x}_i'\boldsymbol{\beta} - \rho\mathbf{x}_{i-1}'\boldsymbol{\beta} + u_i \\ &= (\mathbf{x}_i - \rho\mathbf{x}_{i-1})'\boldsymbol{\beta} + u_i. \end{aligned}$$

This says that $y_i^* = \mathbf{x}_i^{*\prime}\boldsymbol{\beta} + u_i$, where the u_i's have zero expectation, constant variance, and are uncorrelated.

If, as in practice, ρ is unknown, then GLS cannot be implemented. But the following application of FGLS is natural and is commonly used. First run the LS regression of $\mathbf{y}$ on $\mathbf{X}$ to get the residuals e_i. Then regress e_i on e_{i-1} (across $i = 2, \ldots, n$) to estimate ρ as

$$\hat{\rho} = \sum_{i=2}^{n} e_i e_{i-1} \Big/ \sum_{i=2}^{n} e_{i-1}^2 .$$

No intercept is required when the sum of the residuals is zero. Transform the data as above, using $\hat{\rho}$ in place of ρ, then run LS on the transformed data. Under general conditions, this FGLS procedure will give estimators with the same asymptotic distribution as GLS. The rationale for the estimate of ρ is straightforward. In this model,

$$\rho = C(y_i, y_{i-1})/V(y_i) = C(\epsilon_i, \epsilon_{i-1})/V(\epsilon_i),$$

where $\epsilon_i = y_i - \mu_i$, with $\mu_i = \mathbf{x}_i'\boldsymbol{\beta}$. But $e_i = y_i - \hat{\mu}_i$ with $\hat{\mu}_i = \mathbf{x}_i'\mathbf{b}$, so the residuals e_i are "predictors" of the corresponding disturbances ϵ_i. As a consequence, the sample moments of the residuals will consistently estimate the population moments of the disturbances.

An alternative approach may be more attractive when ρ is unknown. Rewrite Eq. (28.3) as

$$y_i = \mathbf{x}_i'\boldsymbol{\gamma}_1 + \mathbf{x}_{i-1}'\boldsymbol{\gamma}_2 + y_{i-1}\gamma_3 + u_i = \mathbf{z}_i'\boldsymbol{\gamma} + u_i, \tag{28.4}$$

where

$$\boldsymbol{\gamma}_1 = \boldsymbol{\beta}, \qquad \boldsymbol{\gamma}_2 = -\boldsymbol{\beta}\rho, \qquad \gamma_3 = \rho,$$

$$\mathbf{z}_i = (\mathbf{x}_i', \mathbf{x}_{i-1}', y_{i-1})', \qquad \boldsymbol{\gamma} = (\boldsymbol{\gamma}_1', \boldsymbol{\gamma}_2', \gamma_3)'.$$

Because $E(y_i|\mathbf{z}_i) = \mathbf{z}_i'\boldsymbol{\gamma}$, with the u_i's homoskedastic and uncorrelated, we may fit Eq. (28.4) by LS, using observations $i = 2, \ldots, n$. Neither the CR nor the NeoCR model is applicable, because the lagged value of the dependent variable appears among the explanatory variables. Indeed the LS estimates of Eq. (28.4) are not unbiased, but under general conditions, they are consistent. The argument is similar to that in Section 26.5.

In fitting Eq. (28.4), one may well want to impose the restriction that $\boldsymbol{\gamma}_2 = -\boldsymbol{\gamma}_1\gamma_3$. If so, the required LS algorithm will be a nonlinear one. More on this in Chapter 29. Under general conditions, the resulting estimator will have the same asymptotic distribution as the GLS estimator. Under normality, another alternative is available: the likelihood function may be maximized jointly with respect to $\boldsymbol{\beta}$, σ^2, and ρ.

28.4. Remarks

- In empirical research, when the AR(1) specification is entertained but ρ is unknown, it is customary first to test the null hypothesis $\rho = 0$. If the null is accepted then LS estimates are used; if it is rejected then FGLS estimates are used. This procedure is a variant of the pretest estimation strategy introduced in Section 24.4.

• The traditional statistic for testing $\rho = 0$ is the *Durbin-Watson statistic* d, which is virtually equal to $2(1 - \hat{\rho})$. Tables of critical values are provided in most econometrics textbooks, along with rather complicated instructions. A considerably simpler test procedure, which has an asymptotic justification, treats $\sqrt{n}\hat{\rho}$ as a $\mathcal{N}(0, 1)$ variable on the null hypothesis $\rho = 0$: see Judge et al. (1988, pp. 394–401).

• A significant value of d or of $\hat{\rho}$ should not be read automatically as evidence in favor of the AR(1) specification. After all, many other stationary population models also generate first-order autocorrelation in the residuals. Examination of higher-order residual autocorrelations may suggest a more appropriate specification for $\boldsymbol{\Sigma}$. Furthermore, as seen in Section 26.1, changing expectations can produce a series that appears to be autocorrelated. So omitting an explanatory variable that is itself autocorrelated may well produce autocorrelation in the residuals.

• The situation changes drastically if the lagged value of the dependent variable appears as one of the original explanatory variables, while an AR(1) process is entertained for the disturbances. Then the FGLS approach described above is inappropriate, as are the above tests for $\rho = 0$. The easiest way to see this is to recognize that the original function no longer has a CEF interpretation. For a simple example, suppose that

$$y_i = \alpha + \beta y_{i-1} + \epsilon_i, \tag{28.5}$$

$$\epsilon_i = \rho\epsilon_{i-1} + u_i, \tag{28.6}$$

where the u_i's are independent and identically distributed with expectation zero and variance σ^2, while $|\beta| < 1$ and $|\rho| < 1$. Then $E(y_i|y_{i-1}) \neq \alpha + \beta y_{i-1}$ because $C(y_{i-1}, \epsilon_i) \neq 0$. To proceed, lag Eq. (28.5) to get $y_{i-1} = \alpha + \beta y_{i-2} + \epsilon_{i-1}$, multiply that by ρ, and subtract from Eq. (28.5), to get

$$y_i - \rho y_{i-1} = \alpha(1 - \rho) + \beta y_{i-1} - \rho\beta y_{i-2} + \epsilon_i - \rho\epsilon_{i-1},$$

whence

$$y_i = \alpha(1 - \rho) + (\beta + \rho)y_{i-1} - \rho\beta y_{i-2} + u_i. \tag{28.7}$$

Because u_i is independent of past y's, this says that

$$E(y_i|y_{i-1}, y_{i-2}) = \alpha(1 - \rho) + (\beta + \rho)y_{i-1} - \rho\beta y_{i-2},$$

which is a special case of a stationary AR(2) specification for the observed variable y. Regressing y_i on $(1, y_{i-1}, y_{i-2})$ will give consistent estimates of $\alpha(1 - \rho)$, $(\beta + \rho)$, and $-\rho\beta$. But a first-step regression of y_i on $(1, y_{i-1})$ is a short regression that will not consistently estimate β (nor $\beta + \rho$, for that matter): see Exercise 28.7. With the first step invalid, the remainder of the FGLS procedure will also be invalid. The residuals from the first step will no longer be valid predictors of the ϵ's, and hence the rationale for using them to estimate ρ will vanish.

Exercises

28.1 Suppose that for $i = 1, 2, \ldots, 20$, the random variables y_i are independent with $E(y_i) = \alpha + \beta x_i$, $V(y_i) = \sigma^2 x_i^2$, where $x_i = i$ and $\sigma^2 = 2$. Set this up as a GCR model with $n = 20$, $k = 2$. Let b and b^* denote the LS and GLS estimators of the slope β. Also let $\hat{\sigma}_b^2$ denote the conventional estimator of the variance of b.

(a) Calculate σ_b^2, $\sigma_{b^*}^2$, and $E(\hat{\sigma}_b^2)$.
(b) Comment on the results.

28.2 A researcher believes that the disturbance variance at each observation is proportional to the square of the third explanatory variable, so she divides each observation through by the third explanatory variable before running an LS regression. However, in reality, there was no heteroskedasticity; the CR model was appropriate for the original data. Will her coefficient estimators be unbiased?

28.3 Suppose that $y_1 = \theta + \epsilon_1$, $y_2 = 2\theta + \epsilon_2$, and $y_3 = 3\theta + \epsilon_3$, where the parameter θ is unknown, while ϵ_1, ϵ_2, and ϵ_3 are independent with zero expectations and variances $\sigma_1^2 = 4$, $\sigma_2^2 = 6$, $\sigma_3^2 = 8$. Find the MVLUE of θ.

28.4 Suppose that for $t = 1, 2, \ldots, 20$, the random variables y_t have the AR(1) disturbance pattern, with $E(y_t) = \alpha + \beta x_t$, $V(y_t) = \sigma^2 = 2$, $\rho = 0.8$, and $x_t = t$. Set this up as a GCR model with $n = 20$, $k = 2$. Let b and b^* denote the LS and GLS estimators of the slope β. Also let $\hat{\sigma}_b^2$ denote the conventional estimator of the variance of b.

(a) Calculate σ_b^2, $\sigma_{b^*}^2$, and $E(\hat{\sigma}_b^2)$.
(b) Comment on the results.

GAUSS Hint:
t = seqm(a,d,n) gives an n × 1 vector whose elements are a, ad, ad^n, . . . , ad^{n-1}.

28.5 Suppose that $y_1 = \theta + \epsilon_1$, $y_2 = \theta + \epsilon_2$, $y_3 = \theta + \epsilon_3$, where $\epsilon_1 = u_0 + u_1$, $\epsilon_2 = u_1 + u_2$, $\epsilon_3 = u_2 + u_3$, while u_0, u_1, u_2, u_3 are independent $\mathcal{N}(0, \sigma^2)$ variables. The parameters θ and σ^2 are unknown. You are given one observation on each of the three y's. Determine which of these two estimators of θ is preferable:

$$\bar{y} = (y_1 + y_2 + y_3)/3, \qquad m = (y_1 + y_3)/2.$$

28.6 In the AR(1) case of the GCR model, the parameter ρ is interpretable as the population first autocorrelation coefficient of the y's (as well as of the ϵ's). So it is proposed to take the sample first autocorrelation coefficient of the y's, namely the sample correlation of y_i and y_{i-1}, as an estimate of ρ. Evaluate the proposal.

28.7 Consider the model of Eqs. (28.5)–(28.6), which involves a lagged dependent variable and autocorrelated disturbances. For convenience, take the u's to be normally distributed. Let $E(y_i|y_{i-1}) = \alpha^* + \beta^* y_{i-1}$.

(a) Find α^* and β^* in terms of the parameters α, β, ρ, σ^2.
(b) What parameters would be consistently estimated by a LS linear regression of y_i on $(1, y_{i-1})$?

29 Nonlinear Regression

29.1. Nonlinear CEF's

We have been concerned with linear CEF's, that is, with populations in which $E(y|\mathbf{x}) = \mathbf{x}'\boldsymbol{\beta}$ is linear in the parameter vector $\boldsymbol{\beta}$. Some "nonlinear" CEF's can be cast in that form, as noted in Section 13.3. But inherently nonlinear CEF's also arise, that is, populations in which $E(y|\mathbf{x}) = h(\mathbf{x}, \boldsymbol{\theta})$, with $h(\cdot, \cdot)$ being nonlinear in the unknown parameter vector $\boldsymbol{\theta}$. In such situations, of course, we may run LS linear regression to estimate the BLP $E^*(y|\mathbf{x}) = \mathbf{x}'\boldsymbol{\beta}$, but we now suppose that we are interested in the CEF itself rather than in the BLP.

For random sampling from the multivariate population, we find that nonlinear least squares estimators are consistent, and sketch their asymptotic distribution. (Random sampling is adopted for convenience; the nonstochastic $\mathbf{X}$ case would be treated similarly.) We also discuss instrumental-variable estimation, and maximum-likelihood estimation for a binary response model. All this is an extension to the multivariate case of material developed for the bivariate case in Sections 13.3 and 13.4.

As background, here are some examples of nonlinear regression models.

Cobb-Douglas Production Function. Let y = output, x_1 = labor input, x_2 = capital input, and suppose that

$$E(y|\mathbf{x}) = \theta_0 x_1^{\theta_1} x_2^{\theta_2}.$$

In this CEF, the parameters θ_0, θ_1, θ_2 enter nonlinearly.

Linear Regression with AR(1) Disturbances. In Section 28.3, we found that a linear regression model with AR(1) disturbances implies

$$E(y_i|\mathbf{z}_i) = \mathbf{z}_i'\boldsymbol{\gamma},$$

where

$$\mathbf{z}_i = (\mathbf{x}_i', \mathbf{x}_{i-1}', y_{i-1})', \qquad \boldsymbol{\gamma} = (\boldsymbol{\gamma}_1', \boldsymbol{\gamma}_2', \gamma_3)',$$

$$\boldsymbol{\gamma}_1 = \boldsymbol{\beta}, \qquad \boldsymbol{\gamma}_2 = -\boldsymbol{\beta}\rho, \qquad \gamma_3 = \rho.$$

In this CEF, the linear regression parameters $\boldsymbol{\gamma}_1$, $\boldsymbol{\gamma}_2$, γ_3 are subject to the nonlinear constraint $\boldsymbol{\gamma}_2 = -\boldsymbol{\gamma}_1\gamma_3$. Equivalently, the CEF is nonlinear in the underlying parameters $\boldsymbol{\beta}$, ρ.

Binary Response. If y is a binary variable taking on only the values 0 and 1, then it is implausible that the CEF be linear in the explanatory variables. A linear function is unbounded, while $E(y|\mathbf{x}) = \Pr(y = 1|\mathbf{x})$ is inherently bounded between 0 and 1. A plausible form for the CEF is $E(y|\mathbf{x}) = F(\mathbf{x}'\boldsymbol{\theta})$, where $F(\cdot)$ is the cdf of some continuous distribution. This CEF is nonlinear in the parameter vector $\boldsymbol{\theta}$.

In the *probit model*, $F(\cdot)$ is taken to be the standard normal cdf. Here is a simple scheme that supports the probit model. Suppose that y^*, the unobserved propensity to own a car, is determined by a normal linear model:

$$y^* = \mathbf{x}'\boldsymbol{\theta} - \epsilon,$$

with $\epsilon \sim \mathcal{N}(0, 1)$ independently of $\mathbf{x}$. (Writing $-\epsilon$ rather than $+\epsilon$ is purely a matter of convenience.) Suppose further that y, the observed binary variable that indicates actual ownership, is determined as

$$y = \begin{cases} 1 & \text{if } y^* \geq 0, \\ 0 & \text{if } y^* < 0. \end{cases}$$

Let A be the event that $y = 1$. Now

$$A = \{y = 1\} = \{y^* \geq 0\} = \{\mathbf{x}'\boldsymbol{\theta} - \epsilon \geq 0\} = \{\epsilon \leq \mathbf{x}'\boldsymbol{\theta}\}.$$

With $\epsilon \sim \mathcal{N}(0, 1)$ independently of $\mathbf{x}$, it follows that

$$E(y|\mathbf{x}) = \Pr(A|\mathbf{x}) = F(\mathbf{x}'\boldsymbol{\theta}),$$

where $F(\cdot)$ is the standard normal cdf.

In the probit model, as in many nonlinear regression models, the parameter vector $\boldsymbol{\theta}$ does not directly give the "effects" of explanatory variables on the conditional expectation of the dependent variable. Because $E(y|\mathbf{x}) = F(\mathbf{x}'\boldsymbol{\theta})$, we have $\partial E(y|\mathbf{x})/\partial x_j = f(\mathbf{x}'\boldsymbol{\theta})\theta_j$, where $f(\cdot)$ is the standard normal pdf. These derivatives vary with $\mathbf{x}$: at any value of $\mathbf{x}$, they are proportional to the coefficients θ_j.

A popular alternative to the probit model arises if the ϵ above has the standard logistic, rather than the standard normal, distribution. The standard logistic cdf is (see Section 2.3)

$$G(\epsilon) = \exp(\epsilon)/[1 + \exp(\epsilon)].$$

So in the *logistic model* for binary response, the CEF is $E(y|\mathbf{x}) = G(\mathbf{x}'\boldsymbol{\theta})$, which differs somewhat from that in the probit model, but is again nonlinear in $\boldsymbol{\theta}$.

Censored Dependent Variable. As the previous example suggests, nonlinear CEF's may arise when an underlying dependent variable has a linear model, but is not fully observable. Let y = dollars spent on purchase of a new car. So $y \geq 0$, and many families will have $y = 0$, features that would be incompatible with a normal linear model for y.

A simple scheme that may be appropriate is the *Tobit model.* Suppose that y^*, the unobserved propensity to spend on a new car, is determined by a normal linear model:

$$y^* = \mathbf{x}'\boldsymbol{\theta} - \sigma\epsilon,$$

where $\epsilon \sim \mathcal{N}(0, 1)$ independently of $\mathbf{x}$. Suppose further that y, the observed continuous variable that measures actual expenditure on a new car, is determined as

$$y = \begin{cases} y^* & \text{if } y^* \geq 0, \\ 0 & \text{if } y^* < 0. \end{cases}$$

Then

$$E(y|\mathbf{x}) = F(\mathbf{x}'\boldsymbol{\theta}/\sigma)\mathbf{x}'\boldsymbol{\theta} + \sigma f(\mathbf{x}'\boldsymbol{\theta}/\sigma),$$

where $f(\cdot)$ and $F(\cdot)$ are the $\mathcal{N}(0, 1)$ pdf and cdf. This CEF is nonlinear in the parameters $\boldsymbol{\theta}$, σ.

Proof. Let A be the event that $y^* \geq 0$. Now

$$A = \{y^* \geq 0\} = \{\mathbf{x}'\boldsymbol{\theta} - \sigma\epsilon \geq 0\} = \{\epsilon \leq \mathbf{x}'\boldsymbol{\theta}/\sigma\} = \{\epsilon \leq \tau\},$$

say, where $\tau \equiv \mathbf{x}'\boldsymbol{\theta}/\sigma$. With $\epsilon \sim \mathcal{N}(0, 1)$ independently of $\mathbf{x}$, it follows that $\Pr(A|\mathbf{x}) = F(\tau)$, and that

$$E(\epsilon|\mathbf{x}, A) = E(\epsilon|\epsilon \leq \tau) = \int_{-\infty}^{\tau} tf(t)\, dt \Big/ \int_{-\infty}^{\tau} f(t)\, dt.$$

Now, for the standard normal density,

$$f(t) = (2\pi)^{-1/2} \exp(-t^2/2) \quad \Rightarrow \quad f'(t) \equiv \partial f(t)/\partial t = -tf(t),$$

and of course $\int f'(t)\, dt = f(t)$. So $E(\epsilon|\mathbf{x}, A) = -f(\tau)/F(\tau)$. Further, if A does occur, then $y = \mathbf{x}'\boldsymbol{\theta} - \sigma\epsilon$, so

$$E(y|\mathbf{x}, A) = \mathbf{x}'\boldsymbol{\theta} - \sigma E(\epsilon|\mathbf{x}, A) = \sigma\tau - \sigma[-f(\tau)/F(\tau)]$$

$$= \sigma[\tau + f(\tau)/F(\tau)].$$

By the Law of Iterated Expectations (T8, Section 5.2),

$$E(y|\mathbf{x}) = \Pr(A|\mathbf{x})E(y|\mathbf{x}, A) + \Pr(\text{not } A|\mathbf{x})E(y|\mathbf{x}, \text{not } A)$$

$$= \Pr(A|\mathbf{x})E(y|\mathbf{x}, A),$$

using the fact that y is identically zero if A does not occur. So

$$E(y|\mathbf{x}) = F(\tau)\sigma[\tau + f(\tau)/F(\tau)] = F(\tau)\sigma\tau + \sigma f(\tau)$$

$$= F(\mathbf{x}'\boldsymbol{\theta}/\sigma)\mathbf{x}'\boldsymbol{\theta} + \sigma f(\mathbf{x}'\boldsymbol{\theta}/\sigma). \quad \blacksquare$$

29.2. Estimation

With that as background, we turn to the general case of nonlinear regression. Consider a multivariate population in which

$$E(y|\mathbf{x}) = h(\mathbf{x}, \boldsymbol{\theta}),$$

the function $h(\cdot, \cdot)$ being known apart from the $k \times 1$ parameter vector $\boldsymbol{\theta}$. (Caution: The number of explanatory variables may differ from the number of parameters, as in the AR(1) and Tobit examples above.)

For estimating this nonlinear CEF, we draw on the analogy principle, as we did for the bivariate case in Chapter 13. In the population, the CEF is the best predictor of y given $\mathbf{x}$. In particular, it is the best predictor in the class $h(\mathbf{x}, \mathbf{c})$, where $\mathbf{c}$ denotes a $k \times 1$ vector. Thus, $\boldsymbol{\theta}$ is the value for $\mathbf{c}$ that minimizes $E(u^2)$ in the population, where $u = y - h(\mathbf{x}, \mathbf{c})$. So, given a random sample of observations $(y_i, \mathbf{x}_i')$ $(i = 1, \ldots, n)$, let us choose, as our estimate of $\boldsymbol{\theta}$, the value of $\mathbf{c}$ that has the corresponding property in the sample, namely minimizing the sample mean squared residual, or equivalently the sample sum of squared residuals.

Proceeding, let $\phi = \phi(\mathbf{c}) = \Sigma_{i=1}^{n} u_i^2$, where $u_i = y_i - h(\mathbf{x}_i, \mathbf{c})$, and choose $\mathbf{c} = (c_1, \ldots, c_k)'$ to minimize ϕ. The derivatives are:

$$\partial\phi/\partial c_j = \sum_i (\partial u_i^2/\partial c_j) = \sum_i 2u_i(\partial u_i/\partial c_j) = 2 \sum_i u_i(-z_{ij}) = -2 \sum_i z_{ij}u_i,$$

where

$$z_{ij} = \partial h(\mathbf{x}_i, \mathbf{c})/\partial c_j.$$

The first-order conditions (FOC's) for a minimum are

$$\sum_i z_{ij}u_i = 0 \qquad (j = 1, \ldots, k),$$

or in matrix form,

$$\mathbf{Z}'\mathbf{u} = \mathbf{0},$$

where $\mathbf{Z} = \{z_{ij}\}$ is the $n \times k$ matrix of derivatives of the regression function h with respect to the c's, and $\mathbf{u}$ is the $n \times 1$ vector of deviations of y from h. This is a system of k nonlinear equations in $c_1, \ldots, c_k$: the c's enter both the u's and the z's nonlinearly. Let $\mathbf{t}$ denote the solution value for $\mathbf{c}$; that is, $\hat{\mathbf{Z}}'\hat{\mathbf{u}} = \mathbf{0}$, where $\hat{\mathbf{Z}}$ and $\hat{\mathbf{u}}$ denote $\mathbf{Z}$ and $\mathbf{u}$ evaluated at $\mathbf{c} = \mathbf{t}$. Provided that this locates the global minimum, we refer to $\mathbf{t}$ as the *nonlinear least squares*, or NLLS, *estimator* of $\boldsymbol{\theta}$.

The FOC's of NLLS, namely $\mathbf{Z}'\mathbf{u} = \mathbf{0}$, have a striking resemblance to the FOC's of linear LS, namely $\mathbf{X}'\mathbf{u} = \mathbf{0}$. Indeed, if $h(\mathbf{x}, \boldsymbol{\theta})$ were linear in $\boldsymbol{\theta}$, that is, if $h(\mathbf{x}, \boldsymbol{\theta}) = \mathbf{x}'\boldsymbol{\theta}$, then $z_{ij} = x_{ij}$ and $\mathbf{Z} = \mathbf{X}$. In that event, $\mathbf{Z}$ would not involve $\mathbf{c}$, and $\mathbf{u} = \mathbf{y} - \mathbf{Xc}$ would be linear in $\mathbf{c}$, so the FOC's would be linear in $\mathbf{c}$, and would be solved analytically to get the familiar $\mathbf{c} = (\mathbf{X}'\mathbf{X})^{-1}\mathbf{X}'\mathbf{y}$.

As in Section 13.3, the analogy principle also offers an alternative to NLLS estimation, namely *instrumental variable*, or IV, *estimation*. In the population, the deviations from the CEF have zero expected cross-product with each conditioning variable. That is, let $u = y - h(\mathbf{x}, \mathbf{c})$; then $\boldsymbol{\theta}$ is the value for $\mathbf{c}$ that makes $E(x_j u) = 0$ for every j. This suggests that we choose, as our estimate of $\boldsymbol{\theta}$, the value of $\mathbf{c}$ that has the corresponding properties in the sample, namely satisfying the equations

$$\sum_i x_{ij}u_i = 0.$$

Provided that the number of conditioning variables is the same as the number of parameters, this is a system of k nonlinear equations in c_1, $\ldots$, c_k: the c's enter the u's nonlinearly. If the number of conditioning variables is less than k, then we may use other functions of the x_j, which also have zero expected cross-product with u in the population, to complete the set of instrumental variables. If the number of conditioning variables exceeds k, then we may use a subset of them, or seek to combine them optimally, as instrumental variables: for discussion and references, see Manski (1988, chap. 6).

Observe that the FOC's of NLLS can be interpreted as choosing $\mathbf{c}$ to make the sample summed cross-products of $z_j = \partial h(\mathbf{x}, \mathbf{c})/\partial c_j$ with u equal to zero. So NLLS has an IV interpretation: z_j is a function of $\mathbf{x}$ (not of y), and we know that in the population, the expected cross-product of every function of $\mathbf{x}$ with $y - h(\mathbf{x}, \boldsymbol{\theta})$ is zero.

29.3. Computation of the Nonlinear Least Squares Estimator

The nonlinearity of the FOC's has implications for computing the solution and also for the properties of the estimator. Because of the nonlinearities, the FOC's are solved numerically rather than analytically.

Here we sketch an NLLS algorithm for the case in which there is a single parameter and a single explanatory variable—for example, $h(x, \theta) = x^\theta$. Let

$$h = h(x, c), \qquad z = \partial h/\partial c = z(x, c), \qquad u = y - h = u(y, x, c).$$

We seek the value of c that makes $\mathbf{z}'\mathbf{u} = 0$. Let c° be an initial guessed value for c and define

$$h^\circ = h(x, c^\circ), \qquad z^\circ = \partial h/\partial c^\circ = z(x, c^\circ), \qquad u^\circ = y - h^\circ = u(y, x, c^\circ).$$

The linear approximation to h at the point c° is

$$h = h^\circ + z^\circ(c - c^\circ),$$

so to that order of approximation,

$$u = y - h = y - [h^\circ + z^\circ(c - c^\circ)] = u^\circ - z^\circ(c - c^\circ).$$

Applied to all n observations, this says

$$\mathbf{u} = \mathbf{u}^\circ - \mathbf{z}^\circ(c - c^\circ).$$

So to the same order of approximation,

$$\phi(c) = \mathbf{u}'\mathbf{u} = \mathbf{u}^{\circ\prime}\mathbf{u}^\circ + (c - c^\circ)^2\mathbf{z}^{\circ\prime}\mathbf{z}^\circ - 2(c - c^\circ)\mathbf{z}^{\circ\prime}\mathbf{u}^\circ,$$

$$\phi'(c) = \partial\phi(c)/\partial c = 2(c - c^\circ)\mathbf{z}^{\circ\prime}\mathbf{z}^\circ - 2\mathbf{z}^{\circ\prime}\mathbf{u}^\circ.$$

Set $\phi'(c) = 0$ and solve for

$$c - c^\circ = \mathbf{z}^{\circ\prime}\mathbf{u}^\circ/\mathbf{z}^{\circ\prime}\mathbf{z}^\circ.$$

Take the resulting c as the new c° and restart the calculation. Continue until convergence, that is until $c - c^\circ \cong 0$, where "$\cong 0$" indicates satisfaction of a convergence criterion such as being less than 0.0001 in absolute value. At that point, $\mathbf{z}'\mathbf{u} \cong 0$, as desired.

A few remarks:

• The derivative $z^\circ = \partial h/\partial c^\circ$ may be approximated numerically as

$$[h(x, c^\circ + p) - h(x, c^\circ - p)]/(2p),$$

where p is a (small) step.

• The expression $\mathbf{z}^{\circ\prime}\mathbf{u}^\circ/\mathbf{z}^{\circ\prime}\mathbf{z}^\circ$ may be interpreted as the coefficient in LS linear regression of $\mathbf{u}^\circ$ on $\mathbf{z}^\circ$.

• The algorithm generalizes in a fairly obvious way to the multiparameter case: see Judge et al. (1988, pp. 501–510) or Greene (1990, chap. 12), and see also Exercise 29.3.

29.4. Asymptotic Properties

Because the NLLS estimator $\mathbf{t}$ is a nonlinear function of $\mathbf{y}$, its sampling properties are not readily obtainable as they would be in the linear case. Indeed exact results are not available, but asymptotic theory is available for random sampling from a multivariate population. For convenience, we sketch the one-parameter case, and proceed quite informally.

For the population, define the random variables

$$u = y - h(x, \theta), \qquad z = \partial h(x, \theta)/\partial\theta, \qquad s = zu.$$

Let

$$w = -\partial s/\partial\theta = -[z(\partial u/\partial\theta) + u(\partial z/\partial\theta)] = z^2 - u(\partial z/\partial\theta),$$

using $\partial u/\partial\theta = -\partial h(x, \theta)/\partial\theta = -z$. We have

$$E(u) = 0, \qquad E(zu) = 0, \qquad E[(\partial z/\partial\theta)u] = 0,$$

because u is the deviation from the CEF while z and $\partial z/\partial\theta$ are functions of x, not of y. So

$$E(s) = 0, \qquad V(s) = E(s^2) = E(z^2u^2), \qquad E(w) = E(z^2).$$

Correspondingly, for the sample, define the variables

$$\hat{u}_i = y_i - h(x_i, t), \qquad \hat{z}_i = \partial h(x_i, t)/\partial t, \qquad \hat{s}_i = \hat{z}_i\hat{u}_i,$$

where t denotes the solution value that makes $\Sigma_i\hat{s}_i = 0$. A linear approximation at the point θ gives

$$\hat{s}_i \cong s_i + (\partial s_i/\partial\theta)(t - \theta) = s_i - w_i(t - \theta),$$

so

$$\sum_i \hat{s}_i = 0 \cong \sum_i s_i - \sum_i w_i(t - \theta).$$

Neglecting the approximation error, we have

$$(t - \theta) = \sum_i s_i \Big/ \sum_i w_i = \bar{s}/\bar{w},$$

say, whence

$$\sqrt{n}(t - \theta) = \sqrt{n}(\bar{s}/\bar{w}).$$

Once again, as in Section 12.3, we see a complicated sample statistic exhibited as a ratio of sample means. Here $\bar{s} = (1/n)\ \Sigma_i s_i$ is the sample mean in random sampling on the random variable s. Because $E(s) = 0$, the CLT implies

$$\sqrt{n}\bar{s} \xrightarrow{D} \mathcal{N}[0, V(s)].$$

Further, $\bar{w} = (1/n)\Sigma_i w_i$ is the sample mean in random sampling on the variable w, so $\bar{w} \xrightarrow{P} E(w)$ by the LLN. Then the Slutsky Theorem S4 (Section 9.5) implies

$$\sqrt{n}(t - \theta) \xrightarrow{D} \mathcal{N}(0, \phi^2),$$

with

$$\phi^2 = V(s)/E^2(w) = E(z^2u^2)/E^2(z^2).$$

(Caution: Do not confuse this ϕ with the $\phi = \phi(\mathbf{c})$ of Section 29.2.) Equivalently, we say that

$$t \overset{A}{\sim} \mathcal{N}(\theta, \phi^2/n).$$

We see that the NLLS estimator is consistent, although not unbiased. In fact, no unbiased estimator of θ exists in nonlinear regression models.

In practice, ϕ^2 will be unknown. The natural estimator is

$$(29.1) \quad \hat{\phi}^2 = \left(\sum_i \hat{z}_i^2 \hat{u}_i^2/n\right) \Big/ \left(\sum_i \hat{z}_i^2/n\right)^2.$$

We can construct confidence intervals and test hypotheses on θ in the usual manner, relying on asymptotic normality. Thus, $t \pm 1.96\hat{\sigma}/\sqrt{n}$ will provide an approximate 95% confidence interval for θ.

For the multiparameter case, the results generalize directly. Let $\mathbf{z} = \{\partial h(\mathbf{x}, \boldsymbol{\theta})/\partial\theta_j\}$ be the $k \times 1$ vector of derivatives of the CEF with respect to its parameters. Then

$$\sqrt{n}(\mathbf{t} - \boldsymbol{\theta}) \overset{D}{\rightarrow} \mathcal{N}(\mathbf{0}, \boldsymbol{\Phi}),$$

with

$$\boldsymbol{\Phi} = [E(\mathbf{zz}')]^{-1}E(\mathbf{zz}'u^2)[E(\mathbf{zz}')]^{-1}.$$

Here the matrix $\Sigma_i\hat{\mathbf{z}}_i\hat{\mathbf{z}}_i'/n$ will serve as the estimator for $E(\mathbf{zz}')$, while the matrix $\Sigma_i\hat{\mathbf{z}}_i\hat{\mathbf{z}}_i'\hat{u}_i^2/n$ will serve as the estimator for $E(\mathbf{zz}'u^2)$.

A few remarks to conclude this discussion of NLLS estimation:

• There is a formal resemblance between the expression for $\boldsymbol{\Phi}$ and the variance matrix of the linear LS estimator in the GCR model (Section 27.2), namely $V(\mathbf{b}) = (\mathbf{X}'\mathbf{X})^{-1}\mathbf{X}'\boldsymbol{\Sigma}\mathbf{X}(\mathbf{X}'\mathbf{X})^{-1}$, and also the formula for the asymptotic variance matrix of the LS estimator of a BLP (Section 25.5).

• Suppose that the population conditional variance function is constant: $V(y|\mathbf{x}) = E(u^2|\mathbf{x}) = \sigma^2$, say. Then $E(\mathbf{zz}'u^2) = \sigma^2E(\mathbf{zz}')$, and $\boldsymbol{\Phi}$ simplifies to $\boldsymbol{\Phi} = \sigma^2[E(\mathbf{zz}')]^{-1}$. There is a formal resemblance between this expression for $\boldsymbol{\Phi}$ and the variance matrix for the linear LS estimator in the CR model, namely $V(\mathbf{b}) = \sigma^2(\mathbf{X}'\mathbf{X})^{-1}$.

In the present homoskedastic case, the natural estimator for $\boldsymbol{\Phi}$ will be

$$\hat{\boldsymbol{\Phi}} = \hat{\sigma}^2\left(\sum_i \hat{\mathbf{z}}_i\hat{\mathbf{z}}_i'/n\right)^{-1},$$

with $\hat{\sigma}^2 = \Sigma_i\hat{u}_i^2/n$. Standard computer programs for NLLS are likely to incorporate this estimator rather than the more general form.

• Suppose that in the population, $y|\mathbf{x} \sim \mathcal{N}[h(\mathbf{x}, \boldsymbol{\theta}), \sigma^2]$. Then, provided that the distribution of $\mathbf{x}$ does not involve $\boldsymbol{\theta}$, the NLLS estimator of $\boldsymbol{\theta}$ is also the ML estimator of $\boldsymbol{\theta}$. The argument here is the same as that for the linear CNR model (Section 19.2).

29.5. Probit Model

For a specific example of nonlinear CEF's, return to the probit model. Here y is a binary variable (equal to 1 or 0), and the CEF is

$$E(y|\mathbf{x}) = \Pr(y = 1|\mathbf{x}) = F(\mathbf{x}'\boldsymbol{\theta}),$$

with $F(\cdot)$ denoting the $\mathcal{N}(0, 1)$ cdf. We can estimate $\boldsymbol{\theta}$ consistently by NLLS. But that is not optimal, because heteroskedasticity is present: the variance of a binary variable depends on its expectation. A version of FGLS might be used, but instead we consider ZES-rule (or ML) estimation, which is operational because the form of the conditional pmf of the binary variable y has been automatically specified.

That conditional pmf of $y|\mathbf{x}$ is

$$[F(\mathbf{x}'\boldsymbol{\theta})]^y[1 - F(\mathbf{x}'\boldsymbol{\theta})]^{1-y},$$

whose logarithm is

$$L = y \log F(\mathbf{x}'\boldsymbol{\theta}) + (1 - y) \log[1 - F(\mathbf{x}'\boldsymbol{\theta})].$$

The scores (derivatives of L with respect to the parameters) are

$$\begin{aligned} s_j = \partial L/\partial\theta_j &= (y/F)fx_j - [(1 - y)/(1 - F)]fx_j \\ &= fx_j(y - F)/[F(1 - F)] \\ &= z_j u, \end{aligned}$$

say, where

$$z_j = fx_j/[F(1 - F)], \qquad u = y - F, \qquad f = f(\mathbf{x}'\boldsymbol{\theta}), \qquad F = F(\mathbf{x}'\boldsymbol{\theta}),$$

with $f(\cdot)$ denoting the $\mathcal{N}(0, 1)$ pdf.

The general rule that score variables have expectation zero is easily verified here: because $u = y - E(y|\mathbf{x})$ and z_j is a function of $\mathbf{x}$ (not of y), it follows that $E(s_j) = E(z_j u) = 0$ $(j = 1, \ldots, k)$. The ZES-rule estimators are the values that make the sample counterparts of the expected scores equate to zero. For a sample of n observations $(y_i, \mathbf{x}_i')$ $(i = 1, \ldots, n)$, let

$$s_{ij} = z_{ij}u_i,$$

where

$$z_{ij} = f_i x_{ij}/[F_i(1 - F_i)], \qquad u_i = y_i - F_i,$$

with

$$f_i = f(\mathbf{x}_i'\mathbf{c}), \qquad F_i = F(\mathbf{x}_i'\mathbf{c}).$$

Choose $\mathbf{c}$ to make $\Sigma_i s_{ij} = 0$, that is, to make

$$\sum_i z_{ij}u_i = 0 \qquad (j = 1, \ldots, k).$$

In matrix form this is

$$\mathbf{Z}'\mathbf{u} = \mathbf{0},$$

where $\mathbf{Z} = \{z_{ij}\}$ is $n \times k$. This is a system of k nonlinear equations in $c_1, \ldots, c_k$: the c's enter both the u's and the z's nonlinearly. Various computer programs are available for numerical solution: see Judge et al. (1988, pp. 786–795), and also see Exercise 29.3. Provided that a global maximum of the log-likelihood is located, the ZES-rule estimator, say $\mathbf{t}$, is the ML estimator of $\boldsymbol{\theta}$.

For statistical inference, we may rely on the general theory for multiparameter maximum likelihood estimation, namely

$$\sqrt{n}(\mathbf{t} - \boldsymbol{\theta}) \xrightarrow{\text{D}} \mathcal{N}(\mathbf{0}, \boldsymbol{\Phi}).$$

Here $\boldsymbol{\Phi} = [V(\mathbf{s})]^{-1} = [E(\mathbf{W})]^{-1}$, with $\mathbf{s} = \{s_j\} = \{\partial L/\partial\theta_j\}$ being the population score vector, and $\mathbf{W} = \{w_{jh}\} = -\{\partial s_j/\partial\theta_h\} = -\{\partial^2 L/(\partial\theta_j\partial\theta_h)\}$ being the population second derivative, or Hessian, matrix. The general rule that ML estimators are BAN can be verified here by showing that the $\boldsymbol{\Phi}$ for ML is less than that for NLLS. The required calculation parallels that which shows that GLS is preferable to LS in a linear GCR model.

In practice, the $\boldsymbol{\Phi}$ of ML may be estimated as the sample second moment matrix of the estimated score vectors, $\Sigma_i \hat{\mathbf{s}}_i \hat{\mathbf{s}}_i'/n$, where the hats denote evaluation at $\mathbf{c} = \mathbf{t}$. Some computer programs (including that in Exercise 29.3) will instead evaluate the sample mean second derivatives at $\mathbf{t}$. For discussion, see Greene (1990, pp. 677–678).

Maximum likelihood estimation is also straightforward for the logistic and Tobit models introduced in Section 29.1: see Maddala (1983, pp. 22–27, 151–158).

Exercises

29.1 Suppose that the logistic, rather than probit, model applies. So $E(y|\mathbf{x}) = G(\mathbf{x}'\boldsymbol{\theta})$, with $G(a) = \exp(a)/[1 + \exp(a)]$. Show that the ZES-rule estimator $\mathbf{c}$ satisfies $\mathbf{X}'\mathbf{u} = \mathbf{0}$, where $\mathbf{u} = \{u_i\}$ with $u_i = y_i - G(\mathbf{x}_i'\mathbf{c})$.

29.2 The Tobit model specifies that $y^* \sim \mathcal{N}(\mathbf{x}'\boldsymbol{\theta}, \sigma^2)$, while the probit model specifies that $y^* \sim \mathcal{N}(\mathbf{x}'\boldsymbol{\theta}, 1)$. Why is the variance set at 1 in the probit model?

29.3 For the SCF data set, let

$$y = \begin{cases} 1 \text{ if earnings} < 9 \text{ (thousand dollars)}, \\ 0 \text{ otherwise}, \end{cases}$$

and let the x variables be defined as in Exercise 17.4. Model A for this quite artificial binary response example is

$$E(y|\mathbf{x}) = F(\mathbf{x}'\boldsymbol{\theta}),$$

where $\mathbf{x}$ consists of $x_1, x_2, x_3, x_6, x_7, x_8$ (that is, the constant, education, experience, and the three regional dummies), and $F(\cdot)$ is the standard normal cdf. Model B is a shortened version, with only x_1, x_2, x_3 included as explanatory variables. Estimate both models by maximum likelihood.

GAUSS Hint:
ASG2903 is a simple program for doing maximum likelihood estimation of the probit model. The "Model-specific Section" of the program is specialized to deal with the present probit Model A. When you have a different model (probit Model B here, say, or in the future), that section has to be changed.

```
/* ASG2903 */
                    /* ----- MODEL-SPECIFIC SECTION ----- */
new; output file = asg2903.out reset;
loadp datestr,timestr; datestr "        " timestr;?;
"ASG2903          EXERCISE 29.3          Name .............. ";
"SCF dataset: 100 family heads from 1963 Survey of Consumer Finances";?;
```

```
"MAXIMUM LIKELIHOOD ESTIMATION";
"Probit Model: E(y|x) = F(x'c), F(.) = N(0,1) cdf.
y = 1 if Earnings > $ 9000; x = 1, ed, exp, 3 regional dummies.";?;
  n = 100; load D[n,12] = scf;
  ys = D[.,9]; y = ys .> 9;
  x1 = ones(n,1); x2 = D[.,3]; x3 = D[.,5];
  x6 = D[.,8] .== 2*x1; x7 = D[.,8] .== 3*x1; x8 = D[.,8] .== 4*x1;
  X = x1~x2~x3~x6~x7~x8;
  let c0 = -5  0.3  0.1  -0.1  -0.6  -0.1;
            /* -- END OF MODEL-SPECIFIC SECTION -- */
c = c0; k = rows(c); df = n - k; dc = 1; tol = 1e-3; iter = 1;
fn logl(p,y) = sumc(y .* ln(p) + (1 - y) .* ln(1 - p));
            /* --------- ITERATION LOOP --------- */
"** Computing ML Iterations: Criterion is -2 Loglikelihood **";
do until abs(dc) < tol;
  w = X * c; p = cdfn(w); f =pdfn(w); lk = logl(p,y);
  xu = X'((f ./ (p .* (1-p))) .* (y-p)); v = f + w .* p;
  r = sqrt(f .* (y .* (v ./ (p .* p)) + (1-y) .* ((v-w)/((1-p) .* (1-p)))));
  Z = X .* r; Q = Z'Z; QI = invpd(Q); dc = QI*xu; cn = c + dc;
gosub PRIT; c =cn; iter = iter + 1; endo;
gosub FINAL;  end;
            /* ----------- SUBROUTINES ----------- */
PRIT:  format 1,0; "     Iteration # " iter  "          ";;
       format 10,4; "     Criterion =   "-2*lk;  "c = " cn'; return;
FINAL: Vc = QI; se = sqrt(diag(Vc)); cls; format 5,0;
       " Sample size:" n;; "     Degrees of freedom:" df;
       " # of iterations:" iter;; format 14,4;" Tolerance" tol;?;
       "-2 Loglikelihood:            " -2*lk;?;
       "Coeff     Estimate        Std Error ";
       j = 1; do until j > k; format 1,0; "c " j;;
       format 14,4; c[j,1];; se[j,1]; j = j + 1; endo; return;
```

29.4 For the setup of Exercise 29.3, estimate Models A and B by nonlinear least squares.

GAUSS Hint:

ASG2904 is a simple program for doing nonlinear least squares regression. The "Model-specific Section" of the program is specialized to deal with the present probit Model A. When you have a different model (probit Model B here, say, or in the future), that section has to be changed.

```
/* ASG2904 */
                /* ----- MODEL-SPECIFIC SECTION ----- */
new; output file = asg2904.out reset;
loadp datestr,timestr; datestr "              " timestr;?;
"ASG2904        EXERCISE 29.4        Name .............. ";
"SCF dataset: 100 family heads from 1963 Survey of Consumer Finances";?;
"NONLINEAR LEAST SQUARES ESTIMATION";
"Probit Model: E(y|x) = F(x'c), F(.) = N(0,1) cdf.
y = 1 if Earnings > $ 9000; x = 1, ed, exp, 3 regional dummies.";?;
  n = 100; load D[n,12] = scf;
  ys = D[.,9]; y = ys .> 9;
  x1 = ones(n,1); x2 = D[.,3]; x3 = D[.,5];
  x6 = D[.,8] .== 2*x1; x7 = D[.,8] .== 3*x1; x8 = D[.,8] .== 4*x1;
  fn h(c) = cdfn( c[1,.] + c[2,.] .* x2 + c[3,.] .* x3 + c[4,.] .* x6
                    + c[5,.] .* x7 + c[6,.] .* x8);
  let c0 = -5  0.3  0.1  -0.1  -0.6  -0.1;
              /* -- END OF MODEL-SPECIFIC SECTION -- */
c = c0; k = rows(c); df = n - k; dc = 1; tol = 1e-3; p = 1e-6; E = eye(k);
mxsrch = 10; fn delc(c,a,j) = c + a*E[.,j] * p;
              /* --------- ITERATION LOOP ---------- */
"** Computing NLLS Iterations: Criterion is Sum of Squared Residuals **";
iter = 0; u = y - h(c); sse = u'u; cn = c; gosub PRIT; iter = 1;
do until abs(dc) < tol;
  Z = zeros(n,k); gosub GRAD; Q = Z'Z; u = y - h(c); zu = Z'u;
  QI = invpd(Q); dc = QI*zu; sc = c; sdc = dc; gosub SEARCH;
gosub PRIT; dc = c - cn; c = cn; iter = iter + 1; endo;
gosub FINAL; end;
              /* ---------- SUBROUTINES ------------ */
GRAD:    /* Columns of partials of h(c) with respect to c */
  j = 1; do until j > k; Z[.,j] = (h(delc(c,1,j)) - h(delc(c,-1,j)))/(2*p);
  j = j + 1; endo; return;
SEARCH: /* Compares sse at c + s*dc and at c + (s/2)*dc, where
          dc = proposed change in c. Initially s is set at 1, then
          s is halved until -sse stops declining */
  sc1 = sc + sdc; sm = 1/2;srch = 1; cc = sc1; gosub COM; e1 = sse;
  do until srch > mxsrch;
  sc2 = sc + sm*sdc;                cc = sc2; gosub COM; e2 = sse;
  if e1 >= e2; sc1=sc2; sm=sm/2; sc2=sc + sm*sdc; e1 = e2; srch=srch + 1;
  else; src = srch - 1; cn = sc1; return; endif; endo;
```

```
    src = 10; cn = sc2; return;
COM: sse = 0; u = y - h(cc); sse = u'u; return;
PRIT:   format 1,0; "     Iteration # " iter "           ";;
        format 10,4; "     Criterion = " sse; "c = " cn'; return;
FINAL: s2 = sse/df; Vc = s2*invpd(Q); se = sqrt(diag(Vc)); cls; format 5,0;
        " Sample size:" n;; "     Degrees of freedom:" df;
        " # of iterations:" iter;; format 14,4; " Tolerance" tol;?;
        " Sum of Squared Residuals:   " sse;?;
        "Coeff      Estimate          Std Error";
        j = 1; do until j > k; format 1,0; "c " j;;
        format 14,4; c[j,1];; se[j,1]; j = j + 1; endo; return;
```

29.5 The standard errors computed in the NLLS program ASG2904 rely on an assumption of homoskedasticity. In binary response models, that assumption is automatically violated because

$$V(y|\mathbf{x}) = E(y|\mathbf{x})[1 - E(y|\mathbf{x})].$$

Modify the program to produce correct standard errors.

30 Regression Systems

30.1. Introduction

Suppose that we have a two-equation linear regression model. The data consist of

$$\mathbf{Y} = (\mathbf{y}_1, \mathbf{y}_2), \qquad \mathbf{Z} = (\mathbf{x}_1, \mathbf{x}_2, \ldots, \mathbf{x}_k),$$

where each of the vectors is $n \times 1$. The matrix $\mathbf{Z}$ is nonstochastic with rank($\mathbf{Z}$) = k. The columns of $\mathbf{Y}$ are random vectors with

$$E(\mathbf{y}_1) = \mathbf{X}_1\boldsymbol{\beta}_1, \qquad E(\mathbf{y}_2) = \mathbf{X}_2\boldsymbol{\beta}_2,$$

$$V(\mathbf{y}_1) = \sigma_{11}\mathbf{I}, \qquad V(\mathbf{y}_2) = \sigma_{22}\mathbf{I}, \qquad C(\mathbf{y}_1, \mathbf{y}_2) = \sigma_{12}\mathbf{I}.$$

Here $\mathbf{X}_1(n \times k_1)$ and $\mathbf{X}_2(n \times k_2)$ are submatrices (possibly overlapping) of $\mathbf{Z}$, and the 2×2 matrix

$$\boldsymbol{\Sigma}^* = \begin{pmatrix} \sigma_{11} & \sigma_{12} \\ \sigma_{21} & \sigma_{22} \end{pmatrix}$$

is positive definite.

A CR model applies to $(\mathbf{y}_1, \mathbf{X}_1)$, and a CR model applies to $(\mathbf{y}_2, \mathbf{X}_2)$. The new feature is the covariance, $\sigma_{12} = \sigma_{21}$, between corresponding elements of $\mathbf{y}_1$ and $\mathbf{y}_2$.

This is the two-equation case of the *regression-system*, or set-of-regressions, or multivariate regression, or SUR ("seemingly unrelated regressions") *model* that plays a central role in contemporary econometrics. For economic settings, consider an input demand system where y_1 and y_2 are the cost shares for labor and capital, while the x's include output and input prices. Or consider the reduced form of a simultaneous supply and demand system where y_1 = quantity, y_2 = price, and the x's include income, input prices, and prices of substitutes. Or consider investment demand y_1 and y_2 by two firms in an industry, where the observations run over time, while $\mathbf{X}_1$ includes variables for the first firm, and $\mathbf{X}_2$ includes variables for the second firm. In all these cases one might expect correlation between the two dependent variables at each observation. In some cases, $\mathbf{X}_1$ and $\mathbf{X}_2$ will coincide (each containing all k of the x's, so $\mathbf{X}_1 = \mathbf{X}_2 = \mathbf{Z}$); in other cases they will differ.

For an underlying framework, we suppose that there is a multivariate population with joint pdf for the random vectors $\mathbf{y} = (y_1, y_2)'$ and $\mathbf{z} = (x_1, \ldots, x_k)'$. With $\mathbf{x}_1$ ($k_1 \times 1$) and $\mathbf{x}_2$ ($k_2 \times 1$) being (possibly overlapping) subvectors of $\mathbf{z}$, suppose further that

$$E(\mathbf{y}|\mathbf{z}) = \begin{pmatrix} \mathbf{x}_1'\boldsymbol{\beta}_1 \\ \mathbf{x}_2'\boldsymbol{\beta}_2 \end{pmatrix}, \qquad V(\mathbf{y}|\mathbf{z}) = \begin{pmatrix} \sigma_{11} & \sigma_{12} \\ \sigma_{21} & \sigma_{22} \end{pmatrix}.$$

So both CEF's are linear and the conditional variances and covariance are constant. If we use the stratified sampling scheme, choosing a set of vectors $\mathbf{z}_i'$ ($i = 1, \ldots, n$), and then drawing independently (over i) from the bivariate conditional distributions of $\mathbf{y}|\mathbf{z}_i'$, then the SUR model will result. There is also a neoclassical variant of the model, in which sampling is random from the joint distribution of $(\mathbf{y}', \mathbf{z}')$.

While our discussion will be confined to the two-equation case, the model and analysis generalize directly to the case where there are more than two equations.

30.2. Stacking

The CR specification applies to each regression separately. So equation-by-equation LS estimation will be unbiased, and the associated variance estimates will be unbiased as well. Those separate LS coefficient vectors are

$$\mathbf{b}_1 = \mathbf{A}_1\mathbf{y}_1, \qquad \mathbf{b}_2 = \mathbf{A}_2\mathbf{y}_2,$$

where

$$\mathbf{A}_1 = (\mathbf{Q}_{11})^{-1}\mathbf{X}_1', \qquad \mathbf{Q}_{11} = \mathbf{X}_1'\mathbf{X}_1,$$
$$\mathbf{A}_2 = (\mathbf{Q}_{22})^{-1}\mathbf{X}_2', \qquad \mathbf{Q}_{22} = \mathbf{X}_2'\mathbf{X}_2.$$

Clearly,

$$E(\mathbf{b}_1) = \boldsymbol{\beta}_1, \qquad E(\mathbf{b}_2) = \boldsymbol{\beta}_2,$$
$$V(\mathbf{b}_1) = \sigma_{11}(\mathbf{Q}_{11})^{-1}, \qquad V(\mathbf{b}_2) = \sigma_{22}(\mathbf{Q}_{22})^{-1}.$$

The fact that $C(\mathbf{y}_1, \mathbf{y}_2)$ is nonzero suggests that better estimates may be obtained by estimating the two regressions jointly. To do so, first *stack* the two regressions into one. Let

$$\mathbf{y} = \begin{pmatrix} \mathbf{y}_1 \\ \mathbf{y}_2 \end{pmatrix}, \qquad \mathbf{X} = \begin{pmatrix} \mathbf{X}_1 & \mathbf{O} \\ \mathbf{O} & \mathbf{X}_2 \end{pmatrix},$$
$$\boldsymbol{\Sigma} = \begin{pmatrix} \sigma_{11}\mathbf{I} & \sigma_{12}\mathbf{I} \\ \sigma_{21}\mathbf{I} & \sigma_{22}\mathbf{I} \end{pmatrix}, \qquad \boldsymbol{\beta} = \begin{pmatrix} \boldsymbol{\beta}_1 \\ \boldsymbol{\beta}_2 \end{pmatrix}.$$

Then $E(\mathbf{y}) = \mathbf{X}\boldsymbol{\beta}$, $V(\mathbf{y}) = \boldsymbol{\Sigma}$ positive definite, $\mathbf{X}$ nonstochastic, and $\mathbf{X}$ has full column rank, so a GCR model applies. The SUR model is just a special case of the GCR model, special in the structure of its $\boldsymbol{\Sigma}$ matrix, and, at this stage, in the pattern of its $\mathbf{X}$ matrix.

For some purposes, a more compact display of $\boldsymbol{\Sigma}$ is convenient. If $\mathbf{D}$ is $m \times m$ and $\mathbf{E}$ is $n \times n$, then the *Kronecker product* of $\mathbf{D}$ and $\mathbf{E}$, denoted $\mathbf{D} \otimes \mathbf{E}$, is the $mn \times mn$ matrix obtained in blocks by multiplying each element of $\mathbf{D}$ into the matrix $\mathbf{E}$. On that definition, we can write the SUR model variance matrix $\boldsymbol{\Sigma}$ as $\boldsymbol{\Sigma}^* \otimes \mathbf{I}$, where the $\mathbf{I}$ is $n \times n$.

In our two-equation case, there are $2n$ observations, $k_1 + k_2$ regression coefficients, and 3 distinct elements in the variance matrix $\boldsymbol{\Sigma}$. Consider LS regression of the $2n \times 1$ vector $\mathbf{y}$ on the $2n \times (k_1 + k_2)$ matrix $\mathbf{X}$. The estimated coefficient vector is $\hat{\boldsymbol{\beta}} = \mathbf{A}\mathbf{y}$, where $\mathbf{A} = \mathbf{Q}^{-1}\mathbf{X}'$. Now

$$\mathbf{X} = \begin{pmatrix} \mathbf{X}_1 & \mathbf{O} \\ \mathbf{O} & \mathbf{X}_2 \end{pmatrix}, \qquad \mathbf{X}' = \begin{pmatrix} \mathbf{X}_1' & \mathbf{O}' \\ \mathbf{O}' & \mathbf{X}_2' \end{pmatrix},$$

$$\mathbf{Q} = \mathbf{X}'\mathbf{X} = \begin{pmatrix} \mathbf{Q}_{11} & \mathbf{O} \\ \mathbf{O} & \mathbf{Q}_{22} \end{pmatrix}, \qquad \mathbf{Q}^{-1} = \begin{pmatrix} (\mathbf{Q}_{11})^{-1} & \mathbf{O} \\ \mathbf{O} & (\mathbf{Q}_{22})^{-1} \end{pmatrix},$$

$$\mathbf{A} = \mathbf{Q}^{-1}\mathbf{X}' = \begin{pmatrix} (\mathbf{Q}_{11})^{-1}\mathbf{X}_1' & \mathbf{O} \\ \mathbf{O} & (\mathbf{Q}_{22})^{-1}\mathbf{X}_2' \end{pmatrix} = \begin{pmatrix} \mathbf{A}_1 & \mathbf{O} \\ \mathbf{O} & \mathbf{A}_2 \end{pmatrix}.$$

So

$$\hat{\boldsymbol{\beta}} = \mathbf{A}\mathbf{y} = \begin{pmatrix} \mathbf{A}_1 & \mathbf{O} \\ \mathbf{O} & \mathbf{A}_2 \end{pmatrix} \begin{pmatrix} \mathbf{y}_1 \\ \mathbf{y}_2 \end{pmatrix} = \begin{pmatrix} \mathbf{A}_1\mathbf{y}_1 \\ \mathbf{A}_2\mathbf{y}_2 \end{pmatrix} = \begin{pmatrix} \mathbf{b}_1 \\ \mathbf{b}_2 \end{pmatrix}.$$

We conclude that LS estimation of the stacked regression reduces to equation-by-equation LS.

30.3. Generalized Least Squares

Because a GCR model applies, GLS estimation of the stacked regression should be preferable to LS. Taking $\mathbf{\Sigma}^*$ and hence $\mathbf{\Sigma}$ to be known, the normal equations for GLS estimation are

$$(\mathbf{X}'\mathbf{\Sigma}^{-1}\mathbf{X})\mathbf{b}^* = \mathbf{X}'\mathbf{\Sigma}^{-1}\mathbf{y},$$

with

$$\mathbf{b}^* = \begin{pmatrix} \mathbf{b}_1^* \\ \mathbf{b}_2^* \end{pmatrix}.$$

For GCR models, a data-transformation device is ordinarily used to convert a GLS problem into an LS problem (see Sections 27.3 and 27.4). But that device is not particularly convenient in the SUR special case, so we continue to work directly with the GLS normal equations.

It is easy to verify that

$$\mathbf{\Sigma}^{-1} = \begin{pmatrix} \sigma^{11}\mathbf{I} & \sigma^{12}\mathbf{I} \\ \sigma^{21}\mathbf{I} & \sigma^{22}\mathbf{I} \end{pmatrix}, \quad \text{with} \begin{pmatrix} \sigma^{11} & \sigma^{12} \\ \sigma^{21} & \sigma^{22} \end{pmatrix} = \mathbf{\Sigma}^{*-1},$$

illustrating a general rule for Kronecker products: if $\mathbf{D}$ and $\mathbf{E}$ are nonsingular, then $(\mathbf{D} \otimes \mathbf{E})^{-1} = \mathbf{D}^{-1} \otimes \mathbf{E}^{-1}$.

Continuing, we have

$$\mathbf{X}'\mathbf{\Sigma}^{-1} = \begin{pmatrix} \sigma^{11}\mathbf{X}_1' & \sigma^{12}\mathbf{X}_1' \\ \sigma^{21}\mathbf{X}_2' & \sigma^{22}\mathbf{X}_2' \end{pmatrix}, \qquad \mathbf{X}'\mathbf{\Sigma}^{-1}\mathbf{y} = \begin{pmatrix} \sigma^{11}\mathbf{X}_1'\mathbf{y}_1 + \sigma^{12}\mathbf{X}_1'\mathbf{y}_2 \\ \sigma^{21}\mathbf{X}_2'\mathbf{y}_1 + \sigma^{22}\mathbf{X}_2'\mathbf{y}_2 \end{pmatrix},$$

$$\mathbf{X}'\mathbf{\Sigma}^{-1}\mathbf{X} = \begin{pmatrix} \sigma^{11}\mathbf{X}_1'\mathbf{X}_1 & \sigma^{12}\mathbf{X}_1'\mathbf{X}_2 \\ \sigma^{21}\mathbf{X}_2'\mathbf{X}_1 & \sigma^{22}\mathbf{X}_2'\mathbf{X}_2 \end{pmatrix}.$$

The normal equations may now be solved for $\mathbf{b}^*$.

We do not give the explicit solution here. Instead, we observe that the following expressions satisfy the GLS normal equations:

$$\text{(30.1)} \quad \mathbf{b}_1^* = \mathbf{b}_1 - \alpha_1\mathbf{A}_1(\mathbf{y}_2 - \mathbf{X}_2\mathbf{b}_2^*),$$

$$\text{(30.2)} \quad \mathbf{b}_2^* = \mathbf{b}_2 - \alpha_2\mathbf{A}_2(\mathbf{y}_1 - \mathbf{X}_1\mathbf{b}_1^*),$$

where

$$\text{(30.3)} \quad \alpha_1 = \sigma_{12}/\sigma_{22}, \qquad \alpha_2 = \sigma_{12}/\sigma_{11}.$$

These expressions show that $\mathbf{b}_1^*$ depends on $\mathbf{y}_2$ as well as on $\mathbf{y}_1$, and similarly for $\mathbf{b}_2^*$. They also show how knowledge of $\mathbf{\Sigma}^*$ up to proportionality suffices to compute the GLS estimator.

Three special cases are instructive.

Orthogonal Explanatory Variables. If $\mathbf{X}_1'\mathbf{X}_2 = \mathbf{O}$, then $\mathbf{A}_2\mathbf{X}_1 = \mathbf{O}$, so

$$\mathbf{A}_2(\mathbf{y}_1 - \mathbf{X}_1\mathbf{b}_1^*) = \mathbf{A}_2\mathbf{y}_1 = \mathbf{A}_2(\mathbf{X}_1\mathbf{b}_1 + \mathbf{e}_1) = \mathbf{A}_2\mathbf{e}_1,$$

and similarly $\mathbf{A}_1\mathbf{y}_2 = \mathbf{A}_1\mathbf{e}_2$, where $\mathbf{e}_1$ and $\mathbf{e}_2$ are the LS residual vectors. So the system (30.1)–(30.2) reduces to

$$\mathbf{b}_1^* = \mathbf{b}_1 - \alpha_1\mathbf{A}_1\mathbf{e}_2, \qquad \mathbf{b}_2^* = \mathbf{b}_2 - \alpha_2\mathbf{A}_2\mathbf{e}_1,$$

which we may write as

$$\mathbf{b}_1^* = \mathbf{A}_1\mathbf{y}_1^\circ, \qquad \mathbf{b}_2^* = \mathbf{A}_2\mathbf{y}_2^\circ,$$

where

$$\mathbf{y}_1^\circ = \mathbf{y}_1 - \alpha_1\mathbf{e}_2, \qquad \mathbf{y}_2^\circ = \mathbf{y}_2 - \alpha_2\mathbf{e}_1.$$

This provides a simple algorithm for GLS in the present case: construct $\mathbf{y}_1^\circ$ and regress it on $\mathbf{X}_1$ to get $\mathbf{b}_1^*$; construct $\mathbf{y}_2^\circ$ and regress it on $\mathbf{X}_2$ to get $\mathbf{b}_2^*$.

Identical Explanatory Variables. If $\mathbf{X}_1 = \mathbf{X}_2$, then $\mathbf{A}_2 = \mathbf{A}_1$, so $\mathbf{A}_2\mathbf{y}_1 = \mathbf{A}_1\mathbf{y}_1 = \mathbf{b}_1$ and $\mathbf{A}_2\mathbf{X}_1 = \mathbf{A}_1\mathbf{X}_1 = \mathbf{I}$, so $\mathbf{A}_2(\mathbf{y}_1 - \mathbf{X}_1\mathbf{b}_1^*) = \mathbf{b}_1 - \mathbf{b}_1^*$. Similarly, $\mathbf{A}_1(\mathbf{y}_2 - \mathbf{X}_2\mathbf{b}_2^*) = \mathbf{b}_2 - \mathbf{b}_2^*$. So the system (30.1)–(30.2) reduces to

$$\mathbf{b}_1^* - \mathbf{b}_1 = -\alpha_1(\mathbf{b}_2 - \mathbf{b}_2^*), \qquad \mathbf{b}_2^* - \mathbf{b}_2 = -\alpha_2(\mathbf{b}_1 - \mathbf{b}_1^*).$$

Here the solution is obvious: $\mathbf{b}_1^* = \mathbf{b}_1$, $\mathbf{b}_2^* = \mathbf{b}_2$. This is a striking result: if the explanatory-variable matrices of the two regressions are identical, then $\mathbf{b}^* = \mathbf{b}$. That is, the GLS and LS estimators will coincide in every sample.

Uncorrelated Disturbances. If $\sigma_{12} = 0$, then $\alpha_1 = \alpha_2 = 0$, and the system (30.1)–(30.2) reduces to $\mathbf{b}_1^* = \mathbf{b}_1$, $\mathbf{b}_2^* = \mathbf{b}_2$. If the population covariance between the two dependent variables is zero, then $\mathbf{b}^* = \mathbf{b}$. That is, the GLS and LS estimators will coincide in every sample.

30.4. Comparison of GLS and LS Estimators

The variance matrix of $\mathbf{b}^*$ is given by $V(\mathbf{b}^*) = (\mathbf{X}'\mathbf{\Sigma}^{-1}\mathbf{X})^{-1}$. Upon inversion, we find that the variance matrix of the GLS estimator of $\boldsymbol{\beta}_2$, which lies in the southeast block of $V(\mathbf{b}^*)$, can be written as

(30.4) $V(\mathbf{b}_2^*) = \sigma_{22}(\mathbf{X}_2'\mathbf{X}_2 + \phi^2\mathbf{X}_2^{*\prime}\mathbf{X}_2^*)^{-1}$,

where

$$\phi^2 = \rho^2/(1 - \rho^2), \qquad \rho = \sigma_{12}/\sqrt{(\sigma_{11}\sigma_{22})}, \qquad \mathbf{X}_2^* = \mathbf{M}_1\mathbf{X}_2.$$

Here ρ (which must lie between -1 and 1) is the population correlation coefficient of y_1 and y_2. The variance matrix of $\mathbf{b}_2^*$ contrasts with the variance matrix of $\mathbf{b}_2$, the LS estimator of $\boldsymbol{\beta}_2$, which is

(30.5) $V(\mathbf{b}_2) = \sigma_{22}(\mathbf{X}_2'\mathbf{X}_2)^{-1}$.

Because $\phi^2 \geq 0$ and the matrix $\mathbf{X}_2^{*\prime}\mathbf{X}_2^*$ is nonnegative definite, we have

$$\mathbf{X}_2'\mathbf{X}_2 + \phi^2\mathbf{X}_2^{*\prime}\mathbf{X}_2^* \geq \mathbf{X}_2'\mathbf{X}_2.$$

Inversion reverses the inequality, so $V(\mathbf{b}_2) \geq V(\mathbf{b}_2^*)$. The difference will be large—that is, GLS will be much more precise than LS—when the matrix $\phi^2\mathbf{X}_2^{*\prime}\mathbf{X}_2^*$ is large relative to the matrix $\mathbf{X}_2'\mathbf{X}_2$. This occurs when ρ^2 is large and/or $\mathbf{X}_2$ is poorly fitted by LS linear regression on $\mathbf{X}_1$.

Extreme cases are again instructive. Suppose that $\mathbf{X}_1$ and $\mathbf{X}_2$ are orthogonal, that is, $\mathbf{X}_1'\mathbf{X}_2 = \mathbf{O}$. Then $\mathbf{X}_2^* = \mathbf{X}_2$, and

$$\mathbf{X}_2'\mathbf{X}_2 + \phi^2\mathbf{X}_2^{*\prime}\mathbf{X}_2^* = (1 + \phi^2)\mathbf{X}_2'\mathbf{X}_2 = \mathbf{X}_2'\mathbf{X}_2/(1 - \rho^2),$$

so

$$V(\mathbf{b}_2^*) = (1 - \rho^2)V(\mathbf{b}_2),$$

and a sharp comparison is seen. For example, if $\rho = 0.8$, then $1 - \rho^2 = 0.36$: the GLS coefficient estimates will have variances that are about one-third as large as those of the LS coefficient estimates. Evidently, GLS is particularly advantageous when $\mathbf{X}_1$ and $\mathbf{X}_2$ are orthogonal and ρ^2 is large.

At the other extreme, if $\mathbf{X}_1$ and $\mathbf{X}_2$ are identical, then $\mathbf{X}_2^* = \mathbf{O}$ and $V(\mathbf{b}_2^*) = V(\mathbf{b}_2)$, as it must because $\mathbf{b}_2^* = \mathbf{b}_2$ in every sample. Or if $\sigma_{12} = 0$, then $\phi^2 = 0$ and again $V(\mathbf{b}_2^*) = V(\mathbf{b}_2)$, as it must because $\mathbf{b}_2^* = \mathbf{b}_2$ in every sample.

In general, the explanatory variables are not identical and the covariance is nonzero, so GLS will be different from, and hence preferable to, LS in the SUR model.

30.5. Feasible Generalized Least Squares

We proceed to the practical situation, where $\mathbf{\Sigma}^*$, and hence $\mathbf{\Sigma}$, is unknown. The natural estimators of the elements of $\mathbf{\Sigma}^*$ come from the residual vectors of equation-by-equation LS regression, $\mathbf{e}_1$ and $\mathbf{e}_2$. Let

$$\hat{\sigma}_{ij} = \mathbf{e}_i'\mathbf{e}_j/n \qquad (i, j = 1, 2),$$

and

$$\hat{\mathbf{\Sigma}}^* = \begin{pmatrix} \hat{\sigma}_{11} & \hat{\sigma}_{12} \\ \hat{\sigma}_{21} & \hat{\sigma}_{22} \end{pmatrix}, \qquad \hat{\mathbf{\Sigma}} = \begin{pmatrix} \hat{\sigma}_{11}\mathbf{I} & \hat{\sigma}_{12}\mathbf{I} \\ \hat{\sigma}_{21}\mathbf{I} & \hat{\sigma}_{22}\mathbf{I} \end{pmatrix}.$$

Then, by the rule $\hat{\mathbf{\Sigma}}^{-1} = (\hat{\mathbf{\Sigma}}^* \otimes \mathbf{I})^{-1} = \hat{\mathbf{\Sigma}}^{*-1} \otimes \mathbf{I}$, we have

$$\hat{\mathbf{\Sigma}}^{-1} = \begin{pmatrix} \hat{\sigma}^{11}\mathbf{I} & \hat{\sigma}^{12}\mathbf{I} \\ \hat{\sigma}^{21}\mathbf{I} & \hat{\sigma}^{22}\mathbf{I} \end{pmatrix}, \quad \text{with} \begin{pmatrix} \hat{\sigma}^{11} & \hat{\sigma}^{12} \\ \hat{\sigma}^{21} & \hat{\sigma}^{22} \end{pmatrix} = \hat{\mathbf{\Sigma}}^{*-1}.$$

The FGLS estimator

$$\hat{\mathbf{b}}^* = \begin{pmatrix} \hat{\mathbf{b}}_1^* \\ \hat{\mathbf{b}}_2^* \end{pmatrix}$$

is the solution to the normal equations $(\mathbf{X}'\hat{\mathbf{\Sigma}}^{-1}\mathbf{X})\hat{\mathbf{b}}^* = \mathbf{X}'\hat{\mathbf{\Sigma}}^{-1}\mathbf{y}$, that is,

$$\hat{\mathbf{b}}^* = (\mathbf{X}'\hat{\mathbf{\Sigma}}^{-1}\mathbf{X})^{-1}\mathbf{X}'\hat{\mathbf{\Sigma}}^{-1}\mathbf{y}.$$

Comparison with Eqs. (30.1)–(30.2) shows that the FGLS normal equations are solved by:

$$\hat{\mathbf{b}}_1^* = \mathbf{b}_1 - \hat{\alpha}_1\mathbf{A}_1(\mathbf{y}_2 - \mathbf{X}_2\hat{\mathbf{b}}_2^*),$$

$$\hat{\mathbf{b}}_2^* = \mathbf{b}_2 - \hat{\alpha}_2\mathbf{A}_2(\mathbf{y}_1 - \mathbf{X}_1\hat{\mathbf{b}}_1^*),$$

where

$$\hat{\alpha}_1 = \hat{\sigma}_{12}/\hat{\sigma}_{22}, \qquad \hat{\alpha}_2 = \hat{\sigma}_{12}/\hat{\sigma}_{11}.$$

So the algebraic analysis of Section 30.3 carries over: FGLS and LS coincide if $\mathbf{X}_1 = \mathbf{X}_2$ or if $\hat{\sigma}_{12} = 0$. Of course, the latter condition will occur only by coincidence even if $\sigma_{12} = 0$.

This FGLS estimator, introduced by Zellner (1962), is sometimes known as ZEF (Zellner efficient). Under general conditions, it has the same asymptotic distribution as the GLS estimator. That conclusion stems from the quality of the estimator of $\mathbf{\Sigma}^*$, which we now explore.

We know from CR theory that $\mathbf{e}_1 = \mathbf{M}_1\mathbf{y}_1$, with $E(\mathbf{e}_1) = \mathbf{0}$ and $V(\mathbf{e}_1) = \sigma_{11}\mathbf{M}_1$. So $E(\mathbf{e}_1'\mathbf{e}_1) = (n - k_1)\sigma_{11}$, whence $E(\hat{\sigma}_{11}) = \sigma_{11}(1 - k_1/n)$. The bias goes to zero as n increases. Equally relevant is the fact that the variance goes to zero. This convergence is especially easy to see under normality. For then $w_1 = \mathbf{e}_1'\mathbf{e}_1/\sigma_{11} \sim \chi^2(n - k_1)$: see D6 in Section 21.1. Now, $\hat{\sigma}_{11} = (\sigma_{11}/n)w_1$. So

$$V(\hat{\sigma}_{11}) = (\sigma_{11}/n)^2V(w_1) = (\sigma_{11}/n)^2 2(n - k_1) = (2\sigma_{11}^2/n)(1 - k_1/n),$$

which goes to zero with n. Thus $\hat{\sigma}_{11}$ converges in mean square to σ_{11}, so it is a consistent estimator of σ_{11}. Similarly for $\hat{\sigma}_{22}$.

Proceeding to the covariance, we have $\mathbf{e}_1 = \mathbf{M}_1\mathbf{y}_1$, and $\mathbf{e}_2 = \mathbf{M}_2\mathbf{y}_2$, so by R6 (Section 15.1),

$$C(\mathbf{e}_1, \mathbf{e}_2) = \mathbf{M}_1 C(\mathbf{y}_1, \mathbf{y}_2)\mathbf{M}_2' = \sigma_{12}\mathbf{M}_1\mathbf{M}_2,$$

whence

$$E(\mathbf{e}_1'\mathbf{e}_2) = \text{tr}[C(\mathbf{e}_1, \mathbf{e}_2)] = \sigma_{12}\text{tr}(\mathbf{M}_1\mathbf{M}_2).$$

So

$$E(\hat{\sigma}_{12}) = \sigma_{12}\text{tr}(\mathbf{M}_1\mathbf{M}_2)/n.$$

Now

$$\text{tr}(\mathbf{M}_1\mathbf{M}_2) = \text{tr}(\mathbf{I} - \mathbf{N}_1 - \mathbf{N}_2 + \mathbf{N}_1\mathbf{N}_2) = n - k_1 - k_2 + \text{tr}(\mathbf{N}_1\mathbf{N}_2).$$

If $\mathbf{X}_1'\mathbf{X}_2 = \mathbf{O}$, then $\mathbf{N}_1\mathbf{N}_2 = \mathbf{O}$ and $\text{tr}(\mathbf{M}_1\mathbf{M}_2) = n - k_1 - k_2$. At the other extreme, if $\mathbf{X}_1 = \mathbf{X}_2$, then $\mathbf{N}_1 = \mathbf{N}_2$ and $\text{tr}(\mathbf{M}_1\mathbf{M}_2) = n - k_1 = n - k_2$. Indeed, it can be shown that $n - k_1 - k_2 \leq \text{tr}(\mathbf{M}_1\mathbf{M}_2) \leq n - \min(k_1, k_2)$: see Theil (1971, pp. 317–322). So $E(\hat{\sigma}_{12})$ goes to σ_{12} as n goes to infinity. Similarly it can be shown that $V(\hat{\sigma}_{12})$ goes to zero, so $\hat{\sigma}_{12}$ is consistent for σ_{12}.

The result is that $\hat{\mathbf{\Sigma}}^*$ is consistent for $\mathbf{\Sigma}^*$. This quality of the variance-matrix estimator suffices to make the asymptotic distribution of FGLS the same as that of GLS. One might divide by the appropriate scalars rather than by n to get an unbiased estimator of $\mathbf{\Sigma}^*$, but there is no advantage in doing so because asymptotic properties are sought.

Having obtained the FGLS estimate $\hat{\mathbf{b}}^*$, one proceeds to estimate its variance matrix as

$$\hat{V}(\hat{\mathbf{b}}^*) = (\mathbf{X}'\hat{\mathbf{\Sigma}}^{-1}\mathbf{X})^{-1},$$

and to use this just as one would use $\hat{V}(\mathbf{b}) = \hat{\sigma}^2\mathbf{Q}^{-1}$ in the CR (or CNR) model. This practice has an asymptotic justification.

Despite the fact that it is a nonlinear function of $\mathbf{y}$, the ZEF estimator is unbiased under quite general conditions, as shown in an elegant argument by Kakwani (1967).

30.6. Restrictions

In empirical applications of the SUR model, cross-equation restrictions are often imposed. For concreteness, suppose that we start with

$$E(\mathbf{y}_1) = \mathbf{x}_1\beta_{11} + \mathbf{x}_2\beta_{21},$$

$$E(\mathbf{y}_2) = \mathbf{x}_1\beta_{12} \qquad\qquad + \mathbf{x}_3\beta_{32}.$$

Upon stacking, this becomes

$$E(\mathbf{y}) = \begin{pmatrix} \mathbf{x}_1 & \mathbf{x}_2 & \mathbf{0} & \mathbf{0} \\ \mathbf{0} & \mathbf{0} & \mathbf{x}_1 & \mathbf{x}_3 \end{pmatrix} \begin{pmatrix} \beta_{11} \\ \beta_{21} \\ \beta_{12} \\ \beta_{32} \end{pmatrix} = \begin{pmatrix} \mathbf{X}_1 & \mathbf{O} \\ \mathbf{O} & \mathbf{X}_2 \end{pmatrix} \begin{pmatrix} \boldsymbol{\beta}_1 \\ \boldsymbol{\beta}_2 \end{pmatrix} = \mathbf{X}\boldsymbol{\beta},$$

say. Consider the cross-equation restriction $\beta_{21} = \beta_{32}$ ($= \beta_0$, say). It implies that

$$E(\mathbf{y}) = \begin{pmatrix} \mathbf{x}_1 & \mathbf{x}_2 & \mathbf{0} \\ \mathbf{0} & \mathbf{x}_3 & \mathbf{x}_1 \end{pmatrix} \begin{pmatrix} \beta_{11} \\ \beta_0 \\ \beta_{12} \end{pmatrix} = \mathbf{X}_0\boldsymbol{\beta}_0,$$

say.

Although this $\mathbf{X}_0$ matrix does not have the characteristic block-diagonal SUR pattern, GLS or FGLS estimation can proceed. Observe that $\mathbf{X}_0 = \mathbf{XT}$, where

$$\mathbf{T} = \begin{pmatrix} 1 & 0 & 0 \\ 0 & 1 & 0 \\ 0 & 0 & 1 \\ 0 & 1 & 0 \end{pmatrix}.$$

So

$$\mathbf{X}_0'\boldsymbol{\Sigma}^{-1}\mathbf{X}_0 = \mathbf{T}'\mathbf{X}'\boldsymbol{\Sigma}^{-1}\mathbf{XT}, \qquad \mathbf{X}_0'\boldsymbol{\Sigma}^{-1}\mathbf{y} = \mathbf{T}'\mathbf{X}'\boldsymbol{\Sigma}^{-1}\mathbf{y},$$

and the restricted GLS estimator of $\boldsymbol{\beta}_0$ is $(\mathbf{T}'\mathbf{X}'\boldsymbol{\Sigma}^{-1}\mathbf{X}\mathbf{T})^{-1}\mathbf{T}'\mathbf{X}'\boldsymbol{\Sigma}^{-1}\mathbf{y}$.

When such restrictions are imposed, GLS and LS can differ even though the original explanatory variables are identical and the disturbances are uncorrelated. The previous conclusion that GLS and LS coincide relied in part on the block-diagonal structure of the $\mathbf{X}$ matrix. For a sharp example, suppose that $\mathbf{X}_1 = \mathbf{X}_2$ $(= \mathbf{X}^\circ$, say), that $\sigma_{12} = 0$, and that the restriction is $\boldsymbol{\beta}_1 = \boldsymbol{\beta}_2$ $(=\boldsymbol{\beta}_0$, say). Set up the restricted model as

$$E(\mathbf{y}) = \begin{pmatrix} \mathbf{X}^\circ \\ \mathbf{X}^\circ \end{pmatrix} \boldsymbol{\beta}_0, \qquad V(\mathbf{y}) = \begin{pmatrix} \sigma_{11}\mathbf{I} & \mathbf{O} \\ \mathbf{O} & \sigma_{22}\mathbf{I} \end{pmatrix}.$$

The LS estimator of $\boldsymbol{\beta}_0$ is

$$\mathbf{b}_0 = (2\mathbf{X}^{\circ\prime}\mathbf{X}^\circ)^{-1}(\mathbf{X}^{\circ\prime}\mathbf{y}_1 + \mathbf{X}^{\circ\prime}\mathbf{y}_2) = (1/2)(\mathbf{b}_1 + \mathbf{b}_2),$$

while the GLS estimator is

$$\begin{aligned} \mathbf{b}_0^* &= (\sigma_{11}^{-1}\mathbf{X}^{\circ\prime}\mathbf{X}^\circ + \sigma_{22}^{-1}\mathbf{X}^{\circ\prime}\mathbf{X}^\circ)^{-1}(\sigma_{11}^{-1}\mathbf{X}^{\circ\prime}\mathbf{y}_1 + \sigma_{22}^{-1}\mathbf{X}^{\circ\prime}\mathbf{y}_2) \\ &= (\sigma_{11}^{-1} + \sigma_{22}^{-1})^{-1}(\mathbf{X}^{\circ\prime}\mathbf{X}^\circ)^{-1}(\sigma_{11}^{-1}\mathbf{X}^{\circ\prime}\mathbf{y}_1 + \sigma_{22}^{-1}\mathbf{X}^{\circ\prime}\mathbf{y}_2) \\ &= \theta\mathbf{b}_1 + (1 - \theta)\mathbf{b}_2, \end{aligned}$$

where

$$\theta = \sigma_{11}^{-1}/(\sigma_{11}^{-1} + \sigma_{22}^{-1}) = \sigma_{22}/(\sigma_{11} + \sigma_{22}).$$

We see that $\mathbf{b}_0$ and $\mathbf{b}_0^*$ are distinct weighted averages of $\mathbf{b}_1$ and $\mathbf{b}_2$.

The rule that identical explanatory variables make GLS coincide with LS is still valid on the understanding that once cross-equation restrictions are imposed, the explanatory variables in the two equations should no longer be considered identical. After all, linear restrictions on coefficients are like zero-null-subvector restrictions, as we saw in the single-equation context of Section 22.2.

30.7. Alternative Estimators

There is a useful way to look at the algebra of the SUR model that may clarify the relations among the various estimators. Let

$$\mathbf{u}_j = \mathbf{y}_j - \mathbf{X}_j\mathbf{c}_j \quad (j = 1, 2), \qquad \mathbf{u} = (\mathbf{u}_1', \mathbf{u}_2')', \qquad \mathbf{U} = (\mathbf{u}_1, \mathbf{u}_2).$$

Then the GLS procedure can be restated as: choose $\mathbf{c} = (\mathbf{c}_1', \mathbf{c}_2')'$ to minimize the criterion

$$\begin{aligned}
\phi(\mathbf{c}) &= \mathbf{u}'\boldsymbol{\Sigma}^{-1}\mathbf{u} \\
&= (\mathbf{u}_1', \mathbf{u}_2') \begin{pmatrix} \sigma^{11}\mathbf{I} & \sigma^{12}\mathbf{I} \\ \sigma^{21}\mathbf{I} & \sigma^{22}\mathbf{I} \end{pmatrix} \begin{pmatrix} \mathbf{u}_1 \\ \mathbf{u}_2 \end{pmatrix} \\
&= (\mathbf{u}_1', \mathbf{u}_2') \begin{pmatrix} \sigma^{11}\mathbf{u}_1 + \sigma^{12}\mathbf{u}_2 \\ \sigma^{21}\mathbf{u}_1 + \sigma^{22}\mathbf{u}_2 \end{pmatrix} \\
&= \sigma^{11}\mathbf{u}_1'\mathbf{u}_1 + \sigma^{22}\mathbf{u}_2'\mathbf{u}_2 + \sigma^{12}\mathbf{u}_1'\mathbf{u}_2 + \sigma^{21}\mathbf{u}_2'\mathbf{u}_1 \\
&= \mathrm{tr}(\boldsymbol{\Sigma}^{*-1}\mathbf{U}'\mathbf{U}).
\end{aligned}$$

Here $\mathbf{U}'\mathbf{U}$ is the 2×2 matrix of sums of squares and cross-products of deviations from the regressions. By the same token, FGLS chooses $\mathbf{c}$ to minimize the criterion $\mathrm{tr}(\hat{\boldsymbol{\Sigma}}^{*-1}\mathbf{U}'\mathbf{U})$, while LS chooses $\mathbf{c}$ to minimize the criterion $\mathrm{tr}(\mathbf{U}'\mathbf{U}) = \mathbf{u}'\mathbf{u}$.

Now suppose that we have the normal version of the SUR model. That is, $\mathbf{y} \sim \mathcal{N}(\boldsymbol{\mu}, \boldsymbol{\Sigma})$, where

$$\mathbf{y} = \begin{pmatrix} \mathbf{y}_1 \\ \mathbf{y}_2 \end{pmatrix}, \qquad \boldsymbol{\mu} = \begin{pmatrix} \boldsymbol{\mu}_1 \\ \boldsymbol{\mu}_2 \end{pmatrix}, \qquad \boldsymbol{\Sigma} = \begin{pmatrix} \sigma_{11}\mathbf{I} & \sigma_{12}\mathbf{I} \\ \sigma_{21}\mathbf{I} & \sigma_{22}\mathbf{I} \end{pmatrix}$$

with $\boldsymbol{\mu}_j = \mathbf{X}_j\boldsymbol{\beta}_j$ $(j = 1, 2)$. The pdf of the $2n \times 1$ random vector $\mathbf{y}$ is

$$f(\mathbf{y}) = (2\pi)^{-n}|\boldsymbol{\Sigma}|^{-1/2}\exp(-w/2),$$

with

$$w = (\mathbf{y} - \boldsymbol{\mu})'\boldsymbol{\Sigma}^{-1}(\mathbf{y} - \boldsymbol{\mu}) = \mathbf{u}'\boldsymbol{\Sigma}^{-1}\mathbf{u} = \mathrm{tr}(\boldsymbol{\Sigma}^{*-1}\mathbf{U}'\mathbf{U}).$$

Further, because $\boldsymbol{\Sigma} = \boldsymbol{\Sigma}^* \otimes \mathbf{I}_n$, we have $|\boldsymbol{\Sigma}| = |\boldsymbol{\Sigma}^*|^n$. So the sample log-likelihood function is, apart from an irrelevant constant,

$$\log \mathcal{L} = (-1/2)\,[n \log|\boldsymbol{\Sigma}^*| + \mathrm{tr}(\boldsymbol{\Sigma}^{*-1}\mathbf{U}'\mathbf{U})].$$

If $\boldsymbol{\Sigma}^*$ is known, then this log-likelihood is evidently maximized by taking, as the estimator of $\boldsymbol{\beta}$, the value of $\mathbf{c}$ that minimizes $\mathrm{tr}(\boldsymbol{\Sigma}^{*-1}\mathbf{U}'\mathbf{U})$. But this is exactly the GLS criterion, so in the SUR model with $\boldsymbol{\Sigma}^*$ known, the normal-ML and GLS estimates of $\boldsymbol{\beta}$ are identical. If $\boldsymbol{\Sigma}^*$ is unknown, normal-ML will choose estimates of the σ's along with estimates of the β's. It is straightforward to show that for any choice of $\mathbf{c}$, the sample log-likelihood function above is maximized with respect to the σ's by taking $\hat{\mathbf{\Sigma}}^* = \mathbf{U}'\mathbf{U}/n$: compare the single-equation case of Section 19.2. Inserting that (conditional) solution into the log-likelihood, we have the "concentrated log-likelihood function"

$$\log \mathscr{L}^* = (-n/2)(\log|\mathbf{U}'\mathbf{U}/n| + 2),$$

which remains to be maximized with respect to $\mathbf{c}$. So in the SUR model with $\mathbf{\Sigma}$ unknown, the normal-ML procedure reduces to minimizing $\log |\mathbf{U}'\mathbf{U}/n|$ or, for that matter, minimizing $|\mathbf{U}'\mathbf{U}|$. Along with the LS, GLS, and FGLS criteria above, the normal-ML criterion is just a scalar measure of the matrix $\mathbf{U}'\mathbf{U}$.

Because the ML and FGLS criteria are different, we should anticipate that the resulting estimates will be different in general. Nevertheless, the estimators have the same asymptotic distribution. If the explanatory variables are identical in the two equations, then normal-ML and FGLS (and LS) estimates do coincide in every sample.

Having obtained the FGLS estimator $\hat{\mathbf{b}}^*$, we might calculate fresh residuals $\mathbf{e}^* = \mathbf{y} - \mathbf{X}\hat{\mathbf{b}}^*$, and use those residuals to re-estimate $\mathbf{\Sigma}^*$. Using that new estimate of $\mathbf{\Sigma}^*$ in place of the original one will generate a new FGLS estimate, say $\hat{\mathbf{b}}^{**}$. If this process is continued until convergence, that is, until the successive estimates of $\boldsymbol{\beta}$ stabilize, then the result is called the *iterative FGLS estimator*, sometimes known as IZEF (iterative Zellner efficient). All the successive estimators, including the terminal one, will share the asymptotic distribution of the GLS estimator $\mathbf{b}^*$. If carried through to convergence, IZEF will solve the FOC's for normal-ML estimation: the iteration procedure turns out to be an algorithm for solving the FOC's for minimization of $|\mathbf{U}'\mathbf{U}|$.

Exercises

30.1 True or false? In the SUR model, if the explanatory variables in the two equations are identical, then the LS residuals from the two equations are uncorrelated with each other.

30.2 True or false? In the SUR model, if the explanatory variables in the two equations are orthogonal to each other, then the LS coefficient estimates for the two equations are uncorrelated with each other.

30.3 Suppose that

$$E(\mathbf{y}_1) = \mathbf{x}_1\beta_1, \qquad E(\mathbf{y}_2) = \mathbf{x}_2\beta_2,$$

$$V(\mathbf{y}_1) = 4\mathbf{I}, \qquad V(\mathbf{y}_2) = 5\mathbf{I}, \qquad C(\mathbf{y}_1, \mathbf{y}_2) = 2\mathbf{I}.$$

Here $\mathbf{y}_1$, $\mathbf{y}_2$, $\mathbf{x}_1$, and $\mathbf{x}_2$ are $n \times 1$, with $\mathbf{x}_1'\mathbf{x}_1 = 5$, $\mathbf{x}_2'\mathbf{x}_2 = 6$, $\mathbf{x}_1'\mathbf{x}_2 = 3$. Let b_2 and b_2^* denote the LS and GLS estimators of β_2. Calculate $V(b_2)$ and $V(b_2^*)$.

30.4 Suppose that y_1 and y_2 are bivariate–normally distributed with unknown expectations μ_1 and μ_2, and known variances and covariance. Consider random sampling, sample size 100, from that population.

(a) Can we improve on the sample means $\bar{y}_1$ and $\bar{y}_2$ as estimators of μ_1 and μ_2? If so, how? If not, why not?
(b) Now suppose that it is known that $\mu_2 = 2\mu_1$. How would your answer in (a) change?

30.5 For the two-equation SUR model, suppose that $\sigma_{12} = 0$. Show that LS is preferable to FGLS, at least for samples of modest size.

30.6 For the two-equation SUR model, derive Eq. (30.4):

$$V(\mathbf{b}_2^*) = \sigma_{22}(\mathbf{X}_2'\mathbf{X}_2 + \phi^2\mathbf{X}_2^{*\prime}\mathbf{X}_2^*)^{-1}.$$

Hint: Adapt the Submatrix of Inverse Theorem in Section 17.5.

30.7 Table A.6 contains annual data on two firms, General Electric (GE) and Westinghouse (WE), for 1935–1954, taken from Theil (1971, p. 296). The variables are: V1 = Year number (1, . . . , 20), V2 = GE investment, V3 = GE market value, V4 = GE lagged capital stock, V5 = WE investment, V6 = WE market value, V7 = WE lagged capital stock. Variables V2–V7 are measured in millions of 1947 dollars. This data set is presumed to be available as an ASCII file labeled GEWE.

Suppose that the SUR model applies to

$$E(\mathbf{y}_1) = \mathbf{z}_1\beta_1 + \mathbf{z}_2\beta_2 + \mathbf{z}_3\beta_3,$$

$$E(\mathbf{y}_2) = \mathbf{z}_1\beta_4 + \mathbf{z}_4\beta_5 + \mathbf{z}_5\beta_6,$$

where y_1 = V2, $z_1 = 1$, z_2 = V3, z_3 = V4, y_2 = V5, z_4 = V6, z_5 = V7.

Using this data set, write and run a program to

(a) Calculate the LS estimates of $\boldsymbol{\beta}_1$ and $\boldsymbol{\beta}_2$, along with their standard errors.
(b) Using the residuals from those LS regressions, estimate $\boldsymbol{\Sigma}^*$.
(c) Calculate the FGLS estimates of $\boldsymbol{\beta}_1$ and $\boldsymbol{\beta}_2$, along with their standard errors.

GAUSS Hints:

(1) If A is m × m, and B is n × n, then the GAUSS command C = A .*. B gives the Kronecker product of A and B, namely the mn × mn matrix C whose typical block consists of the matrix B, multiplied by an element of A.
(2) But when n is large, asking for eye(n) to calculate $\boldsymbol{\Sigma}^{-1} = \boldsymbol{\Sigma}^{*-1} \otimes \mathbf{I}$ may exceed the memory capacity. So, use the expressions for $\mathbf{X}'\boldsymbol{\Sigma}^{-1}\mathbf{X}$ and $\mathbf{X}'\boldsymbol{\Sigma}^{-1}\mathbf{y}$ given in Section 30.3.

30.8 For the setup of Exercise 30.7, calculate the iterative FGLS estimates. That is, get the residuals from FGLS, use them to re-estimate $\boldsymbol{\Sigma}^*$, and thus $\boldsymbol{\Sigma}$ and $\boldsymbol{\beta}$. Continue this process until convergence, say until the successive estimates of $\boldsymbol{\beta}$ coincide to three decimal places.

31 *Structural Equation Models*

31.1. Introduction

Economists find it natural to model economic phenomena as a set of simultaneous equations in which *several* dependent variables are *jointly* determined. It may appear that such models are not only natural, but indeed essential: simultaneity, reciprocal causation, and feedback are ubiquitous in the real world. Suppose that the validity of LS estimation rested on unilateral causation running from the right-hand-side determining variables to a left-hand-side dependent variable in a regression equation. Then LS estimation would be inappropriate for models of joint determination. (For an argument along these lines, see Judge et al. 1988, pp. 599–601.)

But this rationale for special econometric treatment of simultaneous-equation models may be questioned on several counts.

The causal requirement that in regression the x's have to be the variables that actually determine y does not appear in the specification of the CR model: nothing in the CR model requires that the x's cause y. Indeed it is not obvious why the validity of a conditional expectation function (and its estimability by least squares) should depend on the assumption that x causes y. For example, suppose that x = father's height and y = daughter's height are bivariate–normally distributed. Then $E(y|x) = \alpha + \beta x$, so in random sampling, the LS regression of y on x will unbiasedly estimate α and β. But also $E(x|y) = \gamma + \delta y$, so in random sampling, the LS regression of x on y will unbiasedly estimate γ and δ. Neither of those regressions relies on an assumption of causal direction.

The fact that a model contains several dependent variables whose values are determined jointly cannot be an adequate reason to abandon LS. The SUR (regression systems) model of Chapter 30 had several dependent variables whose values might be viewed as jointly determined, and yet its parameters were estimable by LS.

It is sometimes said that the SUR model is not really simultaneous because one does not have to solve any equations to get the explicit equations for the y's. (From that perspective, the algebraic system $y_1 + y_2 = 3x$, $y_1 - y_2 = x$ is simultaneous, while its solution set, $y_1 = 2x$, $y_2 = x$, is not simultaneous.) If so, the notion that a simultaneous-equation model is required to represent an economic system correctly is tenuous. The solution to a system of equations is, after all, logically equivalent to the system itself. So, when an economic system can be represented correctly by a simultaneous-equation model, it can also be represented correctly by the reduced form of that model. If a pair of supply and demand equations (simultaneous) correctly represents a market, then so does the corresponding pair of quantity and price equations (nonsimultaneous).

A sounder case for special treatment of simultaneous-equation models can be made by arguing that those models represent situations in which the parameters of interest are not the parameters of a CEF (or BLP) among observable variables. For such situations, it should be clear that LS, which is inherently designed to estimate CEF's (or BLP's), will not be an appropriate estimation procedure.

31.2. Permanent Income Model

An instructive nonsimultaneous example at this point is Milton Friedman's permanent income model of consumption:

$$y = \alpha + \beta z + v, \qquad x = z + u,$$

where y = consumption, x = income, z = permanent income, v = transitory consumption, u = transitory income. The observed variables are y and x, while the unobserved variables z, u, v are assumed to have expectations μ, 0, 0, variances σ_z^2, σ_u^2, σ_v^2, and zero covariances. The parameters of interest are the slope β (which is called the "marginal propensity to consume out of permanent income") and the intercept α (which is relevant to Friedman's hypothesis that the relation goes through the origin).

For convenience, suppose that z, u, and v are trivariate–normally distributed, so that

$$\begin{pmatrix} x \\ y \end{pmatrix} = \begin{pmatrix} 0 \\ \alpha \end{pmatrix} + \begin{pmatrix} 1 & 1 & 0 \\ \beta & 0 & 1 \end{pmatrix} \begin{pmatrix} z \\ u \\ v \end{pmatrix}$$

is bivariate–normally distributed. Then

$$E(y|x) = \alpha^* + \beta^* x,$$

with

$$\beta^* = \sigma_{xy}/\sigma_x^2, \qquad \alpha^* = \mu_y - \beta^* \mu_x.$$

We calculate

$$\sigma_{xy} = C(z + u, \alpha + \beta z + v) = \beta\sigma_z^2, \qquad \sigma_x^2 = V(z + u) = \sigma_z^2 + \sigma_u^2,$$

$$\mu_y = \alpha + \beta E(z) + E(v) = \alpha + \beta\mu, \qquad \mu_x = E(z) + E(u) = \mu.$$

Let $\theta = \sigma_z^2/(\sigma_z^2 + \sigma_u^2)$. Then

$$\beta^* = \theta\beta, \qquad \alpha^* = \alpha + (1 - \theta)\,\beta\mu,$$

so

$$E(y|x) = [\alpha + (1 - \theta)\beta\mu] + (\theta\beta)x.$$

Clearly the parameters of interest, namely α and β, are not the intercept and slope of this CEF for the observable variables y and x.

If so, it is not surprising that the sample LS regression of y on x, namely $\hat{y} = a + bx$, is inappropriate for estimation of α and β. If we randomly sample from the joint distribution of x and y, then a NeoCR model applies, whence $E(a) = \alpha^*$ and $E(b) = \beta^* = \theta\beta$: see Chapter 25. (The same conclusion follows under a classical, stratified-on-x, sampling scheme.) This result is often described by saying that LS gives *biased estimators of the structural parameters* α and β. But a fairer description is that LS gives *unbiased estimators of the CEF parameters* α^* and β^*, which happen to be different from α and β. The latter description makes it clear that the issue is not one of estimators, but rather of parameters to be estimated.

Normality is not crucial to this argument. If the underlying variables are not joint-normal, then $E(y|x)$ may be nonlinear. But the best linear predictor $E^*(y|x)$ will still be $\alpha^* + \beta^* x$, with α^* and β^* as above. In random sampling, LS linear regression of y on x would consistently estimate α^* and β^*, and for that very reason be inconsistent for α and β.

31.3. Keynesian Model

Next consider a simultaneous example from the same perspective. Take this stochastic version of the simplest Keynesian model:

$$(31.1)\quad y = \alpha + \beta x + u,$$

$$(31.2)\quad x = y + z,$$

where y = consumption, x = income = output, z = investment, and u = "consumption shock." Equation (31.1) represents the demand for consumption, while (31.2) is the equilibrium condition, which says that output is equated to the sum of consumption demand and investment demand. Assume that z and u are random variables with

$$E(z) = \mu, \qquad V(z) = \sigma_z^2, \qquad E(u) = 0, \qquad V(u) = \sigma_u^2, \qquad C(z, u) = 0.$$

The zero-covariance assumption captures the idea that z is *exogenous*.

The understanding is that for given values of the pair (z, u), the model determines the values of the *endogenous* variables x and y. (So, paradoxically, it is the simultaneous-equation model that explicitly incorporates one-way causation.) The parameters of interest are α and β, and our concern is with whether those are estimable by sample LS regression of y on x.

The solution for the endogenous variables is made explicit in the *reduced form*, which expresses each endogenous variable in terms of the exogenous variable and the shock:

$$(31.3)\quad y = (\alpha + \beta z + u)/(1 - \beta),$$

$$(31.4)\quad x = (\alpha + z + u)/(1 - \beta).$$

For convenience suppose that z and u are bivariate normal. Then x and y are bivariate normal, so

$$E(y|x) = \alpha^* + \beta^* x,$$

with

$$\beta^* = \sigma_{xy}/\sigma_x^2, \qquad \alpha^* = \mu_y - \beta^* \mu_x.$$

From Eqs. (31.3)–(31.4) we calculate

$$\sigma_{xy} = (\beta\sigma_z^2 + \sigma_u^2)/(1 - \beta)^2, \qquad \sigma_x^2 = (\sigma_z^2 + \sigma_u^2)/(1 - \beta)^2,$$

$$\mu_y = (\alpha + \beta\mu)/(1 - \beta), \qquad \mu_x = (\alpha + \mu)/(1 - \beta).$$

Let $\theta = \sigma_z^2/(\sigma_z^2 + \sigma_u^2)$. Then

$$\beta^* = \theta\beta + (1 - \theta), \qquad \alpha^* = \theta\alpha - (1 - \theta)\mu,$$

so

$$E(y|x) = [\theta\alpha - (1 - \theta)\mu] + [\theta\beta + (1 - \theta)]x.$$

Clearly the parameters of interest, namely α and β, are not the intercept and slope of this CEF for the observable variables y and x.

If so, it is not surprising that the sample LS regression of y on x, namely $\hat{y} = a + bx$, is inappropriate for estimation of α and β. If we randomly sample from the joint distribution of x and y, then a NeoCR model applies, whence $E(a) = \alpha^*$ and $E(b) = \beta^* = \theta\beta + (1 - \theta)$. (The same conclusion follows if we adopt a classical, stratified-on-x, sampling scheme.) This result is traditionally described by saying that LS gives biased estimators of the structural parameters α and β. But a fairer description is that LS gives unbiased estimators of the CEF parameters α^* and β^*, which happen to be different from α and β. Again, the issue is not one of estimation methods, but rather of the parameters that are the targets of estimation.

Normality is not crucial to this argument. If the underlying variables are not joint-normal, then $E(y|x)$ may be nonlinear. But the best linear predictor $E^*(y|x)$ will still be $\alpha^* + \beta^* x$, with α^* and β^* as above. In random sampling, LS regression of y on x will consistently estimate the parameters α^* and β^*, and for that very reason be inconsistent for α and β.

Here is a direct way to view the situation without relying on normality. If we evaluate $E(y|x)$ from Eq. (31.1), we get

$$E(y|x) = \alpha + \beta x + E(u|x).$$

From Eq. (31.4),

$$C(x, u) = C(\alpha + z + u, u)/(1 - \beta) = \sigma_u^2/(1 - \beta).$$

Because x and u are correlated, we see that $E(u|x) \neq E(u) = 0$. So $E(y|x) \neq \alpha + \beta x$. In the consumption demand equation $y = \alpha + \beta x + u$, the *systematic part*, namely $\alpha + \beta x$, is not the CEF of y conditional on x. In that sense, the structural equation (31.1) is not a regression equation.

31.4. Estimation of the Keynesian Model

To estimate the parameters of Eq. (31.1), we may look for a CEF among observable variables in which α and β appear uncontaminated by θ. Our attention is directed to the reduced-form equation (31.4). We have

$$E(x|z) = [\alpha + z + E(u|z)]/(1 - \beta).$$

But $E(u|z) = E(u) = 0$ since z and u are independent under normality, so

$$E(x|z) = \pi_1 + \pi_2 z,$$

with

$$\text{(31.5)} \qquad \pi_1 = \alpha/(1 - \beta), \qquad \pi_2 = 1/(1 - \beta).$$

So the systematic part, namely $\pi_1 + \pi_2 z$, of the reduced-form income equation $x = \pi_1 + \pi_2 z + v$, where $v = u/(1 - \beta)$, is the CEF of x conditional on z.

Consequently, we should anticipate that LS regression of x on z will estimate the π's. If we randomly sample from the (bivariate-normal) joint distribution of x and z, then a NeoCR model will apply, so the sample LS regression of x on z, namely $\hat{x} = p_1 + p_2 z$, will unbiasedly estimate π_1 and π_2; those estimates will be consistent as well.

How can we convert these estimates of the reduced-form parameters into estimates of the structural parameters? Evidently, if we knew the reduced-form parameters π_1 and π_2, we could solve Eq. (31.5) to deduce the values of the structural parameters as

$$\alpha = \pi_1/\pi_2, \qquad \beta = (\pi_2 - 1)/\pi_2.$$

So it is natural to convert the reduced-form estimates into structural-form estimates via

$$\hat{\alpha} = p_1/p_2, \qquad \hat{\beta} = (p_2 - 1)/p_2.$$

This illustrates the *indirect least squares*, or ILS, *method*: use LS to estimate the reduced-form parameters, and then convert into estimates of the structural-form parameters. The ILS estimates are consistent (via S2, Section 9.5), although not unbiased (because of the nonlinearity).

An alternative approach to estimating α and β runs as follows. In Eq. (31.1), take expectations conditional on z:

$$E(y|z) = \alpha + \beta E(x|z),$$

using $E(u|z) = E(u) = 0$. Let $x^* = E(x|z) = \pi_1 + \pi_2 z$. Because x^* is a one-to-one function of z, we can write

$$E(y|x^*) = \alpha + \beta x^*,$$

which is a CEF in which α and β are the intercept and slope. If x^* were observed in our sample, we could regress y on $(1, x^*)$ to get unbiased estimates of α and β. But x^* is unobservable because π_1 and π_2 are unknown. Still, x^* is estimable as $\hat{x} = p_1 + p_2 z$, so the suggestion is to regress y on $(1, \hat{x})$ to estimate α and β. This illustrates the *two-stage least-squares*, or 2SLS, *method*: the first stage uses LS to estimate the reduced form and obtain fitted values; the second stage uses LS on the structural equation after the fitted values replace the observed values on the right-hand side. We should anticipate that the 2SLS estimates are consistent, though not unbiased.

The analogy principle also suggests a third approach. Consider again the consumption demand equation

$$(31.1) \qquad y = \alpha + \beta x + u.$$

We know that in the population $E(u) = 0$ and $C(z, u) = 0$. That is, α and β are the values for c_1 and c_2 that make $E(u) = 0$ and $E(zu) = 0$, where now $u = y - (c_1 + c_2 x)$. So let us choose as our estimates the values that make the analogous sample quantities zero. That is, take the values of c_1 and c_2 that make $\Sigma_i u_i = 0$ and $\Sigma_i z_i u_i = 0$. This illustrates the *instrumental variable*, or IV, *method*. We should anticipate that the IV estimates are consistent, though not unbiased.

No method for obtaining unbiased estimates of the structural parameters α and β exists, because there is no CEF among the observable variables x, y, and z that has α and β as its coefficients.

31.5. Structure versus Regression

We have just examined two situations where the parameters of interest are not the coefficients of a CEF among observable variables. Why then are the parameters interesting? Following Marschak (1953), we develop a rationale that takes prediction as the ultimate goal of the research.

For the permanent income model, recall that

$$\beta^* = \theta\beta, \qquad \alpha^* = \alpha + (1 - \theta)\beta\mu,$$

where

$$\theta = \sigma_z^2/(\sigma_z^2 + \sigma_u^2).$$

Why is knowledge of α^* and β^* in the CEF $E(y|x) = \alpha^* + \beta^* x$ not adequate? If our objective is to predict y given x, and there has been no change in the population, then knowledge of α^* and β^* would indeed suffice. But suppose that the population has changed, and that the new population is the same as the old one, except for a change in the variance of transitory income. That is, α, β, μ, and σ_z^2 are the same, but σ_u^2 is different. Then θ, α^*, and β^* will all be different. Unless we have estimated the constituent parts of α^* and β^*—those parts being the structural parameters—we will be ignorant of the new CEF parameters.

For the Keynesian model, recall that

$$\beta^* = \theta\beta + (1 - \theta), \qquad \alpha^* = \theta\alpha - (1 - \theta)\mu,$$

where

$$\theta = \sigma_z^2/(\sigma_z^2 + \sigma_u^2).$$

Why is knowledge of α^* and β^* in the CEF $E(y|x) = \alpha^* + \beta^* x$ not adequate? If our objective is to predict y given x, and there has been no change in the population, then knowledge of α^* and β^* would suffice. But suppose that the population has changed, and that the new population is the same as the first, except for a change in the variance of investment. That is, α, β, μ, and σ_u^2 are the same, but σ_z^2 is different. Then θ, α^*, and β^* will be different. Unless we have estimated the constituent parts of α^* and β^*—those parts being the structural parameters—we will be ignorant of the new CEF parameters.

To say that a set of parameters is structural is to claim that it is plausible that one of them will change while the rest of them remain invariant. It is a claim about the world rather than about the algebra or econometrics.

From that perspective, a conditional expectation function may or may not be structural. The relevance of this remark is not confined to multi-equation models.

Consider a perennial question: Which is the correct regression in a bivariate distribution? We have argued (Section 16.1) that both $E(y|x)$ and $E(x|y)$ may be legitimate targets of interest. But one target may be more interesting than the other. Suppose, for example, that $f_1(x)$, the marginal pdf of x, will change with no change in $g_2(y|x)$, the conditional pdf of y given x. Then $E(y|x)$ will remain the same while $E(x|y)$ will not. If this is how the world works, then $E(y|x)$ will be structural, while $E(x|y)$ will not. For example, suppose that y = daughter's height, and x = father's height. Suppose that you are asked to adapt results for the population at large to the subpopulation consisting of families in which the father has played professional basketball. Which CEF do you anticipate will be unchanged from the population to the subpopulation?

Consider another question: What is the objection to running a short regression—won't that estimate a CEF in its own right? Suppose that

$$E(y|x_1, x_2) = \beta_0 + \beta_1 x_1 + \beta_2 x_2.$$

It is often said that it is wrong to omit x_2 and run the short regression of y on x_1 alone. But it is quite possible that

$$E(y|x_1) = \beta_0^* + \beta_1^* x_1$$

is also correct. It is certainly correct, with $\beta_1^* = \beta_1 + \phi\beta_2$, if the CEF of x_2 on x_1 in the population is linear, with slope ϕ. Nevertheless, considerations of structure may dictate a preference for the long CEF. Suppose that the world changes because the joint pdf of x_1 and x_2 changes, with no change in the conditional pdf of y given x_1 and x_2. Then the conditional pdf of y given x_1 will change. In particular, a change in ϕ will, with β_1 and β_2 invariant, produce a change in β_1^*. Unless we have estimated the constituent parts of β_1^*—the structural parameters—we will be ignorant of the new CEF for y given x_1. In a sense, the original short regression is not wrong; it is just inadequate.

It should not be presumed that a long regression is inevitably more structural than a short regression. Suppose that a SUR model applies to

$$(31.6)\qquad E(y_1|\mathbf{z}) = \mathbf{z}'\boldsymbol{\beta}_1, \qquad E(y_2|\mathbf{z}) = \mathbf{z}'\boldsymbol{\beta}_2,$$
$$V(y_1|\mathbf{z}) = \omega_{11}, \qquad V(y_2|\mathbf{z}) = \omega_{22}, \qquad C(y_1, y_2|\mathbf{z}) = \omega_{12}.$$

If the distribution of the random vector $\mathbf{y} = (y_1, y_2)'$ conditional on $\mathbf{z}$ is bivariate normal, then

$$(31.7)\qquad E(y_2|\mathbf{z}, y_1) = \mathbf{z}'\boldsymbol{\beta}_2^* + \theta y_1,$$

with

$$\theta = \omega_{12}/\omega_{11}, \qquad \boldsymbol{\beta}_2^* = \boldsymbol{\beta}_2 - \theta\boldsymbol{\beta}_1.$$

If the conditional variance of y_1 given $\mathbf{z}$ changes with no change in the conditional expectations of y_1 and y_2 given $\mathbf{z}$, then the long regression (Eq. 31.7) will change while the short regression in Eq. (31.6) remains invariant.

To recapitulate: even if we are interested only in prediction (that is, in CEF's), if our interest includes predictions for populations other than the one from which our data came, we may well need to isolate structural parameters, those that may change individually. The argument may be reinforced if we recall an introductory microeconomics course, and ask why the determination of price and quantity is modeled in terms of separate demand and supply functions, rather than directly in terms of the exogenous variables (income, input prices, prices of substitutes). The answer is that we want to study what happens when the demand function shifts, while the supply function remains the same (or vice versa).

Finally, we might concede some validity to the notion that causality is a requirement for regression models. To the extent that causality is interpreted as "structural-ness," we may well agree that causality is needed to support interest in the parameters of a regression, while maintaining that it is not needed to support estimation of the parameters of a regression.

Exercises

31.1 Consider the permanent income model of Section 31.2.

(a) Suppose it is known that consumption is proportional to permanent income, in the sense that $\alpha = 0$. Propose a simple estimator of β that is consistent under random sampling.

(b) Alternatively, suppose that we observe not only y and x, but also $x' = z + u'$, where u' has zero expectation and is uncorrelated with z, u, and v. Propose a simple estimator of β that is consistent under random sampling. Hint: Find $C(x', x)$.

31.2 In the Keynesian model of Section 31.3, show that the ILS, 2SLS, and IV estimates of β are identical. Hint: $x = y + z$ at every observation.

31.3 Suppose that the endogenous variables q = quantity and p = price are jointly determined by this simultaneous-equation model:

Demand. $q = 30 - 2p + u,$

Supply. $q = 20 + p + v,$

in which the disturbances u and v are independent normal variables with zero expectations and variances $\sigma_u^2 = 5$, $\sigma_v^2 = 10$. Consider random sampling from the joint probability distribution of quantity and price. Let b denote the slope in the LS regression of quantity on price (with a constant term included). Calculate $E(b)$.

31.4 Consider the two-equation model:

$$y_1 = \alpha_1 x + u_1, \qquad y_2 = \alpha_2 y_1 + u_2,$$

where x, u_1, u_2 are independent $\mathcal{N}(0, 1)$ variables.

(a) Calculate $E(y_2|x)$ and $V(y_2|x)$.
(b) Calculate $E(y_2|x, y_1)$.
(c) In random sampling, will LS regression of y_2 on y_1 give an unbiased estimator of α_2? Explain.

31.5 Suppose that $y_1 = x + u_1$, $y_2 = 2y_1 + u_2$, where x, u_1, and u_2 are trivariate normal with zero expectations, unit variances, and $C(u_1, u_2) = 1/2$, $C(x, u_1) = 0 = C(x, u_2)$. Consider random sampling, sample size 50, from the joint distribution of x, y_1, y_2.

(a) Let p be the slope in the LS regression of y_2 on x. Find $E(p)$.
(b) Let b be the slope in the LS regression of y_2 on y_1. Find $E(b)$.

31.6 Suppose that

$$y_2 = \alpha_1 y_1 + \alpha_2 x + u_1, \qquad y_2 = \alpha_3 y_1 + u_2,$$

where x, u_1, u_2 have a trivariate normal distribution, with

$$E(x) = E(u_1) = E(u_2) = 0,$$

$$V(x) = 3, \qquad V(u_1) = V(u_2) = 2,$$

$$C(x, u_1) = C(x, u_2) = 0, \qquad C(u_1, u_2) = 1,$$

$$\alpha_1 = 1, \qquad \alpha_2 = 2, \qquad \alpha_3 = -3.$$

Consider random sampling, sample size 50, from the joint distribution of x, y_1, y_2. Let b be the slope in the sample LS linear regression of y_1 on y_2. Find $E(b)$.

31.7 The structural form of a model is

$$y_1 = \alpha_1 y_2 + \alpha_2 x_1 \qquad\qquad\qquad + u_1,$$

$$y_2 = \alpha_3 y_1 \qquad\qquad + \alpha_4 x_2 + \alpha_5 x_3 + u_2,$$

where x_1, x_2, x_3 are independent $\mathcal{N}(0, 1)$ variables, u_1, u_2 are independent $\mathcal{N}(0, 3)$ variables, and $\alpha_1 = -2$, $\alpha_2 = 2$, $\alpha_3 = 2$, $\alpha_4 = 4$, $\alpha_5 = 5$. The exogenous variables (the x's) are independent of the structural shocks (the u's).

(a) Find the pair of reduced-form equations. Include expressions for the reduced-form disturbances in terms of the structural disturbances.

(b) Find the variances and covariances for the x's and the y's. Display your results as a matrix $V(\mathbf{z})$, where $\mathbf{z} = (x_1, x_2, x_3, y_1, y_2)'$.

(c) Let $E(y_1 | y_2, x_1) = \alpha_1^* y_2 + \alpha_2^* x_1$. Find the α^*'s.

(d) Discuss the qualitative relation between the α^*'s and the α's.

32 Simultaneous-Equation Model

32.1. A Supply-Demand Model

In this chapter, we develop a specification that may be appropriate for linear simultaneous-equation models. But first, to fix ideas and introduce notation, we consider a two-equation system in which the endogenous variables y_1 = quantity and y_2 = price are determined by the exogenous variables x_1 = income, x_2 = wage rate, and x_3 = interest rate, and the disturbances u_1 = demand shock, u_2 = supply shock. For convenience we suppress intercepts in both equations. The *structural form* of the model is

$$\text{(32.1) Demand.} \quad y_1 = \alpha_1 y_2 + \alpha_2 x_1 + u_1,$$

$$\text{(32.2) Supply.} \quad y_2 = \alpha_3 y_1 + \alpha_4 x_2 + \alpha_5 x_3 + u_2,$$

Taking the terms in y_1 and y_2 over to the left-hand side and adopting a matrix representation we have

$$(y_1, y_2)\begin{pmatrix} 1 & -\alpha_3 \\ -\alpha_1 & 1 \end{pmatrix} = (x_1, x_2, x_3)\begin{pmatrix} \alpha_2 & 0 \\ 0 & \alpha_4 \\ 0 & \alpha_5 \end{pmatrix} + (u_1, u_2),$$

or

$$\mathbf{y}'\mathbf{\Gamma} = \mathbf{x}'\mathbf{B} + \mathbf{u}'.$$

In the structural-form coefficient matrices $\mathbf{\Gamma}$ and $\mathbf{B}$, the columns refer to equations, while the rows refer to variables.

Solve for each endogenous variable in terms of exogenous variables and structural shocks to get the *reduced form* of the model:

(32.3) Quantity. $y_1 = \pi_{11}x_1 + \pi_{21}x_2 + \pi_{31}x_3 + v_1$,

(32.4) Price. $y_2 = \pi_{12}x_1 + \pi_{22}x_2 + \pi_{32}x_3 + v_2$.

In matrix form, we have

$$(y_1, y_2) = (x_1, x_2, x_3)\begin{pmatrix}\pi_{11} & \pi_{12}\\ \pi_{21} & \pi_{22}\\ \pi_{31} & \pi_{32}\end{pmatrix} + (v_1, v_2),$$

or

$$\mathbf{y}' = \mathbf{x}'\mathbf{\Pi} + \mathbf{v}'.$$

To be more explicit: we solved by post-multiplying the structural form through by $\mathbf{\Gamma}^{-1}$ to get

$$\mathbf{y}' = \mathbf{x}'\mathbf{B}\mathbf{\Gamma}^{-1} + \mathbf{u}'\mathbf{\Gamma}^{-1} = \mathbf{x}'\mathbf{\Pi} + \mathbf{v}'.$$

Here $\mathbf{\Pi} = \mathbf{B}\mathbf{\Gamma}^{-1}$ is the reduced-form coefficient matrix, and $\mathbf{v}' = \mathbf{u}'\mathbf{\Gamma}^{-1}$ is the reduced-form disturbance vector. In $\mathbf{\Pi}$, the columns refer to equations; the rows refer to variables.

For our supply-demand model we have

$$\mathbf{\Gamma}^{-1} = (1/\Delta)\begin{pmatrix}1 & \alpha_3\\ \alpha_1 & 1\end{pmatrix}, \quad \text{with } \Delta = 1 - \alpha_1\alpha_3.$$

So

$$\mathbf{\Pi} = \begin{pmatrix}\pi_{11} & \pi_{12}\\ \pi_{21} & \pi_{22}\\ \pi_{31} & \pi_{32}\end{pmatrix} = (1/\Delta)\begin{pmatrix}\alpha_2 & \alpha_2\alpha_3\\ \alpha_1\alpha_4 & \alpha_4\\ \alpha_1\alpha_5 & \alpha_5\end{pmatrix},$$

$$\mathbf{v}' = (v_1, v_2) = (u_1 + \alpha_1 u_2, \alpha_3 u_1 + u_2)/\Delta.$$

To recapitulate: we began with a structural form that consisted of two linear equations relating the two endogenous variables, three exogenous variables, and two structural disturbances. We derived the reduced form, which consists of two linear equations, each of which expresses one endogenous variable as a linear function of the exogenous variables and a reduced-form disturbance (which in turn is a linear function of the structural disturbances). All exogenous variables and structural disturbances appeared in each reduced-form equation, although they did not all appear in each structural-form equation.

32.2. Specification of the Simultaneous-Equation Model

Our statistical specification for the linear *simultaneous-equation model*, or SEM, starts with a multivariate population. Suppose that the joint distribution of the $m \times 1$ endogenous-variable vector $\mathbf{y}$, the $k \times 1$ exogenous-variable vector $\mathbf{x}$, and the $m \times 1$ structural-disturbance vector $\mathbf{u}$, has these properties:

(A1) $\mathbf{y}'\boldsymbol{\Gamma} = \mathbf{x}'\mathbf{B} + \mathbf{u}'$,

(A2) $\boldsymbol{\Gamma}$ nonsingular,

(A3) $E(\mathbf{u}|\mathbf{x}) = \mathbf{0}$,

(A4) $V(\mathbf{u}|\mathbf{x}) = \boldsymbol{\Sigma}^*$ positive definite.

Here $\boldsymbol{\Gamma}$ is $m \times m$, $\mathbf{B}$ is $k \times m$, $\boldsymbol{\Sigma}^*$ is $m \times m$. Assumption (A1) gives the system of m structural equations in m endogenous variables. Assumption (A2) says that the system is complete, in the sense that $\mathbf{y}$ is uniquely determined by $\mathbf{x}$ and $\mathbf{u}$. Assumption (A3) says that $\mathbf{x}$ is exogenous, in the sense that the conditional expectation of the structural shock vector is the same for all values of $\mathbf{x}$. Assumption (A4) is a homoskedasticity requirement; positive definiteness simply rules out situations where there is an exact linear dependency among the structural disturbances.

In some variants of the SEM it is assumed that $\mathbf{u}|\mathbf{x}$ is multinormal, in others that $\mathbf{u}$ and $\mathbf{x}$ are stochastically independent. For some purposes, a weaker exogeneity condition, namely $C(\mathbf{x}, \mathbf{u}) = \mathbf{O}$, suffices.

The specification in (A1)–(A4), when supplemented by a sampling scheme, will constitute our SEM. The following implications are immediate:

(B1) $\mathbf{y}' = \mathbf{x}'\boldsymbol{\Pi} + \mathbf{v}'$,

with

(B2) $\boldsymbol{\Pi} = \mathbf{B}\boldsymbol{\Gamma}^{-1}, \qquad \mathbf{v}' = \mathbf{u}'\boldsymbol{\Gamma}^{-1}$.

Here Eq. (B1) is the reduced form, in which each endogenous variable is expressed as a linear function of the exogenous-variable vector $\mathbf{x}$ and the reduced-form disturbance vector $\mathbf{v}$, the latter being a linear function of $\mathbf{u}$. Linear function rules applied to Eqs. (A3)–(A4) imply

(B3) $E(\mathbf{v}|\mathbf{x}) = \mathbf{0}$,

(B4) $V(\mathbf{v}|\mathbf{x}) = \mathbf{\Omega}^*$,

with

(B5) $\mathbf{\Omega}^* = (\mathbf{\Gamma}^{-1})'\mathbf{\Sigma}^*\mathbf{\Gamma}^{-1}$ positive definite.

So the reduced-form disturbance vector $\mathbf{v}$ is mean-independent of, and homoskedastic with respect to, the exogenous variable vector $\mathbf{x}$. Then it follows from Eq. (B1) that

(B6) $E(\mathbf{y}'|\mathbf{x}) = \mathbf{x}'\mathbf{\Pi}$,

(B7) $V(\mathbf{y}|\mathbf{x}) = \mathbf{\Omega}^*$.

Equation (B6) says that the systematic part of the reduced form constitutes a set of conditional expectation functions, and (B7) says that the conditional variance function is constant. If $\mathbf{u}|\mathbf{x}$ is multinormal, then $\mathbf{y}|\mathbf{x}$ will also be multinormal.

Mean-independence implies uncorrelatedness, so

(B8) $C(\mathbf{x}, \mathbf{u}) = \mathbf{O}$,

(B9) $C(\mathbf{x}, \mathbf{v}) = \mathbf{O}$.

In conjunction with Eqs. (B1), (B4), and (B5), these imply that

(B10) $C(\mathbf{y}, \mathbf{v}) = C(\mathbf{\Pi}'\mathbf{x} + \mathbf{v}, \mathbf{v}) = V(\mathbf{v}) = \mathbf{\Omega}^*$,

(B11) $C(\mathbf{y}, \mathbf{u}) = C(\mathbf{y}, \mathbf{\Gamma}'\mathbf{v}) = C(\mathbf{y}, \mathbf{v})\mathbf{\Gamma} = \mathbf{\Omega}^*\mathbf{\Gamma} = (\mathbf{\Gamma}^{-1})'\mathbf{\Sigma}^*$.

In Eq. (B11) all elements of $(\mathbf{\Gamma}^{-1})'\mathbf{\Sigma}^*$ may be nonzero, so each endogenous variable may be correlated with every structural disturbance. And with $C(\mathbf{y}, \mathbf{u}) \neq \mathbf{O}$, we have

(B12) $E(\mathbf{u}|\mathbf{y}) \neq E(\mathbf{u}) = \mathbf{0}$.

The contrast between Eq. (A3) and Eq. (B12) is critical: the structural disturbances are mean-independent of the exogenous variables, but not of the endogenous variables. (In weaker form, the contrast is between Eq. B8 and Eq. B11: the structural disturbances are uncorrelated with $\mathbf{x}$, but not with $\mathbf{y}$.)

To illustrate, suppose that the SEM applies to our supply-demand model. The structural form is

(32.1) Demand. $y_1 = \alpha_1 y_2 + \alpha_2 x_1 \qquad\qquad + u_1,$

(32.2) Supply. $y_2 = \alpha_3 y_1 \qquad\qquad + \alpha_4 x_2 + \alpha_5 x_3 + u_2.$

Taking expectations conditional on y_2 and x_1 in Eq. (32.1), we get

$$E(y_1|y_2, x_1) = \alpha_1 y_2 + \alpha_2 x_1 + E(u_1|y_2, x_1),$$

in which the last term is not equal to $E(u_1) = 0$. If it were, then $C(y_2, u_1)$ would be 0. But

$$C(y_2, u_1) = C(v_2, u_1) = (1/\Delta)C(\alpha_3 u_1 + u_2, u_1) = (\alpha_3\sigma_{11} + \sigma_{12})/\Delta$$

is (coincidence apart) nonzero. While u_1 is uncorrelated with x_1, it is correlated with y_2. Consequently the systematic part of the structural demand equation, namely $\alpha_1 y_2 + \alpha_2 x_1$, is not the CEF (or BLP) for y_1 given y_2 and x_1. If $\mathbf{u}|\mathbf{x}$ and $\mathbf{x}$ were multinormal, then we would be assured that all CEF's were linear. In that case $E(y_1|y_2, x_1) = \alpha_1^* y_2 + \alpha_2^* x_1$, say, where α_1^* and α_2^* are deducible from the structural parameters and variances and covariances. That assurance is not available in general, but the negative conclusion remains: $E(y_1|y_2, x_1) \neq \alpha_1 y_2 + \alpha_2 x_1$.

Similarly in the supply equation (32.2), we have

$$E(y_2|y_1, x_2, x_3) = \alpha_3 y_1 + \alpha_4 x_2 + \alpha_5 x_3 + E(u_2|y_1, x_2, x_3)$$
$$\neq \alpha_3 y_1 + \alpha_4 x_2 + \alpha_5 x_3.$$

This analysis may be summarized by saying that in the SEM, the systematic parts of the structural equations are not regression functions. It follows that they ought not to be estimated by least squares.

In contrast, consider the reduced form of our supply-demand model:

(32.3) Quantity. $y_1 = \pi_{11} x_1 + \pi_{21} x_2 + \pi_{31} x_3 + v_1,$

(32.4) Price. $y_2 = \pi_{12} x_1 + \pi_{22} x_2 + \pi_{32} x_3 + v_2.$

Because $E(v_1|x_1, x_2, x_3) = E(v_2|x_1, x_2, x_3) = 0$, we have

$$E(y_1|x_1, x_2, x_3) = \pi_{11} x_1 + \pi_{21} x_2 + \pi_{31} x_3,$$

$$E(y_2|x_1, x_2, x_3) = \pi_{12} x_1 + \pi_{22} x_2 + \pi_{32} x_3.$$

The systematic part of each reduced-form equation is the CEF of a y given the x's. So the reduced-form equations are regression equations, and hence their parameters are presumably estimable by least squares.

32.3. Sampling

Let us now turn from the population to the sample. Suppose that we obtain a sample of n observations from the multivariate distribution of $\mathbf{x}$ and $\mathbf{y}$ by stratified sampling: n values of $\mathbf{x}'$, namely $\mathbf{x}_i'$ $(i = 1, \ldots, n)$, are selected, forming the rows of the $n \times k$ observed matrix $\mathbf{X}$, with $\text{rank}(\mathbf{X}) = k$. For each observation, a random drawing is made from the relevant joint-conditional distribution $g_2(\mathbf{y}|\mathbf{x})$, giving the $\mathbf{y}_i'$ $(i = 1, \ldots, n)$, which form the rows of the $n \times m$ observed matrix $\mathbf{Y}$. Successive drawings are independent. This completes the specification of our SEM.

For convenience we will confine our attention to the two-equation case, where

$$\mathbf{Y} = (\mathbf{y}_1, \mathbf{y}_2), \qquad \mathbf{\Pi} = (\boldsymbol{\pi}_1, \boldsymbol{\pi}_2), \qquad \mathbf{\Omega}^* = \begin{pmatrix} \omega_{11} & \omega_{12} \\ \omega_{21} & \omega_{22} \end{pmatrix}.$$

We have

$$E(\mathbf{y}_1) = \mathbf{X}\boldsymbol{\pi}_1, \qquad E(\mathbf{y}_2) = \mathbf{X}\boldsymbol{\pi}_2,$$

$$V(\mathbf{y}_1) = \omega_{11}\mathbf{I}, \qquad V(\mathbf{y}_2) = \omega_{22}\mathbf{I}, \qquad C(\mathbf{y}_1, \mathbf{y}_2) = \omega_{12}\mathbf{I}.$$

Except for notation, this is just the two-equation SUR (regression systems) model of Chapter 30. The conclusion is that in the SEM, a SUR model applies to the reduced form. If the data were obtained by random sampling from the joint distribution of $(\mathbf{x}', \mathbf{y}')$, then we would have a neoclassical version of the SUR model.

32.4. Remarks

The SEM specification turns out to be a roundabout, rather exotic, way of specifying a SUR model for the reduced form. We have already discussed methods for estimating the parameters of a SUR model, including LS, GLS, FGLS, and normal-ML. Why is a separate discussion needed for this special case of a SUR model? Indeed, why not just estimate the reduced form by LS? After all, the reduced form appears to have identical explanatory variables, a SUR situation in which LS coincides with GLS: see Section 30.3.

What does justify special econometric consideration of the SEM?

In simultaneous-equation models, the targets of research are the structural parameters (the α's in our supply-demand example, the elements of $\mathbf{\Gamma}$ and $\mathbf{B}$ in the general case), rather than the reduced-form parameters (the π's, or $\mathbf{\Pi}$). So the SEM is a situation in which the parameters of interest are not those of the CEF's among observable variables. As a consequence, rules are needed for converting estimates of the π's into estimates of the α's, that is, for converting estimates of $\mathbf{\Pi}$ into estimates of $\mathbf{\Gamma}$ and $\mathbf{B}$.

Why not just get LS estimates of the π's and convert them into estimates of α's in the obvious way, as was done for the Keynesian model in Section 31.4?

The answer here is two-fold. First, an SEM may imply restrictions on the π's, in which case the best estimates of the π's are not obtained by equation-by-equation LS on the reduced form. Second, there may be no way to convert estimates of π's into estimates of α's, because the parameters α may not be uniquely deducible from the parameters π. These two items are interrelated: *identification* deals with the issue of whether $\mathbf{\Pi}$ uniquely determines $\mathbf{\Gamma}$ and $\mathbf{B}$; *restrictions* deal with the issue of whether the prior knowledge of certain elements of $\mathbf{\Gamma}$ and $\mathbf{B}$ implies restrictions on $\mathbf{\Pi}$.

33 *Identification and Restrictions*

33.1. Introduction

We now investigate whether the structural parameters are uniquely determined by the reduced-form parameters. If we know the elements of the $\mathbf{\Pi}$ matrix, can we uniquely deduce the values of the elements of $\mathbf{\Gamma}$ and $\mathbf{B}$? At first glance, this task seems hopeless, because there are only km elements in $\mathbf{\Pi}$, while there are m^2 elements in $\mathbf{\Gamma}$ and km elements in $\mathbf{B}$. It appears that the number of unknowns, $m(m + k)$, must exceed the number of equations, mk. But a structural model typically will include prior knowledge of certain elements of $\mathbf{\Gamma}$ and $\mathbf{B}$. Such knowledge reduces the number of unknowns, and hence opens up the possibility of a unique solution for the remaining structural parameters. The prior knowledge may even be rich enough to constrain the values of the reduced-form coefficients.

The key to our investigation is the matrix equation that relates the reduced-form coefficients to the structural-form coefficients, namely $\mathbf{\Pi} = \mathbf{B}\mathbf{\Gamma}^{-1}$, which we may rewrite as

$$\mathbf{\Pi}\mathbf{\Gamma} = \mathbf{B}.$$

We will treat $\mathbf{\Pi}$ as known along with particular elements of $\mathbf{\Gamma}$ and $\mathbf{B}$. The question will be whether we can solve $\mathbf{\Pi}\mathbf{\Gamma} = \mathbf{B}$ uniquely for the remaining unknown elements of $\mathbf{\Gamma}$ and $\mathbf{B}$. When a structural parameter is uniquely determined in that manner, then we say that the parameter is *identified in terms of* $\mathbf{\Pi}$, or more simply, that it is identified. Although the focus is on identification, a few preliminary remarks about estimation are also included.

33.2. Supply-Demand Models

We explore identification via three variants of a supply-demand system. Each variant is a two-equation model in which the endogenous variables y_1 = quantity and y_2 = price are determined by the exogenous variables x_1 = income, x_2 = wage rate, and x_3 = interest rate, and the structural disturbances u_1 = demand shock, and u_2 = supply shock.

Model A. First take the example of Section 32.1. The structural form is:

(33.1A) Demand. $y_1 = \alpha_1 y_2 + \alpha_2 x_1 + u_1,$

(33.2A) Supply. $y_2 = \alpha_3 y_1 + \alpha_4 x_2 + \alpha_5 x_3 + u_2.$

Observe that in this economic model, the values 1 and 0 have been preassigned to certain elements of $\mathbf{\Gamma}$ and $\mathbf{B}$.

The reduced form is:

(33.3) Quantity. $y_1 = \pi_{11}x_1 + \pi_{21}x_2 + \pi_{31}x_3 + v_1,$

(33.4) Price. $y_2 = \pi_{12}x_1 + \pi_{22}x_2 + \pi_{32}x_3 + v_2.$

As previously shown, the reduced-form coefficients relate to the structural-form coefficients via:

$$\mathbf{\Pi} = \begin{pmatrix} \pi_{11} & \pi_{12} \\ \pi_{21} & \pi_{22} \\ \pi_{31} & \pi_{32} \end{pmatrix} = (1/\Delta) \begin{pmatrix} \alpha_2 & \alpha_2\alpha_3 \\ \alpha_1\alpha_4 & \alpha_4 \\ \alpha_1\alpha_5 & \alpha_5 \end{pmatrix}, \text{ with } \Delta = 1 - \alpha_1\alpha_3.$$

Reading this as a system of six equations in five unknowns, we see that if we were given the π's, then we could solve uniquely for the α's:

$$\alpha_3 = \pi_{12}/\pi_{11}, \qquad \alpha_1 = \pi_{21}/\pi_{22} = \pi_{31}/\pi_{32},$$

$$\Delta = 1 - \alpha_1\alpha_3, \qquad \alpha_2 = \Delta\pi_{11}, \qquad \alpha_4 = \Delta\pi_{22}, \qquad \alpha_5 = \Delta\pi_{32}.$$

We conclude that all the structural coefficients are identified in terms of the reduced-form coefficients. Furthermore, there is a restriction on the reduced-form coefficients, namely

$$\pi_{21}/\pi_{22} = \pi_{31}/\pi_{32}.$$

It is not surprising to find one restriction on the π's, because all six π's are functions of only five α's.

Actually, it is more convenient to analyze identification via $\mathbf{\Pi\Gamma} = \mathbf{B}$, which we write out here as:

$$\begin{pmatrix} \pi_{11} & \pi_{12} \\ \pi_{21} & \pi_{22} \\ \pi_{31} & \pi_{32} \end{pmatrix} \begin{pmatrix} 1 & -\alpha_3 \\ -\alpha_1 & 1 \end{pmatrix} = \begin{pmatrix} \alpha_2 & 0 \\ 0 & \alpha_4 \\ 0 & \alpha_5 \end{pmatrix}.$$

Reading off, we see these six equations in five unknowns:

$$\text{(33.5A)} \quad \begin{aligned} \pi_{11} - \alpha_1\pi_{12} &= \alpha_2 & \qquad \pi_{12} - \alpha_3\pi_{11} &= 0 \\ \pi_{21} - \alpha_1\pi_{22} &= 0 & \qquad \pi_{22} - \alpha_3\pi_{21} &= \alpha_4 \\ \pi_{31} - \alpha_1\pi_{32} &= 0 & \qquad \pi_{32} - \alpha_3\pi_{31} &= \alpha_5. \end{aligned}$$

On the left, which refers to the demand equation, we see three equations in two unknowns; on the right, which refers to the supply equation, we see three equations in three unknowns.

It is easy to solve the system on the left of (33.5A). First solve either of the equations that has 0 on its right-hand side. Because of the restriction on $\mathbf{\Pi}$, they give the same answer, namely

$$\alpha_1 = \pi_{21}/\pi_{22} = \pi_{31}/\pi_{32}.$$

Insert that value for α_1 into the remaining equation to get

$$\alpha_2 = \pi_{11} - \alpha_1\pi_{12}.$$

We conclude that the coefficients of the demand equation are identified in terms of $\mathbf{\Pi}$.

It is also easy to solve the system on the right of (33.5A). First solve the equation that has 0 on its right-hand side to get

$$\alpha_3 = \pi_{12}/\pi_{11}.$$

Insert that value of α_3 into the remaining equations to get

$$\alpha_4 = \pi_{22} - \alpha_3\pi_{21}, \qquad \alpha_5 = \pi_{32} - \alpha_3\pi_{31}.$$

We conclude that the coefficients of the supply equation are also identified in terms of $\mathbf{\Pi}$.

With respect to estimation, because there is a restriction on $\mathbf{\Pi}$, equation-by-equation LS estimation of the reduced form will not be optimal: see Section 30.6. But if we estimate the reduced form subject to that restriction, then estimates of the π's can be converted into estimates of the α's using the sample counterpart of the system (33.5A).

Model B. Modify the structural model by allowing the wage rate x_2 to enter the demand equation:

(33.1B) Demand. $y_1 = \alpha_1 y_2 + \alpha_2 x_1 + \alpha_6 x_2 \qquad\qquad + u_1,$

(33.2B) Supply. $y_2 = \alpha_3 y_1 \qquad\qquad + \alpha_4 x_2 + \alpha_5 x_3 + u_2.$

The reduced-form equations are again (33.3)–(33.4) but now, in the $\mathbf{\Pi\Gamma} = \mathbf{B}$ format, the relation between reduced-form and structural coefficients is:

$$\begin{pmatrix} \pi_{11} & \pi_{12} \\ \pi_{21} & \pi_{22} \\ \pi_{31} & \pi_{32} \end{pmatrix} \begin{pmatrix} 1 & -\alpha_3 \\ -\alpha_1 & 1 \end{pmatrix} = \begin{pmatrix} \alpha_2 & 0 \\ \alpha_6 & \alpha_4 \\ 0 & \alpha_5 \end{pmatrix}.$$

Reading off, we see these six equations in six unknowns:

$$\text{(33.5B)} \quad \begin{aligned} \pi_{11} - \alpha_1\pi_{12} &= \alpha_2 & \qquad \pi_{12} - \alpha_3\pi_{11} &= 0 \\ \pi_{21} - \alpha_1\pi_{22} &= \alpha_6 & \qquad \pi_{22} - \alpha_3\pi_{21} &= \alpha_4 \\ \pi_{31} - \alpha_1\pi_{32} &= 0 & \qquad \pi_{32} - \alpha_3\pi_{31} &= \alpha_5. \end{aligned}$$

On each side we see three equations in three unknowns. On the left of (33.5B), solve the last equation for $\alpha_1 = \pi_{31}/\pi_{32}$; insert that value into the first and second equations to get α_2 and α_6. We conclude that the coefficients of the demand equation are identified in terms of $\mathbf{\Pi}$. On the right of (33.5B), solve the first equation for $\alpha_3 = \pi_{12}/\pi_{11}$; insert that value into the other two equations to get α_4 and α_5. We conclude that the coefficients of the supply equation are also identified in terms of $\mathbf{\Pi}$. There are no restrictions on the π's, which is not surprising because the six π's are functions of six α's.

With respect to estimation, because there are no restrictions on $\mathbf{\Pi}$, the reduced form is a SUR model with identical explanatory variables, so equation-by-equation LS will coincide with GLS and hence be optimal. The LS estimates of the π's can be converted into estimates of the α's by using the sample counterpart of the system (33.5B).

Model C. Modify the original structural model by allowing income x_1 to enter the supply equation:

(33.1C) Demand. $y_1 = \alpha_1 y_2 + \alpha_2 x_1 \qquad\qquad + u_1,$

(33.2C) Supply. $y_2 = \alpha_3 y_1 + \alpha_7 x_1 + \alpha_4 x_2 + \alpha_5 x_3 + u_2.$

The reduced-form equations are again (33.3)–(33.4) but now, in the $\mathbf{\Pi\Gamma} = \mathbf{B}$ format, the relation between reduced-form and structural coefficients is:

$$\begin{pmatrix} \pi_{11} & \pi_{12} \\ \pi_{21} & \pi_{22} \\ \pi_{31} & \pi_{32} \end{pmatrix} \begin{pmatrix} 1 & -\alpha_3 \\ -\alpha_1 & 1 \end{pmatrix} = \begin{pmatrix} \alpha_2 & \alpha_7 \\ 0 & \alpha_4 \\ 0 & \alpha_5 \end{pmatrix}.$$

Reading off, we see these six equations in six unknowns:

$$\text{(33.5C)} \quad \begin{aligned} \pi_{11} - \alpha_1\pi_{12} &= \alpha_2 & \qquad \pi_{12} - \alpha_3\pi_{11} &= \alpha_7 \\ \pi_{21} - \alpha_1\pi_{22} &= 0 & \qquad \pi_{22} - \alpha_3\pi_{21} &= \alpha_4 \\ \pi_{31} - \alpha_1\pi_{32} &= 0 & \qquad \pi_{32} - \alpha_3\pi_{31} &= \alpha_5. \end{aligned}$$

It is clear how to solve the system on the left of (33.5C), that is, to determine the parameters of the demand equation. First solve either of the equations that has a 0 on its right-hand side for $\alpha_1 = \pi_{31}/\pi_{32} = \pi_{21}/\pi_{22}$, and then get α_2 from the remaining equation. So the coefficients of the demand equation are identified in terms of $\mathbf{\Pi}$. And there is a restriction on the π's, namely $\pi_{31}/\pi_{32} = \pi_{21}/\pi_{22}$, which is not surprising because on the left of (33.5C) there are three equations in two unknowns.

However, the system on the right of (33.5C) consists of three equations in four unknowns. We can assign any value to α_3 and then solve for α_4, α_5, α_7. A different arbitrary value for α_3 would generate different values for α_4, α_5, α_7. The solution is not unique. Evidently, there are an infinity of alternative sets of values for the supply-equation coefficients that, in conjunction with the appropriate set of values for the demand-equation coefficients, produce the same set of values for the π's. Consequently, $\mathbf{\Pi}$ does not contain enough information to uniquely deduce the $\mathbf{\Gamma}$ and $\mathbf{B}$ that produced it. We conclude that the coefficients of the supply equation are not identified in terms of $\mathbf{\Pi}$.

With respect to estimation, because there is a restriction on $\mathbf{\Pi}$, LS estimation of the reduced form will not be optimal. If the reduced form is estimated subject to that restriction, then estimates of the demand equation can be derived. But in Model C, there is no way to estimate the supply equation: to seek estimates of its coefficients is not a meaningful task.

To recapitulate: we have answered the question "Are the α's uniquely determined by the π's?" for three variants of the supply-demand model. Because certain elements of $\mathbf{\Gamma}$ and $\mathbf{B}$ were known a priori, the answer was sometimes yes. Indeed sometimes the knowledge was sufficient enough to restrict the set of admissible π's. We have seen that not only the number of pieces of prior information but also their location is crucial to the answers. In our examples, the prior information consisted only of *exclusions* (zero coefficients) and *normalizations* (a 1 in each column of $\mathbf{\Gamma}$). In other simultaneous-equation models there may be additional pieces of prior information—for example, two structural coefficients may be known to be equal. Such information also serves to aid identification and may even constrain the reduced-form coefficient matrix $\mathbf{\Pi}$.

33.3. Uncorrelated Disturbances

We have focused on getting $\mathbf{\Gamma}$ and $\mathbf{B}$ from $\mathbf{\Pi\Gamma} = \mathbf{B}$, but there is another relation between structural and reduced-form parameters that may assist identification, namely $\mathbf{\Omega}^* = (\mathbf{\Gamma}^{-1})'\mathbf{\Sigma}^*\mathbf{\Gamma}^{-1}$, obtained as Eq. (B5) in Chapter 32. We may rewrite this as

$$(33.6) \qquad \mathbf{\Sigma}^* = \mathbf{\Gamma}'\mathbf{\Omega}^*\mathbf{\Gamma}.$$

Like the coefficient matrix $\mathbf{\Pi}$, the disturbance variance matrix $\mathbf{\Omega}$ is estimable from LS regression on the reduced form. Suppose that both $\mathbf{\Pi}$ and $\mathbf{\Omega}^*$ are known. Can we exploit Eq. (33.6) to help deduce $\mathbf{\Gamma}$? In general, the answer is no. With $\mathbf{\Sigma}^*$ unknown, Eq. (33.6) merely suffices to deduce $\mathbf{\Sigma}^*$ once $\mathbf{\Omega}^*$ and $\mathbf{\Gamma}$ are known. However, there may be prior information on $\mathbf{\Sigma}^*$ that reduces the number of unknowns in Eq. (33.6) and thus frees it to help in identifying $\mathbf{\Gamma}$. For example, suppose that the structural disturbances are known to be uncorrelated with one another. Then $\mathbf{\Sigma}^*$ is diagonal, so there are only m unknown elements of $\mathbf{\Sigma}^*$, while Eq. (33.6) has $m(m + 1)/2$ distinct equations.

To illustrate: for our supply-demand Model C, Eq. (33.6) is

$$\begin{pmatrix} \sigma_{11} & \sigma_{12} \\ \sigma_{21} & \sigma_{22} \end{pmatrix} = \begin{pmatrix} 1 & -\alpha_1 \\ -\alpha_3 & 1 \end{pmatrix} \begin{pmatrix} \omega_{11} & \omega_{12} \\ \omega_{21} & \omega_{22} \end{pmatrix} \begin{pmatrix} 1 & -\alpha_3 \\ -\alpha_1 & 1 \end{pmatrix}.$$

The off-diagonal element is

$$\sigma_{12} = -\alpha_3\omega_{11} + \omega_{12} + \alpha_1\alpha_3\omega_{21} - \alpha_1\omega_{22}.$$

Suppose that the structural disturbances are known to be uncorrelated, so $\sigma_{12} = 0$. Then

$$(33.7) \qquad \alpha_3 = (\omega_{12} - \alpha_1\omega_{22})/(\omega_{11} - \alpha_1\omega_{21}),$$

so α_3 is uniquely determined by the ω's and α_1. Referring back to the analysis of Model C in Section 33.2, we see that this will suffice to complete identification of the supply equation. So in this case, all the structural coefficients will be *identified in terms of* $\mathbf{\Pi}$ *and* $\mathbf{\Omega}^*$.

With respect to estimation, because the ω's are estimable from the residuals of LS regression on the reduced form, the sample counterpart of Eq. (33.7) will in this situation be usable along with the sample counterpart of system (33.5C).

A leading special case, known as the *fully recursive model*, arises when $\mathbf{\Sigma}^*$ is diagonal *and* $\mathbf{\Gamma}$ is triangular. Here, all off-diagonal elements in $\mathbf{\Sigma}^*$ are zero, and in $\mathbf{\Gamma}$ all elements below the diagonal are zero. Then not only are all the structural parameters identified, but in fact they are estimable by LS regression on the structural equations. If $\mathbf{\Gamma}$ is upper-triangular, then $\mathbf{\Gamma}^{-1}$ will also be upper-triangular, so $(\mathbf{\Gamma}^{-1})'$ will be lower-triangular. In conjunction with the diagonality of $\mathbf{\Sigma}^*$, this will imply that

$$C(\mathbf{y}, \mathbf{u}) = (\mathbf{\Gamma}^{-1})'\mathbf{\Sigma}^*$$

is lower-triangular. Any endogenous variable on the right-hand side of a structural equation will be uncorrelated with the disturbance in that equation. If so, each structural equation is a CEF (or at least a BLP), hence identified, and indeed estimable by LS.

33.4. Other Sources of Identification

We have seen that the reduced-form coefficient matrix $\mathbf{\Pi}$ and disturbance variance matrix $\mathbf{\Omega}^*$ may both be used to identify structural parameters in the SEM. Can anything else be used? The answer is effectively no. After all, the most one can hope to learn from stratified sampling is $g_2(\mathbf{y}|\mathbf{x})$, the joint-conditional pdf (or pmf) of the endogenous variables given the exogenous variables. If we learn that distribution, then we will know $E(\mathbf{y}'|\mathbf{x}) = \mathbf{x}'\mathbf{\Pi}$ and $V(\mathbf{y}|\mathbf{x}) = \mathbf{\Omega}^*$. If $\mathbf{y}|\mathbf{x}$ is multinormal, then there is nothing more to learn: knowledge of both $\mathbf{\Pi}$ and $\mathbf{\Omega}^*$ is equivalent to knowledge of $g_2(\mathbf{y}|\mathbf{x})$. So if a structural parameter is not identified in terms of $\mathbf{\Pi}$ and $\mathbf{\Omega}^*$, then it is not identifed in terms of $g_2(\mathbf{y}|\mathbf{x})$; that is, it is not identified. To be sure, if $g_2(\mathbf{y}|\mathbf{x})$ were known to be nonnormal, then there might be more information available, but that situation is rare indeed, and we ignore it here. In random sampling, we can also learn $f_1(\mathbf{x})$, the joint-marginal pdf (or pmf) of the exogenous variables, but the structural parameters do not enter that function.

In the next chapter, we will proceed on the presumption that the only prior information consists of exclusions and normalizations on $\mathbf{\Gamma}$ and $\mathbf{B}$. Then $\mathbf{\Pi}$ is the sole source of information available for identifying the unknown structural coefficients. The remaining task will be to obtain estimates of $\mathbf{\Pi}$ that are convertible into estimates of the unknown, but identified, elements of $\mathbf{\Gamma}$ and $\mathbf{B}$.

Exercises

33.1 Consider the simultaneous-equation model

$$y_1 = \alpha_1 y_2 + \alpha_2 x_1 \qquad\qquad + u_1,$$

$$y_2 = \alpha_3 y_1 \qquad\qquad + \alpha_4 x_2 + u_2,$$

where the exogenous variables x_1 and x_2 are independent of the disturbances u_1 and u_2. The reduced form of the model is

$$y_1 = \pi_1 x_1 + \pi_2 x_2 + v_1,$$

$$y_2 = \pi_3 x_1 + \pi_4 x_2 + v_2.$$

(a) You are told that $\pi_1 = 1$, $\pi_2 = 4$, $\pi_3 = -2$, $\pi_4 = 2$. Determine the values of α_1, α_2, α_3, α_4.

(b) You are also told that x_1, x_2, u_1, u_2 are independent $\mathcal{N}(0, 1)$ variables. Predict the value of y_1 that will occur if $y_2 = x_1 = 1$.

33.2 A simple theoretical model for the labor market consists of the supply function $H = S(W, N)$ and the demand function $W = D(H, X)$, where the endogenous variables are H = hours worked and W = wage rate, while the exogenous variables are N = family size and X = worker characteristics. The economic presumptions are that $\partial S/\partial W > 0$ and $\partial D/\partial H < 0$. A linear version is

$$y_1 = \alpha_1 y_2 + \alpha_2 x_1 + \alpha_3 x_2 + u_1,$$

$$y_2 = \alpha_4 y_1 + \alpha_5 x_1 + \alpha_6 x_3 + \alpha_7 x_4 + \alpha_8 x_5 + u_2,$$

where y_1 = months worked, y_2 = wage rate, $x_1 = 1$, x_2 = family size, x_3 = education, x_4 = age, x_5 = race (= 1 if black, = 0 if white), u_1 = supply shock, u_2 = demand shock. Suppose that the SEM model applies. Write out the system $\mathbf{\Pi\Gamma} = \mathbf{B}$, and analyze identification and restrictions.

34 Estimation in the Simultaneous-Equation Model

34.1. Introduction

We proceed to methods for estimating the structural parameters in the SEM. We continue to confine attention to the two-equation case, with some specialization to our supply-demand models. The only types of prior information that we allow for are exclusions and normalizations.

For the population, the model has

$$E(\mathbf{y}'|\mathbf{x}) = \mathbf{x}'\mathbf{\Pi}, \qquad V(\mathbf{y}|\mathbf{x}) = \mathbf{\Omega}^*,$$

where $\mathbf{\Pi} = (\boldsymbol{\pi}_1, \boldsymbol{\pi}_2)$ is $k \times 2$, and $\mathbf{\Omega}^* = \{\omega_{hi}\}$ is 2×2 and positive definite. Reading off, we have

$$E(y_1|\mathbf{x}) = \mathbf{x}'\boldsymbol{\pi}_1, \qquad E(y_2|\mathbf{x}) = \mathbf{x}'\boldsymbol{\pi}_2,$$

$$V(y_1|\mathbf{x}) = \omega_{11}, \qquad V(y_2|\mathbf{x}) = \omega_{22}, \qquad C(y_1, y_2|\mathbf{x}) = \omega_{12}.$$

We sample by the stratified scheme, so that the $n \times k$ matrix $\mathbf{X}$ is nonstochastic and has rank k, while the $n \times 2$ matrix $\mathbf{Y} = (\mathbf{y}_1, \mathbf{y}_2)$ is random. We have

$$E(\mathbf{y}_1) = \mathbf{X}\boldsymbol{\pi}_1, \qquad E(\mathbf{y}_2) = \mathbf{X}\boldsymbol{\pi}_2,$$

$$V(\mathbf{y}_1) = \omega_{11}\mathbf{I}, \qquad V(\mathbf{y}_2) = \omega_{22}\mathbf{I}, \qquad C(\mathbf{y}_1, \mathbf{y}_2) = \omega_{12}\mathbf{I}.$$

Except for notation, this is precisely the SUR (regression systems) model of Chapter 30. To stack the two equations, let

$$\mathbf{y} = \begin{pmatrix} \mathbf{y}_1 \\ \mathbf{y}_2 \end{pmatrix}, \qquad \mathbf{X}^\circ = \begin{pmatrix} \mathbf{X} & \mathbf{O} \\ \mathbf{O} & \mathbf{X} \end{pmatrix},$$

$$\boldsymbol{\Omega} = \begin{pmatrix} \omega_{11}\mathbf{I} & \omega_{12}\mathbf{I} \\ \omega_{21}\mathbf{I} & \omega_{22}\mathbf{I} \end{pmatrix}, \qquad \boldsymbol{\pi} = \begin{pmatrix} \boldsymbol{\pi}_1 \\ \boldsymbol{\pi}_2 \end{pmatrix}.$$

Then $E(\mathbf{y}) = \mathbf{X}^\circ\boldsymbol{\pi}$, $V(\mathbf{y}) = \boldsymbol{\Omega} = \boldsymbol{\Omega}^* \otimes \mathbf{I}$ is positive definite, and $\mathbf{X}^\circ$ is nonstochastic with full column rank. (Caution: $\mathbf{y}$ now denotes the $2n \times 1$ vector of observations, rather than the original 2×1 random vector.)

34.2. Indirect Feasible Generalized Least Squares

Because the SUR model applies, GLS is the natural estimation procedure. First, suppose that $\boldsymbol{\Omega}^*$, and hence $\boldsymbol{\Omega}$, is known. As the estimator of $\boldsymbol{\pi}$, we would choose the vector $\mathbf{c}$ that minimizes the GLS criterion $\phi(\mathbf{c}) = \mathbf{v}'\boldsymbol{\Omega}^{-1}\mathbf{v}$, where $\mathbf{v} = \mathbf{y} - \mathbf{X}^\circ\mathbf{c}$.

If there are no restrictions on $\boldsymbol{\pi}$, then we have a SUR model with identical explanatory variables, and the solution is obvious: GLS reduces to equation-by-equation LS, as shown in Section 30.3. In the present notation, this means that the GLS estimator of $\boldsymbol{\pi}$ is

$$\mathbf{p} = \begin{pmatrix} \mathbf{p}_1 \\ \mathbf{p}_2 \end{pmatrix} = \begin{pmatrix} \mathbf{A}\mathbf{y}_1 \\ \mathbf{A}\mathbf{y}_2 \end{pmatrix},$$

with $\mathbf{A} = \mathbf{Q}^{-1}\mathbf{X}'$, $\mathbf{Q} = \mathbf{X}'\mathbf{X}$. Reassembling, the GLS estimator of $\boldsymbol{\Pi}$ is

$$\mathbf{P} = (\mathbf{p}_1, \mathbf{p}_2) = (\mathbf{A}\mathbf{y}_1, \mathbf{A}\mathbf{y}_2) = \mathbf{A}(\mathbf{y}_1, \mathbf{y}_2) = \mathbf{A}\mathbf{Y}.$$

The implied estimates of the structural parameters follow by solving the sample counterpart of $\boldsymbol{\Pi}\boldsymbol{\Gamma} = \mathbf{B}$, namely

$$(34.1) \qquad \mathbf{P}\hat{\boldsymbol{\Gamma}} = \hat{\mathbf{B}},$$

for the unknown elements of $\hat{\boldsymbol{\Gamma}}$ and $\hat{\mathbf{B}}$. That is to say, do in the sample what we did in the population for Model B in Section 33.2.

If there are restrictions on $\boldsymbol{\pi}$, then those should be imposed in the minimization. The most convenient way to impose them is to solve them out, which amounts to expressing the π's in terms of α's and choosing estimates of the (unrestricted) α's. Consider, for example, Model A of Chapter 33. Here

$$\boldsymbol{\Pi} = \begin{pmatrix} \pi_{11} & \pi_{12} \\ \pi_{21} & \pi_{22} \\ \pi_{31} & \pi_{32} \end{pmatrix} = (1/\Delta) \begin{pmatrix} \alpha_2 & \alpha_2\alpha_3 \\ \alpha_1\alpha_4 & \alpha_4 \\ \alpha_1\alpha_5 & \alpha_5 \end{pmatrix}, \quad \text{with } \Delta = 1 - \alpha_1\alpha_3.$$

Write $\boldsymbol{\pi} = (\pi_{11}, \pi_{21}, \pi_{31}, \pi_{12}, \pi_{22}, \pi_{32})'$ and $\boldsymbol{\alpha} = (\alpha_1, \alpha_2, \alpha_3, \alpha_4, \alpha_5)'$. Let $\boldsymbol{\pi} = g(\boldsymbol{\alpha})$ be the mapping from the true structural coefficients to the true reduced-form coefficients. Correspondingly, write $\mathbf{c} = (c_1, \ldots, c_6)'$ and $\mathbf{a} = (a_1, \ldots, a_5)'$. Then $\mathbf{c} = g(\mathbf{a})$ is the mapping from the choice vector (estimator) for the structural coefficients to the choice vector (estimator) for the reduced-form coefficients. Referring to the display above, this mapping is

$$\begin{aligned} c_1 &= a_2/(1 - a_1a_3) & c_4 &= a_2a_3/(1 - a_1a_3) \\ c_2 &= a_1a_4/(1 - a_1a_3) & c_5 &= a_4/(1 - a_1a_3) \\ c_3 &= a_1a_5/(1 - a_1a_3) & c_6 &= a_5/(1 - a_1a_3). \end{aligned}$$

We propose to estimate $\boldsymbol{\alpha}$ by the vector $\mathbf{a}$ that minimizes the GLS criterion

$$\psi(\mathbf{a}) = \phi[g(\mathbf{a})] = \mathbf{v}'\boldsymbol{\Omega}^{-1}\mathbf{v},$$

with

$$\mathbf{v} = \mathbf{y} - \mathbf{X}^{\circ}g(\mathbf{a}).$$

The associated estimate of $\boldsymbol{\pi}$ will be $\mathbf{c} = g(\mathbf{a})$.

Computationally, it may be convenient to transform this into an LS problem. To do so, let $\mathbf{H}^*$ be the 2×2 matrix such that $\mathbf{H}^{*\prime}\mathbf{H}^* = \boldsymbol{\Omega}^{*-1}$, and let $\mathbf{H} = \mathbf{H}^* \otimes \mathbf{I}_n$. Then, as is easily verified, $\mathbf{H}'\mathbf{H} = \boldsymbol{\Omega}^{-1}$. With such an $\mathbf{H}$ matrix in hand, we can rewrite the GLS criterion as

$$\psi(\mathbf{a}) = \mathbf{v}^{*\prime}\mathbf{v}^*,$$

with

$$\mathbf{v}^* = \mathbf{H}\mathbf{v} = \mathbf{H}\mathbf{y} - \mathbf{H}\mathbf{X}^{\circ}g(\mathbf{a}) = \mathbf{y}^* - \mathbf{X}^{\circ *}g(\mathbf{a}),$$

say. We take as our estimates of the α's the values of the a's that minimize $\psi(\mathbf{a})$. In view of the form of $g(\mathbf{a})$, this is a nonlinear least squares problem, so the algorithm discussed in Sections 29.2 and 29.3 may be used.

Next suppose that, as in practice, $\boldsymbol{\Omega}^*$, and hence $\boldsymbol{\Omega}$, is unknown. The natural procedure is feasible generalized least squares. The FGLS algorithm will parallel the GLS algorithm, except that an estimator $\hat{\boldsymbol{\Omega}}$ is used in place of $\mathbf{\Omega}$. The estimator comes from the residuals of the LS reduced-form regressions. More explicitly, let

$$\hat{\mathbf{v}}_j = \mathbf{y}_j - \mathbf{X}\mathbf{p}_j = \mathbf{M}\mathbf{y}_j \qquad (j = 1, 2),$$

where $\mathbf{M} = \mathbf{I} - \mathbf{X}\mathbf{A}$, and let

$$\hat{\mathbf{V}} = \mathbf{Y} - \mathbf{X}\mathbf{P} = \mathbf{M}\mathbf{Y} = (\hat{\mathbf{v}}_1, \hat{\mathbf{v}}_2).$$

Then

$$\hat{\mathbf{\Omega}}^* = (1/n)\hat{\mathbf{V}}'\hat{\mathbf{V}}$$

is the estimator for $\mathbf{\Omega}^*$, and $\hat{\mathbf{\Omega}} = \hat{\mathbf{\Omega}}^* \otimes \mathbf{I}$ is the estimator for $\mathbf{\Omega}$.

The rest of the computational algorithm can then track that for GLS. If there are no restrictions on $\boldsymbol{\pi}$, then taking $\mathbf{c} = \mathbf{p}$ (that is, $\hat{\mathbf{\Pi}} = \mathbf{P}$) will solve the minimization problem. (In a SUR model with identical explanatory variables, FGLS, like GLS, coincides with LS.) If there are restrictions on $\boldsymbol{\pi}$, an NLLS algorithm is usable. We refer to the resulting estimates of the α's as *indirect feasible generalized least squares*, or indirect-FGLS, *estimates*, because we are in effect estimating the π's by FGLS and converting them into estimates of the α's. (If there are no restrictions on $\boldsymbol{\pi}$, then indirect-FGLS coincides with indirect least squares.) In the literature, the indirect-FGLS procedure is sometimes referred to as a *minimum-distance* procedure.

With respect to sampling properties: because the FGLS estimates of $\mathbf{\Pi}$ are consistent and BAN, the indirect-FGLS estimates of $\mathbf{B}$ and $\mathbf{\Gamma}$ are also consistent and BAN. So indirect-FGLS is one preferred way to estimate structural parameters in the SEM.

Several remarks about the indirect-FGLS procedure:

• In our algorithm, we use the relation between π's and α's to solve out the restrictions, reducing the problem to unconstrained, but nonlinear, minimization. It is easy to see that the resulting estimates satisfy the sample counterpart of $\mathbf{\Pi}\mathbf{\Gamma} = \mathbf{B}$.

• If one or more of the structural equations is not identified in terms of $\mathbf{\Pi}$, then the indirect-FGLS procedure will break down, as it should.

• If in the population, the conditional distribution of the $m \times 1$ random vector $\mathbf{y}$, given the $k \times 1$ vector $\mathbf{x}$, is multinormal, then maximum-likelihood estimation is available. For historical reasons, this method is known as *full-information maximum likelihood*, or FIML. From the discussion in Section 30.7, we can verify several facts. If $\mathbf{\Omega}^*$ is known, then FIML coincides with GLS. If $\mathbf{\Omega}^*$ is unknown, then FIML minimizes

$|\mathbf{V}'\mathbf{V}|$, which differs from the FGLS criterion $\text{tr}(\hat{\mathbf{\Omega}}^{*-1}\mathbf{V}'\mathbf{V})$, but the estimators have the same asymptotic distribution. Iterating indirect-FGLS until convergence is an algorithm for solving the FOC's of FIML estimation.

The indirect-FGLS and FIML procedures are computationally complex when there are restrictions, because the restrictions are characteristically nonlinear in the π's. As a consequence, structural estimation procedures have been developed that use the unrestricted reduced-form estimate, $\mathbf{P} = (\mathbf{p}_1, \mathbf{p}_2)$, in a quite different way. Nowadays the complexity is less of a concern, but the simpler methods are widely used, and therefore we will sketch two of them.

34.3. Two-Stage Least Squares

The *two-stage least squares*, or 2SLS, *method* is the most popular procedure for estimating a simultaneous-equation model. Its mechanics can be described very simply. In the first stage, each endogenous variable is regressed on all the exogenous variables, and fitted values are obtained. In the second stage, each structural equation is taken in turn, right-hand-side endogenous variables are replaced by their fitted values, and LS is run. The 2SLS algorithm does not involve nonlinear optimization, which accounts for its popularity.

We describe the procedure explicitly in terms of the supply-demand models of Chapter 33. The data consist of the $n \times 2$ matrix $\mathbf{Y} = (\mathbf{y}_1, \mathbf{y}_2)$ and the $n \times k$ matrix $\mathbf{X} = (\mathbf{x}_1, \ldots, \mathbf{x}_k)$. The familiar regression matrices are

$$\mathbf{Q} = \mathbf{X}'\mathbf{X}, \qquad \mathbf{A} = \mathbf{Q}^{-1}\mathbf{X}', \qquad \mathbf{N} = \mathbf{X}\mathbf{A}, \qquad \mathbf{M} = \mathbf{I} - \mathbf{N}.$$

We have

$$\mathbf{AY} = (\mathbf{Ay}_1, \mathbf{Ay}_2) = (\mathbf{p}_1, \mathbf{p}_2) = \mathbf{P}, \qquad \mathbf{AX} = \mathbf{I},$$

$$\mathbf{NY} = (\mathbf{Ny}_1, \mathbf{Ny}_2) = (\hat{\mathbf{y}}_1, \hat{\mathbf{y}}_2) = \hat{\mathbf{Y}}, \qquad \mathbf{NX} = \mathbf{X}.$$

Focus on one of the structural equations, say the demand equation in Model B. In population terms this is

$$y_1 = \alpha_1 y_2 + \alpha_2 x_1 + \alpha_6 x_2 + u_1.$$

For the sample of size n it is

$$\mathbf{y}_1 = \mathbf{y}_2\alpha_1 + \mathbf{x}_1\alpha_2 + \mathbf{x}_2\alpha_6 + \mathbf{u}_1$$

$$= (\mathbf{y}_2, \mathbf{x}_1, \mathbf{x}_2)\begin{pmatrix}\alpha_1 \\ \alpha_2 \\ \alpha_6\end{pmatrix} + \mathbf{u}_1$$

$$= \mathbf{Z}_1\boldsymbol{\alpha}_1 + \mathbf{u}_1,$$

say, where $\mathbf{Z}_1 = (\mathbf{y}_2, \mathbf{x}_1, \mathbf{x}_2)$ is $n \times 3$, and $\boldsymbol{\alpha}_1 = (\alpha_1, \alpha_2, \alpha_6)'$ is 3×1. Regressing $\mathbf{y}_1$ on $\mathbf{Z}_1$ would give the normal equations

$$\mathbf{Z}_1'\mathbf{Z}_1\mathbf{a}_1 = \mathbf{Z}_1'\mathbf{y}_1,$$

the solution to which is the LS coefficient vector $\mathbf{a}_1 = (\mathbf{Z}_1'\mathbf{Z}_1)^{-1}\mathbf{Z}_1'\mathbf{y}_1$. As we know, this is not a sensible estimator of $\boldsymbol{\alpha}_1$.

Instead, replace $\mathbf{Z}_1$ by

$$\hat{\mathbf{Z}}_1 = \mathbf{N}\mathbf{Z}_1 = \mathbf{N}(\mathbf{y}_2, \mathbf{x}_1, \mathbf{x}_2) = (\mathbf{N}\mathbf{y}_2, \mathbf{N}\mathbf{x}_1, \mathbf{N}\mathbf{x}_2) = (\hat{\mathbf{y}}_2, \mathbf{x}_1, \mathbf{x}_2),$$

and regress $\mathbf{y}_1$ on $\hat{\mathbf{Z}}_1$. This gives the normal equations

$$\hat{\mathbf{Z}}_1'\hat{\mathbf{Z}}_1\mathbf{a}_1^* = \hat{\mathbf{Z}}_1'\mathbf{y}_1,$$

the solution to which is the 2SLS estimator

$$\mathbf{a}_1^* = (\hat{\mathbf{Z}}_1'\hat{\mathbf{Z}}_1)^{-1}\hat{\mathbf{Z}}_1'\mathbf{y}_1.$$

The (asymptotic) variance matrix of $\mathbf{a}_1^*$ is estimated as

$$\hat{V}(\mathbf{a}_1^*) = \hat{\sigma}_{11}(\hat{\mathbf{Z}}_1'\hat{\mathbf{Z}}_1)^{-1},$$

where

$$\hat{\sigma}_{11} = \mathbf{e}_1^{*\prime}\mathbf{e}_1^*/(n - k^*),$$

with k^* being the number of right-hand-side variables (columns of $\hat{\mathbf{Z}}_1$), and

$$\mathbf{e}_1^* = \mathbf{y}_1 - \mathbf{Z}_1\mathbf{a}_1^*.$$

Observe that the original values $\mathbf{Z}_1$ are used in calculating residuals, even though the fitted values $\hat{\mathbf{Z}}_1$ were used in calculating coefficients. (In calculating $\hat{\sigma}_{11}$, division by n rather than $n - k^*$ is also acceptable in view of the fact that asymptotic theory is being relied on.)

There are at least two heuristic rationales for the 2SLS procedure, which we exposit in the context of the demand equation of Model B. First, observe that

$$E(y_1|\mathbf{x}) = \alpha_1 E(y_2|\mathbf{x}) + \alpha_2 x_1 + \alpha_6 x_2$$
$$= \alpha_1 y_2^* + \alpha_2 x_1 + \alpha_6 x_2,$$

where

$$y_2^* = \mathbf{x}'\boldsymbol{\pi}_2 = E(y_2|\mathbf{x}).$$

So a sample LS regression of $\mathbf{y}_1$ on $\mathbf{Z}_1^* = (\mathbf{y}_2^*, \mathbf{x}_1, \mathbf{x}_2)$ would give unbiased estimates of $\boldsymbol{\alpha}_1$. That regression cannot be run because y_2^* is unobserved. Still, $\mathbf{p}_2$ unbiasedly and consistently estimates $\boldsymbol{\pi}_2$, whence the fitted-value vector $\hat{\mathbf{y}}_2$ unbiasedly and consistently estimates the conditional expectation vector $\mathbf{y}_2^*$. Making the natural replacement, $\hat{\mathbf{y}}_2$ for $\mathbf{y}_2^*$, suffices to produce consistent estimates of the $\boldsymbol{\alpha}_1$. The second rationale is simpler. The 2SLS normal equations are equivalent to a set of orthogonality conditions:

$$\hat{\mathbf{Z}}_1'\mathbf{u}_1 = \mathbf{0},$$

where $\mathbf{u}_1 = \mathbf{y}_1 - \mathbf{Z}_1\mathbf{a}_1^*$. To show equivalence, use the algebraic fact:

$$\hat{\mathbf{Z}}_1'\mathbf{Z}_1 = \mathbf{Z}_1'\mathbf{N}'\mathbf{Z}_1 = \mathbf{Z}_1'\mathbf{N}'\mathbf{N}\mathbf{Z}_1 = \hat{\mathbf{Z}}_1'\hat{\mathbf{Z}}_1.$$

So 2SLS has an instrumental-variable interpretation. The variables in $\hat{\mathbf{Z}}_1$ are legitimate instruments because they are, at least asymptotically, uncorrelated with the disturbance.

The 2SLS procedure may be applied to each of the structural equations in turn. The fact that it relies on the unrestricted estimator $\mathbf{P}$ as the estimator of $\boldsymbol{\Pi}$ suggests that when there are restrictions on $\boldsymbol{\Pi}$, the 2SLS estimates will not be optimal.

Here are several remarks about the mechanical aspects of 2SLS estimation:

• If a structural equation is not identified in terms of $\boldsymbol{\Pi}$, then the 2SLS procedure will break down, as it should, for that equation. For example, consider the supply equation in Model C. The second stage of 2SLS calls for regressing $\mathbf{y}_2$ on $(\hat{\mathbf{y}}_1, \mathbf{x}_1, \mathbf{x}_2, \mathbf{x}_3)$, but there is exact multicollinearity among those four explanatory variables:

$$\hat{\mathbf{y}}_1 = \mathbf{X}\mathbf{p}_1 = \mathbf{x}_1 p_{11} + \mathbf{x}_2 p_{21} + \mathbf{x}_3 p_{31}.$$

So the solution to the second-stage normal equations is not unique, and the 2SLS estimates are not defined.

• Standard errors for 2SLS cannot be smaller than the conventional standard errors for LS obtained from

$$s_{11}(\mathbf{Z}_1'\mathbf{Z}_1)^{-1}, \quad \text{with } s_{11} = (\mathbf{y}_1 - \mathbf{Z}_1\mathbf{a}_1)'(\mathbf{y}_1 - \mathbf{Z}_1\mathbf{a}_1)/(n - k^*).$$

First, $\hat{\sigma}_{11} > s_{11}$ because LS minimizes the sum of squared residuals of $\mathbf{y}_1$ from a linear combination of the columns of $\mathbf{Z}_1$. Second, $\mathbf{Z}_1'\mathbf{Z}_1 \geq \hat{\mathbf{Z}}_1'\hat{\mathbf{Z}}_1$ because $\hat{\mathbf{Z}}_1 = \mathbf{N}\mathbf{Z}_1$.

• One should not report R^2 for equations estimated by 2SLS. If one uses the conventional sum of squared residuals, then one is measuring the proportion of variation in the dependent variable that is accounted for by the fitted explanatory variables. Alternatively, if one uses the sum of squared residuals that enters the 2SLS estimate of σ_{11}, then there is no guarantee that the resulting R^2 will lie between 0 and 1.

34.4. Relation between 2SLS and Indirect-FGLS

Relying on $\mathbf{P}$, the unrestricted estimator of $\mathbf{\Pi}$, how does 2SLS succeed in producing estimates of $\mathbf{\Gamma}$ and $\mathbf{B}$ even when there are restrictions on $\mathbf{\Pi}$? To explore that issue, we study the algebraic relation between the 2SLS and indirect-FGLS estimators. Consider first the demand equation of Model B. We have

$$\hat{\mathbf{Z}}_1'\mathbf{y}_1 = \mathbf{Z}_1'\mathbf{N}\mathbf{y}_1 = \mathbf{Z}_1'(\mathbf{X}\mathbf{A})\mathbf{y}_1 = \mathbf{Z}_1'\mathbf{X}\mathbf{A}\mathbf{y}_1 = \mathbf{Z}_1'\mathbf{X}\mathbf{p}_1,$$

$$\hat{\mathbf{Z}}_1'\hat{\mathbf{Z}}_1 = \mathbf{Z}_1'\mathbf{N}\mathbf{Z}_1 = \mathbf{Z}_1'(\mathbf{X}\mathbf{A})\mathbf{Z}_1 = \mathbf{Z}_1'\mathbf{X}\mathbf{A}\mathbf{Z}_1 = \mathbf{Z}_1'\mathbf{X}(\mathbf{p}_2, \mathbf{D}),$$

where $\mathbf{D}$ consists of the first two columns of $\mathbf{A}\mathbf{X} = \mathbf{I}$:

$$\mathbf{D} = \mathbf{A}(\mathbf{x}_1, \mathbf{x}_2) = \begin{pmatrix} 1 & 0 \\ 0 & 1 \\ 0 & 0 \end{pmatrix} = (\mathbf{d}_1, \mathbf{d}_2),$$

say. So the normal equations of 2SLS, namely

$$\hat{\mathbf{Z}}_1'\mathbf{y}_1 = \hat{\mathbf{Z}}_1'\hat{\mathbf{Z}}_1\mathbf{a}_1^*,$$

can be read as

$$\mathbf{Z}_1'\mathbf{X}\mathbf{p}_1 = \mathbf{Z}_1'\mathbf{X}(\mathbf{p}_2, \mathbf{D})\begin{pmatrix} a_1^* \\ a_2^* \\ a_6^* \end{pmatrix} = \mathbf{Z}_1'\mathbf{X}(\mathbf{p}_2 a_1^* + \mathbf{d}_1 a_2^* + \mathbf{d}_2 a_6^*),$$

which may be rearranged into

$$\mathbf{Z}_1'\mathbf{X}(\mathbf{p}_1 - \mathbf{p}_2 a_1^*) = \mathbf{Z}_1'\mathbf{X}(\mathbf{d}_1 a_2^* + \mathbf{d}_2 a_6^*) = \mathbf{Z}_1'\mathbf{X}\begin{pmatrix} a_2^* \\ a_6^* \\ 0 \end{pmatrix}.$$

Here $\mathbf{Z}_1'\mathbf{X}$ is square and (coincidence apart) nonsingular, so the 2SLS normal equations are equivalent to

$$\text{(34.2)} \quad \mathbf{p}_1 - \mathbf{p}_2 a_1^* = \begin{pmatrix} a_2^* \\ a_6^* \\ 0 \end{pmatrix}.$$

Similarly, for the supply equation of Model B, the 2SLS normal equations are equivalent to

$$\text{(34.3)} \quad \mathbf{p}_2 - \mathbf{p}_1 a_3^* = \begin{pmatrix} 0 \\ a_4^* \\ a_5^* \end{pmatrix}.$$

Assembled together, Eqs. (34.2) and (34.3) say

$$\mathbf{P}\hat{\boldsymbol{\Gamma}} = \hat{\mathbf{B}},$$

which is Eq. (34.1), the indirect-FGLS (and ILS) estimating equations when there are no restrictions to be imposed. We conclude that for Model B, which has no restriction on $\boldsymbol{\Pi}$, 2SLS coincides with indirect-FGLS.

When restrictions are present, this coincidence will not prevail. For example, consider the demand equation in Model A. Suppose that we tried to use the p's instead of the $\hat{\pi}$'s in the ILS estimating equations, writing

$$\mathbf{p}_1 - \mathbf{p}_2\hat{\alpha}_1 = \begin{pmatrix} \hat{\alpha}_2 \\ 0 \\ 0 \end{pmatrix},$$

that is,

$$\mathbf{p}_1 = (\mathbf{p}_2, \mathbf{d}_1)\begin{pmatrix} \hat{\alpha}_1 \\ \hat{\alpha}_2 \end{pmatrix}.$$

This system of three equations in two unknowns is overdetermined: it has no solution. We might combine the equations by premultiplying through by any 2×3 matrix. One such matrix is $\mathbf{Z}_1'\mathbf{X}$. Premultiplying through by it gives

$$\mathbf{Z}_1'\mathbf{X}\mathbf{p}_1 = \mathbf{Z}_1'\mathbf{X}(\mathbf{p}_2, \mathbf{d}_1)\begin{pmatrix}\hat{\alpha}_1\\ \hat{\alpha}_2\end{pmatrix},$$

which is a system of two equations in two unknowns. Now

$$\mathbf{Z}_1'\mathbf{X}\mathbf{p}_1 = \hat{\mathbf{Z}}_1'\mathbf{y}_1, \qquad \mathbf{Z}_1'\mathbf{X}(\mathbf{p}_2, \mathbf{d}_1) = \hat{\mathbf{Z}}_1'\hat{\mathbf{Z}}_1,$$

so we have arrived at the normal equations for 2SLS. From this perspective, when restrictions are present, there is a surplus of estimating equations. The 2SLS estimates can be viewed as the solution to a collapsed set of those equations. It turns out that collapsing via $\mathbf{Z}_1'\mathbf{X}$ is optimal: see Amemiya (1985, pp. 239–240).

34.5. Three-Stage Least Squares

In 2SLS, we estimate each structural equation separately, acting *as if* classical regression models applied to

$$\mathbf{y}_1 = \hat{\mathbf{Z}}_1\boldsymbol{\alpha}_1 + \mathbf{u}_1, \qquad \mathbf{y}_2 = \hat{\mathbf{Z}}_2\boldsymbol{\alpha}_2 + \mathbf{u}_2.$$

There would seem to be an advantage to estimating the pair of structural equations jointly. If we stack into

$$\mathbf{y} = \begin{pmatrix}\mathbf{y}_1\\ \mathbf{y}_2\end{pmatrix}, \qquad \hat{\mathbf{Z}}^{\circ} = \begin{pmatrix}\hat{\mathbf{Z}}_1 & \mathbf{O}\\ \mathbf{O} & \hat{\mathbf{Z}}_2\end{pmatrix}, \qquad \boldsymbol{\alpha} = \begin{pmatrix}\boldsymbol{\alpha}_1\\ \boldsymbol{\alpha}_2\end{pmatrix}, \qquad \mathbf{u} = \begin{pmatrix}\mathbf{u}_1\\ \mathbf{u}_2\end{pmatrix},$$

then $\mathbf{y} = \hat{\mathbf{Z}}^{\circ}\boldsymbol{\alpha} + \mathbf{u}$ has the appearance of a SUR model. If $\boldsymbol{\Sigma}^*$ (hence $\boldsymbol{\Sigma}$) were known, we might calculate an estimator by the GLS rule, namely

$$\hat{\boldsymbol{\alpha}} = (\hat{\mathbf{Z}}^{\circ\prime}\boldsymbol{\Sigma}^{-1}\hat{\mathbf{Z}}^{\circ})^{-1}\hat{\mathbf{Z}}^{\circ\prime}\boldsymbol{\Sigma}^{-1}\mathbf{y}.$$

Lacking that knowledge, we may adopt an FGLS rule. Estimate $\boldsymbol{\Sigma}^*$ from the proper residuals of 2SLS:

$$\hat{\boldsymbol{\Sigma}}^* = (1/n)\begin{pmatrix}\mathbf{e}_1^{*\prime}\mathbf{e}_1^* & \mathbf{e}_1^{*\prime}\mathbf{e}_2^*\\ \mathbf{e}_2^{*\prime}\mathbf{e}_1^* & \mathbf{e}_2^{*\prime}\mathbf{e}_2^*\end{pmatrix},$$

and construct $\hat{\boldsymbol{\Sigma}}$ from $\hat{\boldsymbol{\Sigma}}^*$. Then calculate

$$\mathbf{a}^{**} = (\hat{\mathbf{Z}}^{\circ\prime}\hat{\boldsymbol{\Sigma}}^{-1}\hat{\mathbf{Z}}^{\circ})^{-1}\hat{\mathbf{Z}}^{\circ\prime}\hat{\boldsymbol{\Sigma}}^{-1}\mathbf{y}.$$

This defines the *three-stage least squares*, or 3SLS, *estimator* of the structural parameters. The LS rule is used three times—first on the reduced form (to get the $\hat{\mathbf{Z}}$'s), next on individual structural equations (to get the

$\mathbf{e}^*$'s), and finally on the structural equations jointly (to get the $\mathbf{a}^{**}$). Clearly, 3SLS will break down (as it should) if the system contains a nonidentified structural equation.

It can be shown that the 3SLS estimator is consistent and BAN, like indirect-FGLS. The computations for 3SLS, like those for 2SLS, do not involve nonlinear optimization, even when restrictions are present. The inverse matrix in the formula above for $\mathbf{a}^{**}$ serves as the estimate of the (asymptotic) variance matrix of $\mathbf{a}^{**}$.

34.6. Remarks

We conclude with some remarks on estimation in the SEM.

- If all the structural equations are identified, and there are no restrictions on $\mathbf{\Pi}$, then indirect least squares, indirect-FGLS, 2SLS, 3SLS, and FIML all produce the same estimates.
- If a parameter is not identified, then there is no method to estimate it consistently.
- In the SEM, there is in general no unbiased estimator of the structural parameters.
- Throughout this chapter, we have confined attention to an SEM in which the only prior information consists of normalizations and exclusions. If other information is available (for example, $\mathbf{\Sigma}^*$ is diagonal, or a coefficient in one structural equation is equal to a coefficient in another), then some modifications are needed in the description of the estimators and their statistical properties.
- We have relied on stratified (nonstochastic $\mathbf{X}$) sampling in this chapter. The statements about asymptotic properties rely on an additional assumption about how additional observations are generated, namely that the matrix $\mathbf{X}'\mathbf{X}/n$ has a positive definite limit: see Section 22.7. If instead sampling is random from the joint distribution of $(\mathbf{x}', \mathbf{y}')$, no substantial change in the results would be required: see Chapter 25.

Exercises

34.1 You are given the following sums of squares and cross-products on the variables y_1 = quantity, y_2 = price, x = income, obtained in a sample of 60 observations:

	x	y_1	y_2
x	360	120	120
y_1	120	110	5
y_2	120	5	80

You are told that the sample was produced by this simultaneous-equation model:

Demand. $y_1 = \alpha_1 y_2 + \alpha_2 x + u_1,$

Supply. $y_2 = \alpha_3 y_1 \qquad\qquad + u_2,$

in which the exogenous variable x was independent of the disturbances u_1 and u_2, while those two disturbances had zero expectations and were correlated with each other.

(a) From the sample data, calculate the LS "estimate" of α_3, and the 2SLS estimate of α_3.
(b) Would you use 2SLS, or some other method, to estimate α_1 and α_2? Justify your answer.
(c) From the information in hand, you are asked to predict the value of y_2 that will prevail when $y_1 = 55$. Would your prediction be 2.5, or 55, or some other number? Justify your answer.

34.2 The usual simultaneous-equation model applies to

$$y_1 = \alpha_1 y_2 + \alpha_2 x_1 \qquad\qquad + u_1,$$

$$y_2 = \alpha_3 y_1 \qquad\qquad + \alpha_4 x_2 + u_2.$$

Here y_1 = quantity, y_2 = price, x_1 = input price, and x_2 = income. These two LS regressions were obtained in a sample of 100 observations:

$$\hat{y}_1 = -6x_1 + 2x_2,$$

$$\hat{y}_2 = 3x_1 + x_2.$$

Calculate estimates of the α's.

34.3 In Exercise 33.2, you considered this supply-demand model for labor:

$$y_1 = \alpha_1 y_2 + \alpha_2 x_1 + \alpha_3 x_2 \qquad\qquad\qquad + u_1,$$

$$y_2 = \alpha_4 y_1 + \alpha_5 x_1 \qquad\qquad + \alpha_6 x_3 + \alpha_7 x_4 + \alpha_8 x_5 + u_2,$$

where y_1 = months worked, y_2 = wage rate, $x_1 = 1$, x_2 = family size, x_3 = education, x_4 = age, x_5 = race (= 1 if black, = 0 if white), u_1 = supply shock, u_2 = demand shock. Now suppose, rather artificially, that this SEM applies to the SCF data set of Exercise 17.4. Take y_2 = wage rate = earnings/months worked.

Write and run a program to:

(a) Calculate the LS "estimates" of the structural coefficients, along with their conventional standard errors.
(b) Calculate the 2SLS estimates of the structural coefficients, along with their standard errors.
(c) Discuss your results from an economic perspective.

34.4 For the setup of Exercise 34.3, write and run programs to:

(a) Calculate the 3SLS estimates of the structural coefficients.
(b) Calculate the indirect-FGLS estimates of the structural coefficients.
(c) Assuming normality, calculate the FIML estimates of the structural coefficients, by iterating the indirect-FGLS algorithm until convergence.
(d) Also comment on the relation among your alternative estimates.

34.5 For the setup of Exercise 34.3:

(a) Use your 2SLS estimates to derive an estimate of the reduced-form coefficient matrix $\mathbf{\Pi}$.
(b) Does this estimate of $\mathbf{\Pi}$ satisfy the restrictions that you found in Exercise 33.2? Explain.
(c) Compare your estimated $\mathbf{\Pi}$ with the unrestricted estimate $\mathbf{P}$.

Appendixes
References
Index

Appendix A
Statistical and Data Tables

Table A.1 Standard normal cumulative distribution function.

	0.00	0.01	0.02	0.03	0.04	0.05	0.06	0.07	0.08	0.09
0.00	0.500	0.504	0.508	0.512	0.516	0.520	0.524	0.528	0.532	0.536
0.10	0.540	0.544	0.548	0.552	0.556	0.560	0.564	0.567	0.571	0.575
0.20	0.579	0.583	0.587	0.591	0.595	0.599	0.603	0.606	0.610	0.614
0.30	0.618	0.622	0.626	0.629	0.633	0.637	0.641	0.644	0.648	0.652
0.40	0.655	0.659	0.663	0.666	0.670	0.674	0.677	0.681	0.684	0.688
0.50	0.691	0.695	0.698	0.702	0.705	0.709	0.712	0.716	0.719	0.722
0.60	0.726	0.729	0.732	0.736	0.739	0.742	0.745	0.749	0.752	0.755
0.70	0.758	0.761	0.764	0.767	0.770	0.773	0.776	0.779	0.782	0.785
0.80	0.788	0.791	0.794	0.797	0.800	0.802	0.805	0.808	0.811	0.813
0.90	0.816	0.819	0.821	0.824	0.826	0.829	0.831	0.834	0.836	0.839
1.00	0.841	0.844	0.846	0.848	0.851	0.853	0.855	0.858	0.860	0.862
1.10	0.864	0.867	0.869	0.871	0.873	0.875	0.877	0.879	0.881	0.883
1.20	0.885	0.887	0.889	0.891	0.893	0.894	0.896	0.898	0.900	0.901
1.30	0.903	0.905	0.907	0.908	0.910	0.911	0.913	0.915	0.916	0.918
1.40	0.919	0.921	0.922	0.924	0.925	0.926	0.928	0.929	0.931	0.932
1.50	0.933	0.934	0.936	0.937	0.938	0.939	0.941	0.942	0.943	0.944
1.60	0.945	0.946	0.947	0.948	0.949	0.951	0.952	0.953	0.954	0.954
1.70	0.955	0.956	0.957	0.958	0.959	0.960	0.961	0.962	0.962	0.963
1.80	0.964	0.965	0.966	0.966	0.967	0.968	0.969	0.969	0.970	0.971
1.90	0.971	0.972	0.973	0.973	0.974	0.974	0.975	0.976	0.976	0.977
2.00	0.977	0.978	0.978	0.979	0.979	0.980	0.980	0.981	0.981	0.982
2.10	0.982	0.983	0.983	0.983	0.984	0.984	0.985	0.985	0.985	0.986
2.20	0.986	0.986	0.987	0.987	0.987	0.988	0.988	0.988	0.989	0.989
2.30	0.989	0.990	0.990	0.990	0.990	0.991	0.991	0.991	0.991	0.992
2.40	0.992	0.992	0.992	0.992	0.993	0.993	0.993	0.993	0.993	0.994
2.50	0.994	0.994	0.994	0.994	0.994	0.995	0.995	0.995	0.995	0.995
2.60	0.995	0.995	0.996	0.996	0.996	0.996	0.996	0.996	0.996	0.996
2.70	0.997	0.997	0.997	0.997	0.997	0.997	0.997	0.997	0.997	0.997
2.80	0.997	0.998	0.998	0.998	0.998	0.998	0.998	0.998	0.998	0.998
2.90	0.998	0.998	0.998	0.998	0.998	0.998	0.998	0.999	0.999	0.999
3.00	0.999	0.999	0.999	0.999	0.999	0.999	0.999	0.999	0.999	0.999

Example: If $Z \sim \mathcal{N}(0, 1)$, then $\Pr(Z \leq 1.15) = F(1.15) = 0.875$.

Table A.2 Chi-square cumulative distribution function.

	$G_k(\cdot)$										
k	0.05	0.10	0.15	0.20	0.25	0.30	0.35	0.40	0.45	0.50	0.55
1	0.00	0.02	0.04	0.06	0.10	0.15	0.21	0.27	0.36	0.45	0.57
2	0.10	0.21	0.33	0.45	0.58	0.71	0.86	1.02	1.20	1.39	1.60
3	0.35	0.58	0.80	1.01	1.21	1.42	1.64	1.87	2.11	2.37	2.64
4	0.71	1.06	1.37	1.65	1.92	2.19	2.47	2.75	3.05	3.36	3.69
5	1.15	1.61	1.99	2.34	2.67	3.00	3.33	3.66	4.00	4.35	4.73
6	1.64	2.20	2.66	3.07	3.45	3.83	4.20	4.57	4.95	5.35	5.77
7	2.17	2.83	3.36	3.82	4.25	4.67	5.08	5.49	5.91	6.35	6.80
8	2.73	3.49	4.08	4.59	5.07	5.53	5.98	6.42	6.88	7.34	7.83
9	3.33	4.17	4.82	5.38	5.90	6.39	6.88	7.36	7.84	8.34	8.86
10	3.94	4.87	5.57	6.18	6.74	7.27	7.78	8.30	8.81	9.34	9.89
11	4.57	5.58	6.34	6.99	7.58	8.15	8.70	9.24	9.78	10.34	10.92
12	5.23	6.30	7.11	7.81	8.44	9.03	9.61	10.18	10.76	11.34	11.95
13	5.89	7.04	7.90	8.63	9.30	9.93	10.53	11.13	11.73	12.34	12.97
14	6.57	7.79	8.70	9.47	10.17	10.82	11.45	12.08	12.70	13.34	14.00
15	7.26	8.55	9.50	10.31	11.04	11.72	12.38	13.03	13.68	14.34	15.02
16	7.96	9.31	10.31	11.15	11.91	12.62	13.31	13.98	14.66	15.34	16.04
17	8.67	10.09	11.12	12.00	12.79	13.53	14.24	14.94	15.63	16.34	17.06
18	9.39	10.86	11.95	12.86	13.68	14.44	15.17	15.89	16.61	17.34	18.09
19	10.12	11.65	12.77	13.72	14.56	15.35	16.11	16.85	17.59	18.34	19.11
20	10.85	12.44	13.60	14.58	15.45	16.27	17.05	17.81	18.57	19.34	20.13
25	14.61	16.47	17.82	18.94	19.94	20.87	21.75	22.62	23.47	24.34	25.22
30	18.49	20.60	22.11	23.36	24.48	25.51	26.49	27.44	28.39	29.34	30.31
35	22.47	24.80	26.46	27.84	29.05	30.18	31.25	32.28	33.31	34.34	35.39
40	26.51	29.05	30.86	32.34	33.66	34.87	36.02	37.13	38.23	39.34	40.46
45	30.61	33.35	35.29	36.88	38.29	39.58	40.81	42.00	43.16	44.34	45.53
50	34.76	37.69	39.75	41.45	42.94	44.31	45.61	46.86	48.10	49.33	50.59
55	38.96	42.06	44.24	46.04	47.61	49.06	50.42	51.74	53.04	54.33	55.65
60	43.19	46.46	48.76	50.64	52.29	53.81	55.24	56.62	57.98	59.33	60.71
65	47.45	50.88	53.29	55.26	56.99	58.57	60.07	61.51	62.92	64.33	65.77
70	51.74	55.33	57.84	59.90	61.70	63.35	64.90	66.40	67.87	69.33	70.82
75	56.05	59.79	62.41	64.55	66.42	68.13	69.74	71.29	72.81	74.33	75.88
80	60.39	64.28	66.99	69.21	71.14	72.92	74.58	76.19	77.76	79.33	80.93
85	64.75	68.78	71.59	73.88	75.88	77.71	79.43	81.09	82.71	84.33	85.98
90	69.13	73.29	76.20	78.56	80.62	82.51	84.29	85.99	87.67	89.33	91.02
95	73.52	77.82	80.81	83.25	85.38	87.32	89.14	90.90	92.62	94.33	96.07
100	77.93	82.36	85.44	87.95	90.13	92.13	94.00	95.81	97.57	99.33	101.11

Example: If $W \sim \chi^2(6)$, then $\Pr(W \leq 4.20) = G_6(4.20) = 0.35$.

Table A.2 (continued)

	$G_k(\cdot)$										
k	0.60	0.65	0.70	0.75	0.80	0.85	0.90	0.95	0.975	0.990	0.995
1	0.71	0.87	1.07	1.32	1.64	2.07	2.71	3.84	5.02	6.63	7.88
2	1.83	2.10	2.41	2.77	3.22	3.79	4.61	5.99	7.38	9.21	10.60
3	2.95	3.28	3.66	4.11	4.64	5.32	6.25	7.81	9.35	11.34	12.84
4	4.04	4.44	4.88	5.39	5.99	6.74	7.78	9.49	11.14	13.28	14.86
5	5.13	5.57	6.06	6.63	7.29	8.12	9.24	11.07	12.83	15.09	16.75
6	6.21	6.69	7.23	7.84	8.56	9.45	10.64	12.59	14.45	16.81	18.55
7	7.28	7.81	8.38	9.04	9.80	10.75	12.02	14.07	16.01	18.48	20.28
8	8.35	8.91	9.52	10.22	11.03	12.03	13.36	15.51	17.53	20.09	21.95
9	9.41	10.01	10.66	11.39	12.24	13.29	14.68	16.92	19.02	21.67	23.59
10	10.47	11.10	11.78	12.55	13.44	14.53	15.99	18.31	20.48	23.21	25.19
11	11.53	12.18	12.90	13.70	14.63	15.77	17.28	19.68	21.92	24.72	26.76
12	12.58	13.27	14.01	14.85	15.81	16.99	18.55	21.03	23.34	26.22	28.30
13	13.64	14.35	15.12	15.98	16.98	18.20	19.81	22.36	24.74	27.69	29.82
14	14.69	15.42	16.22	17.12	18.15	19.41	21.06	23.68	26.12	29.14	31.32
15	15.73	16.49	17.32	18.25	19.31	20.60	22.31	25.00	27.49	30.58	32.80
16	16.78	17.56	18.42	19.37	20.47	21.79	23.54	26.30	28.85	32.00	34.27
17	17.82	18.63	19.51	20.49	21.61	22.98	24.77	27.59	30.19	33.41	35.72
18	18.87	19.70	20.60	21.60	22.76	24.16	25.99	28.87	31.53	34.81	37.16
19	19.91	20.76	21.69	22.72	23.90	25.33	27.20	30.14	32.85	36.19	38.58
20	20.95	21.83	22.77	23.83	25.04	26.50	28.41	31.41	34.17	37.57	40.00
25	26.14	27.12	28.17	29.34	30.68	32.28	34.38	37.65	40.65	44.31	46.93
30	31.32	32.38	33.53	34.80	36.25	37.99	40.26	43.77	46.98	50.89	53.67
35	36.47	37.62	38.86	40.22	41.78	43.64	46.06	49.80	53.20	57.34	60.27
40	41.62	42.85	44.16	45.62	47.27	49.24	51.81	55.76	59.34	63.69	66.77
45	46.76	48.06	49.45	50.98	52.73	54.81	57.51	61.66	65.41	69.96	73.17
50	51.89	53.26	54.72	56.33	58.16	60.35	63.17	67.50	71.42	76.15	79.49
55	57.02	58.45	59.98	61.66	63.58	65.86	68.80	73.31	77.38	82.29	85.75
60	62.13	63.63	65.23	66.98	68.97	71.34	74.40	79.08	83.30	88.38	91.95
65	67.25	68.80	70.46	72.28	74.35	76.81	79.97	84.82	89.18	94.42	98.11
70	72.36	73.97	75.69	77.58	79.71	82.26	85.53	90.53	95.02	100.43	104.21
75	77.46	79.13	80.91	82.86	85.07	87.69	91.06	96.22	100.84	106.39	110.29
80	82.57	84.28	86.12	88.13	90.41	93.11	96.58	101.88	106.63	112.33	116.32
85	87.67	89.43	91.32	93.39	95.73	98.51	102.08	107.52	112.39	118.24	122.32
90	92.76	94.58	96.52	98.65	101.05	103.90	107.57	113.15	118.14	124.12	128.30
95	97.85	99.72	101.72	103.90	106.36	109.29	113.04	118.75	123.86	129.97	134.25
100	102.95	104.86	106.91	109.14	111.67	114.66	118.50	124.34	129.56	135.81	140.17

Table A.3 SCF data set.

V1 = ID number
V2 = Family size
V3 = Education
V4 = Age
V5 = Experience
V6 = Months worked
V7 = Race
V8 = Region
V9 = Earnings
V10 = Income
V11 = Wealth
V12 = Savings

V1	V2	V3	V4	V5	V6	V7	V8	V9	V10	V11	V12
1	4	2	40	33	12	2	3	1.920	1.920	0.470	0.030
2	4	9	33	19	12	1	1	12.403	12.403	3.035	0.874
3	2	17	31	9	12	1	4	5.926	6.396	2.200	0.370
4	3	9	50	36	12	1	2	7.000	7.005	11.600	1.200
5	4	12	28	11	12	1	3	6.990	6.990	0.300	0.275
6	4	13	33	15	12	1	1	6.500	6.500	2.200	1.400
7	5	17	36	14	12	1	3	26.000	26.007	11.991	31.599
8	5	16	44	23	12	1	1	15.000	15.363	17.341	1.766
9	5	9	48	34	12	2	3	5.699	14.999	9.852	3.984
10	5	16	31	10	12	1	3	8.820	9.185	8.722	1.017
11	10	9	41	27	12	1	4	7.000	10.600	0.616	1.004
12	4	10	41	26	12	1	1	6.176	12.089	23.418	0.687
13	7	11	36	20	12	1	2	6.200	6.254	7.600	−0.034
14	5	14	31	12	12	1	3	5.800	9.010	0.358	−1.389
15	5	7	27	15	12	1	2	6.217	6.217	0.108	1.000
16	5	8	42	29	12	1	2	5.500	5.912	5.560	1.831
17	4	12	28	11	11	1	1	4.800	4.800	0.970	0.613
18	2	6	46	35	12	2	3	1.820	2.340	2.600	0.050
19	3	12	47	30	12	1	4	4.558	7.832	31.867	0.013
20	7	8	35	22	12	1	2	7.468	9.563	1.704	1.389
21	3	9	41	27	9	1	1	6.600	7.600	4.820	0.602
22	4	17	30	8	12	1	1	12.850	13.858	32.807	2.221
23	6	12	38	21	12	1	1	5.800	5.802	10.305	1.588
24	3	11	48	32	12	1	3	7.479	19.362	12.652	5.082
25	3	10	36	21	12	1	1	5.700	8.000	7.631	1.846
26	3	12	45	28	12	1	1	12.000	17.200	14.392	0.914
27	6	8	44	31	6	1	1	3.578	4.091	6.649	2.483
28	4	10	44	29	12	1	3	9.600	9.600	6.995	0.837
29	3	3	46	38	12	1	3	3.686	10.425	9.138	1.274
30	4	12	26	9	12	1	3	6.480	6.512	2.933	−0.275
31	5	12	50	33	12	1	4	6.383	7.675	38.260	1.092
32	4	8	46	33	11	1	1	5.610	12.418	12.661	1.157
33	5	8	33	20	12	1	1	6.000	6.079	0.820	0.340
34	4	12	41	24	12	1	2	6.300	6.979	21.286	0.373
35	5	17	33	11	12	1	1	10.513	10.517	9.723	3.307
36	4	12	41	24	12	1	2	30.000	30.996	95.187	10.668
37	3	12	29	12	11	2	1	3.427	5.283	0.171	1.105
38	9	11	27	11	12	1	2	8.500	8.511	3.105	3.500
39	5	12	42	25	12	1	1	11.300	12.700	7.385	0.541
40	5	16	39	18	12	1	3	16.960	16.770	16.049	3.020

Note: V3, V4, V5 are in years; for V7, 1 = white, 2 = black; for V8, 1 = northeast, 2 = northcentral, 3 = south, 4 = west; V9, V10, V11, V12 are in thousands of current dollars.

V1	V2	V3	V4	V5	V6	V7	V8	V9	V10	V11	V12
41	6	12	36	19	12	1	1	8.300	8.300	0.050	0.650
42	4	8	34	21	12	1	2	5.375	5.375	4.464	0.989
43	4	12	40	23	12	1	4	4.770	6.265	7.203	2.532
44	4	12	37	20	12	1	2	4.320	8.520	9.145	6.120
45	5	17	44	22	12	1	4	10.720	24.226	54.524	−2.749
46	2	4	49	40	12	1	3	0.750	0.750	4.000	0.000
47	5	12	33	16	12	1	4	7.310	7.356	6.800	−1.036
48	6	14	36	17	12	1	3	9.000	9.000	6.890	1.351
49	4	15	51	31	12	1	1	14.000	14.660	13.500	−1.150
50	5	12	37	20	12	1	2	3.900	5.593	9.837	−0.248
51	4	19	33	9	12	1	2	10.000	11.841	10.384	0.388
52	4	14	39	20	12	1	3	7.200	7.700	6.842	1.157
53	3	12	44	27	12	1	3	6.500	10.550	4.929	1.656
54	4	7	50	38	12	1	2	8.000	13.700	34.124	3.959
55	4	12	39	22	12	1	2	9.500	12.242	11.731	5.369
56	6	7	46	34	12	1	2	6.000	7.803	5.695	1.405
57	4	12	43	26	12	1	3	6.400	9.879	25.029	0.220
58	6	11	40	24	12	2	3	5.190	9.154	0.600	−0.298
59	2	9	40	26	12	1	3	4.548	7.067	45.105	−0.276
60	8	7	39	27	12	1	2	4.860	4.496	8.511	−0.578
61	6	10	34	19	6	2	4	2.736	4.636	20.205	−1.360
62	4	10	32	17	12	2	4	6.000	9.003	4.727	5.277
63	3	16	42	21	12	1	1	7.800	13.820	2.270	0.980
64	2	8	52	39	12	1	4	6.163	8.891	18.916	2.637
65	6	12	29	12	12	1	1	8.600	8.632	14.194	0.984
66	2	12	27	10	12	1	3	7.899	8.385	13.662	−0.076
67	5	10	37	22	12	1	4	5.048	5.403	0.159	0.902
68	2	12	52	35	12	1	2	4.133	8.573	21.700	10.733
69	3	12	32	15	12	1	3	6.500	6.516	1.180	0.716
70	4	12	35	18	12	1	2	6.000	6.000	5.900	0.200
71	3	13	31	13	12	1	4	10.116	16.778	2.531	0.006
72	5	9	36	22	10	1	1	6.000	9.504	44.461	1.464
73	6	16	34	13	12	1	4	8.950	8.953	4.863	0.948
74	3	12	54	37	12	1	4	4.952	8.703	8.534	0.835
75	4	12	52	35	10	1	1	8.681	12.667	26.085	−2.883
76	6	9	28	14	12	1	2	6.500	6.504	3.775	0.298
77	6	12	44	27	12	1	4	7.668	8.180	3.032	0.481
78	4	17	29	7	12	1	2	11.600	11.600	2.167	5.033
79	4	9	50	36	7	1	3	3.100	5.602	5.072	−0.111
80	4	8	50	37	12	1	3	4.586	10.390	4.100	0.000

(continued)

Table A.3 (continued)

V1	V2	V3	V4	V5	V6	V7	V8	V9	V10	V11	V12
81	4	16	44	23	12	1	2	27.000	30.610	51.892	4.115
82	4	9	34	20	9	1	1	1.500	3.941	1.260	2.575
83	7	10	39	24	12	1	3	1.789	2.936	17.128	−0.112
84	5	12	39	22	12	1	4	11.068	11.068	11.542	−5.577
85	4	14	29	10	12	1	4	8.338	8.338	2.272	2.750
86	3	8	38	25	12	1	3	2.943	6.683	6.100	0.095
87	5	10	30	15	12	1	1	7.212	7.212	0.857	1.348
88	3	10	50	35	12	1	1	7.500	10.411	3.678	0.178
89	2	8	33	20	12	1	3	5.250	8.850	1.650	−0.695
90	4	9	35	21	12	1	1	5.066	8.334	2.143	0.787
91	3	16	36	15	12	1	2	12.848	13.923	18.182	4.642
92	4	12	33	16	12	1	2	6.214	6.214	0.275	1.260
93	6	20	38	13	12	1	1	12.202	12.323	28.953	2.687
94	4	12	46	29	12	1	2	8.190	14.963	11.230	0.720
95	4	16	50	29	12	1	2	7.200	10.060	25.462	5.109
96	2	16	54	33	12	1	1	30.000	32.080	98.033	1.800
97	5	12	31	14	12	1	2	9.190	9.260	5.539	1.684
98	2	18	27	4	12	1	2	7.500	10.450	2.860	1.475
99	5	12	40	23	12	1	3	7.852	9.138	11.197	0.566
100	6	18	34	11	12	1	1	12.000	12.350	30.906	25.405

Source: T. W. Mirer, *Economic Statistics and Econometrics*, 2d ed. (New York: Macmillan, 1988), pp. 18–23.

Table A.4 Noncentral chi-square: complement of cumulative distribution function.

λ^2	$k = 1$	$k = 2$	$k = 3$
0.000	0.050	0.050	0.050
0.500	0.109	0.090	0.081
1.000	0.170	0.133	0.116
1.500	0.232	0.178	0.153
2.000	0.293	0.226	0.192
2.500	0.353	0.274	0.233
3.000	0.410	0.322	0.275
3.500	0.465	0.369	0.317
4.000	0.516	0.415	0.359
4.500	0.564	0.460	0.400
5.000	0.609	0.504	0.440

Note: The entries are the values of $1 - G_k^*(c_k; \lambda^2)$, where $G_k^*(\cdot; \lambda^2)$ is the cdf of the noncentral chi-square distribution with degrees of freedom parameter k and noncentrality parameter λ^2, and c_k satisfies $G_k^*(c_k; 0) = 0.95$. Thus $c_1 = 3.841$, $c_2 = 5.991$, $c_3 = 7.815$. The table was constructed by using the GAUSS command "cdfchinc(c,k,m)", which gives $G_k^*(c; m^2)$.

Table A.5 TIM data set.

V1 = ID number
V2 = Year − 1900
V3 = GNP price index
V4 = Real GNP
V5 = Real gross private domestic investment
V6 = Real personal consumption
V7 = Real disposable personal income
V8 = Change in GNP price index
V9 = Change in consumer price index
V10 = Unemployment rate
V11 = Money stock (M1)
V12 = Treasury bill rate
V13 = Corporate bond rate (Moody's Aaa)

Note: V3 is equal to 100 in 1972; V4, V5, V6, V7 are in billions of 1972 dollars; V11 is in billions of current dollars; V8, V9, V12, V13 are in percent per year; V10 is in percent.

V1	V2	V3	V4	V5	V6	V7	V8	V9	V10	V11	V12	V13
1	56	62.79	671.6	102.6	405.4	446.2	3.205	1.496	4.1	135.0	2.658	3.36
2	57	64.93	683.8	97.0	413.8	455.5	3.408	3.563	4.3	133.8	3.267	3.89
3	58	66.04	680.9	87.5	418.0	460.7	1.710	2.728	6.8	138.9	1.839	3.79
4	59	67.60	721.7	108.0	440.4	479.7	2.362	0.808	5.5	141.2	3.405	4.38
5	60	68.70	737.2	104.7	452.0	489.7	1.627	1.604	5.5	142.2	2.928	4.41
6	61	69.33	756.6	103.9	461.4	503.8	0.917	1.015	6.7	146.7	2.378	4.35
7	62	70.61	800.3	117.6	482.0	524.9	1.846	1.116	5.5	149.4	2.778	4.33
8	63	71.67	832.5	125.1	500.5	542.3	1.501	1.214	5.7	154.9	3.157	4.26
9	64	72.77	876.4	133.0	528.0	580.8	1.535	1.309	5.2	162.0	3.549	4.40
10	65	74.36	929.3	151.9	557.5	616.3	2.185	1.722	4.5	169.6	3.954	4.49
11	66	76.76	984.8	163.0	585.7	646.8	3.228	2.857	3.8	173.8	4.881	5.13
12	67	79.06	1011.4	154.9	602.7	673.5	2.996	2.881	3.8	185.2	4.321	5.51
13	68	82.54	1058.1	161.6	634.4	701.3	4.402	4.200	3.6	199.5	5.339	6.18
14	69	86.79	1087.6	171.4	657.9	722.5	5.149	5.374	3.5	205.9	6.677	7.03
15	70	91.45	1085.6	158.5	672.1	751.6	5.369	5.920	4.9	216.8	6.458	8.04
16	71	96.01	1122.4	173.9	696.8	779.2	4.986	4.299	5.9	231.0	4.348	7.39
17	72	100.00	1185.9	195.0	737.1	810.3	4.156	3.298	5.6	252.4	4.071	7.21
18	73	105.69	1255.0	217.5	768.5	865.3	5.690	6.225	4.9	266.4	7.041	7.44
19	74	114.92	1248.0	195.5	763.6	858.4	8.733	10.969	5.6	278.0	7.886	8.57
20	75	125.56	1233.9	154.8	780.2	875.8	9.259	9.140	8.5	291.8	5.838	8.83
21	76	132.11	1300.4	184.5	823.7	907.4	5.217	5.769	7.7	311.1	4.989	8.43
22	77	139.83	1371.7	213.5	863.9	939.8	5.844	6.452	7.1	336.4	5.265	8.02
23	78	150.05	1436.9	229.7	904.8	981.5	7.309	7.658	6.1	364.2	7.221	8.73
24	79	162.77	1483.0	232.6	930.9	1011.5	8.477	11.259	5.8	390.5	10.041	9.63
25	80	177.36	1480.7	203.6	935.1	1018.4	8.964	13.523	7.1	415.6	11.506	11.94

Source: T. W. Mirer, *Economic Statistics and Econometrics,* 2d ed. (New York: Macmillan, 1988), pp. 24–25.

Table A.6 GEWE data set (V2–V7 in millions of 1947 dollars).

V1 = ID number
V2 = GE investment
V3 = GE market value
V4 = GE lagged capital stock
V5 = WE investment
V6 = WE market value
V7 = WE lagged capital stock

V1	V2	V3	V4	V5	V6	V7
1	33.1	1170.6	97.8	12.9	191.5	1.8
2	45.0	2015.8	104.4	25.9	516.0	0.8
3	77.2	2803.3	118.0	35.1	729.0	7.4
4	44.6	2039.7	156.2	22.9	560.4	18.1
5	48.1	2256.2	172.6	18.8	519.9	23.5
6	74.4	2132.2	186.6	28.6	628.5	26.5
7	113.0	1834.1	220.9	48.5	537.1	36.2
8	91.9	1588.0	287.8	43.3	561.2	60.8
9	61.3	1749.4	319.9	37.0	617.2	84.4
10	56.8	1687.2	321.3	37.8	626.7	91.2
11	93.6	2007.7	319.6	39.3	737.2	92.4
12	159.9	2208.3	346.0	53.5	760.5	86.0
13	147.2	1656.7	456.4	55.6	581.4	111.1
14	146.3	1604.4	543.4	49.6	662.3	130.6
15	98.3	1431.8	618.3	32.0	583.8	141.8
16	93.5	1610.5	647.4	32.2	635.2	136.7
17	135.2	1819.4	671.3	54.4	723.8	129.7
18	157.3	2079.7	726.1	71.8	864.1	145.5
19	179.5	2371.6	800.3	90.1	1193.5	174.8
20	189.6	2759.9	888.9	68.6	1188.9	213.5

Source: H. Theil, *Principles of Econometrics* (New York: John Wiley & Sons, 1971), p. 296.

Appendix B
Getting Started in GAUSS

These notes, adapted from material provided by Aptech Systems Inc., provide some important information about getting started using GAUSS. They refer to Version 1.49B. They do not in any way provide complete documentation, even for the topics covered.

Notation

>	Denotes the DOS prompt.
⟨ ⟩	Denotes a key on the keyboard. For example, ⟨F2⟩ denotes Function Key 2, while ⟨ – ⟩ denotes two keys pressed simultaneously, for example, ⟨Ctrl–F1⟩.
≫	Denotes the GAUSS prompt.
≪	Denotes the GAUSS program terminator.

Editing and Running Programs

1. To get into GAUSS from the operating system: > gauss ⟨Enter⟩ There may be a special command provided in your system.
2. To get out of GAUSS into the operating system: ⟨Esc⟩
3. GAUSS has two modes of operation: COMMAND MODE and EDIT MODE. When you first get into GAUSS you are in COMMAND MODE, as indicated by FILE=COMMAND at the bottom of the screen.
4. In COMMAND MODE you can write and run interactive programs. After the GAUSS prompt ≫, start writing GAUSS statements. End them with semicolons. After the last statement in the program, press ⟨F4⟩, and then ⟨F2⟩. For example:

    ≫ x1=rndu(100,3); x2=rndu(100,1); x=x1~x2; x; ⟨F4⟩ ⟨F2⟩

creates two random matrices and then concatenates them horizontally. The result, x, is a 100 × 4 matrix.

5. GAUSS does not care about blank spaces (with only a few exceptions), or about blank lines in programs. It does not distinguish uppercase and lowercase letters.
6. To get into EDIT MODE, use the command "edit" followed by the name of the file you want to edit. For example:

   ≫ edit myprog; ⟨F4⟩ ⟨F2⟩

   To get out of EDIT MODE and back into COMMAND MODE, use a function key:

   ⟨F1⟩ SAVE: saves the file you are editing, without running it.
   ⟨F2⟩ RUN: saves the file you are editing, and will try to run it.
   ⟨F4⟩ QUIT: drops you back to COMMAND MODE without saving file but clearing screen.

7. After a program file has been run from EDIT MODE, and you are back in COMMAND MODE, you can return to EDIT MODE to re-edit that file by pressing ⟨CTRL–F1⟩.
8. Programs written in EDIT MODE are just like those in COMMAND MODE, except that they do not contain the GAUSS prompt ≫ and program terminator ≪. To run a program from EDIT MODE, press ⟨F2⟩. Programs in EDIT MODE are automatically saved in a file when they are run.
9. To print output on the screen, write the name of the matrix. To print matrix x, for example, write: ≫ x;
10. The format of the output can be changed by using format w,p; here w is the width of the field for each number, and p is the number of decimal places. For example: ≫ format 8,2;
11. To send output to a file, use a command such as:

    ≫ output file = myoutput.out reset;

    After this command is executed, anything printed to the screen will also be sent to the file myoutput.out (which is first cleared).
12. Enclose comments with a combination of slashes and asterisks: /* */. For example:

    /*This is a comment; it will not be executed in a program.*/

13. To use a DOS command, precede it with the word "dos". For example:

    ≫ dos dir; ≫ dos del myfile; ≫ dos copy a:myfile c:;

14. Mathematical operators:

+ − * /	Perform in the standard way on scalars (add, subtract, multiply, divide). On matrices, +, −, and * have the standard definitions.
.* ./	Perform element-by-element multiplication and division, respectively, of matrices.
^	Performs element-by-element exponentiation (raising to a power) of the elements of a matrix.

15. Matrix operators:

~	Concatenates matrices horizontally.
|	Concatenates matrices vertically.
′	Transposes a matrix (interchanges rows and columns).

16. Mathematical functions:

cols(x)	Gives the number of columns in matrix x.
exp(x)	Raises e to powers given by elements of x.
ln(x)	Natural logs (base e) of elements of x.
log(x)	Common logs (base 10) of elements of x.
meanc(x)	Means of the columns of x.
rows(x)	The number of rows in matrix x.
sqrt(x)	Square roots of elements in x.
sumc(x)	Sums of the columns of x.

17. Defining matrices:

eye(k)	k × k identity matrix.
let	Allows matrices to be defined explicitly: let x[2,2] = 1 8 −12 15; [a 2 × 2 matrix] let x = 1 5 7 9; [a 4 × 1 vector] let x = dog cat; [matrix with character elements]
ones(n,k)	n × k matrix of 1's.
rndn(n,k)	n × k matrix of standard normal random variables.
zeros(n,k)	n × k matrix of 0's.

18. Indexing elements of a matrix:

x[i,j] The i,j element of x.
x[.,j] jth column of x.
x[i,.] ith row of x.
x[m:n,j] Rows m through n of column j of x.
x[rv,cv] Rows of x specified in the vector rv, columns of x specified in the vector cv.

19. Loading and saving matrices:

save x; Save matrix x as file named x.fmt.
load x; Load matrix x from file named x.fmt.

20. Flow control:

do until . . . ; . . . ; endo; Do loop. For example:
    i=1; do until i > 10; i=i+1; endo;
if . . . ; . . . ; endif; If statement. For example:
    if age[i,1] < 10; dage[i,1]=1;
    elseif age[i,1] > 10 and age [i,1] <= 20;
        dage[i,1]=2;
    else; dage[i,1]=3; endif;

Simple Exercises

To do these exercises, first get into GAUSS. After GAUSS is loaded into memory, the GAUSS prompt will appear on the screen, preceded by the letter of the default disk drive. Type the exercises below, exactly as they are written. The exercises are all indented and begin with the GAUSS prompt ≫. They all end with ⟨F4⟩ ⟨F2⟩, the keystrokes that tell GAUSS that you are done with the program, and that you want it run. When you press ⟨F4⟩ ⟨F2⟩, the GAUSS program terminator symbol ≪ appears on the screen.

The exercises are sequential, in that each uses results in memory that have been created by the preceding ones.

1. Generate a matrix of random numbers, x, and print it out:

    ≫ x=rndn(2,2); x; ⟨F4⟩ ⟨F2⟩

2. Generate another matrix, y, multiply x and y, and print the result:

    ≫ y=rndu(2,2); y; z=x*y; z; ⟨F4⟩ ⟨F2⟩

3. Define a 2 × 2 matrix using a specified set of numbers:

   ≫ let w[2,2]=1 2 3 4; w; ⟨F4⟩ ⟨F2⟩

4. Sum the elements of each column of w, and find the means of each column:

   ≫ sumc(w); meanc(w); ⟨F4⟩ ⟨F2⟩

5. Recover the screen as it was before the last exercise by pressing ⟨F1⟩. Edit the code by placing a transpose operator symbol before each semicolon:

   ≫ sumc(w)′; meanc(w)′; ⟨F4⟩ ⟨F2⟩

6. Sum the elements of each column of w, assign the result to sw, and print:

   ≫ sw=sumc(w); sw; ⟨F4⟩ ⟨F2⟩

7. Define a 4 × 1 vector using a specified set of numbers. The new vector will be given the name w, so that the old w vanishes:

   ≫ let w=1 2 3 4; w; ⟨F4⟩ ⟨F2⟩

8. Generate two matrices with specified elements, and sum them:

   ≫ let x[2,2]=10 9 8 7; let y[2,2]=−1 −2 −3 −4; x; y; w=x+y;w; ⟨F4⟩ ⟨F2⟩

9. Multiply each element in x by each element in y; divide each element in y by each element in x:

   ≫ x .* y; y ./ x; ⟨F4⟩ ⟨F2⟩

10. Pull out the first row of x, and assign it to w; then pull out the first column of x, and assign it to z; then print both w and z:

    ≫ w=x[1,.]; z=x[.,1]; w; z; ⟨F4⟩ ⟨F2⟩

11. Concatenate x and y. First do it horizontally, then vertically:

    ≫ z1 = x~y; z2 = x|y; z1; z2; ⟨F4⟩ ⟨F2⟩

12. Generate and print a 2 × 2 identity matrix, a 2 × 2 matrix of 1's, and a 2 × 2 matrix of 0's:

    ≫ k=2; i=eye(k); u=ones(k,k); z=zeros(k,k); i; u; z;⟨F4⟩ ⟨F2⟩

13. Save a matrix to a file:

    ≫ save u; ⟨F4⟩ ⟨F2⟩

14. Look for the file u.fmt on the default drive:

    ≫ dos dir u.fmt; ⟨F4⟩ ⟨F2⟩

15. Set the matrix u equal to scalar 0, print it, then load u from memory and print it again:

    ≫ u=0; u; load u; u; ⟨F4⟩ ⟨F2⟩

16. Find out how many rows and columns a matrix has, and print:

    ≫ format 1,0; "The matrix x has " rows(x) " rows, and "
    cols(x) " columns."; ⟨F4⟩ ⟨F2⟩

17. Using a loop, generate a sequence of numbers, and print the numbers divided by 10:

    ≫ i=0; format 2,1;
    do until i>33; print i/10;; i=i+1; endo; ⟨F4⟩ ⟨F2⟩

References

Amemiya, T. 1985. *Advanced Econometrics.* Cambridge, Mass.: Harvard University Press.

Conlisk, J. 1971. "When collinearity is desirable." *Western Economic Journal* 9:393–407.

DeGroot, M. H. 1975. *Probability and Statistics.* Reading, Mass.: Addison-Wesley.

Frisch, R., and F. V. Waugh. 1933. "Partial time regressions as compared with individual trends." *Econometrica* 1:387–401.

Goldberger, A. S. 1964. *Econometric Theory.* New York: John Wiley & Sons.

Gouriéroux, C., A. Holly, and A. Monfort. 1982. "Likelihood ratio test, Wald test, and Kuhn-Tucker test in linear models with inequality constraints on the regression parameters." *Econometrica* 50:63–80.

Greene, W. H. 1990. *Econometric Analysis.* New York: Macmillan.

Intriligator, M. D. 1978. *Econometric Models, Techniques, and Applications.* Englewood Cliffs, N.J.: Prentice-Hall.

Johnston, J. J. 1984. *Econometric Methods.* 3d ed. New York: McGraw-Hill.

Judge, G. G., R. C. Hill, W. E. Griffiths, H. Lütkepohl, and T.-C. Lee. 1988. *Introduction to the Theory and Practice of Econometrics.* 2d ed. New York: John Wiley & Sons.

Kakwani, N. C. 1967. "The unbiasedness of Zellner's seemingly unrelated regression equations estimators." *Journal of the American Statistical Association* 62:141–142.

Kosobud, R., and J. N. Morgan, eds. 1964. *Consumer Behavior of Individual Families over Two and Three Years.* Ann Arbor: Institute for Social Research, The University of Michigan.

Leamer, E. E. 1983. "Let's take the con out of econometrics." *American Economic Review* 73:31–43.

Lovell, M. 1983. "Data mining." *Review of Economics and Statistics* 65:1–12.

McCloskey, D. N. 1985. "The loss function has been mislaid: the rhetoric of significance tests." *American Economic Review* 75:201–205.

Maddala, G. S. 1983. *Limited-Dependent and Qualitative Variables in Econometrics.* London: Cambridge University Press.

Manski, C. F. 1988. *Analog Estimation Methods in Econometrics.* New York: Chapman and Hall.

Marschak, J. 1953. "Economic measurements for policy and prediction," pp. 1–26 in W. C. Hood and T. C. Koopmans, eds., *Studies in Econometric Method.* New York: John Wiley & Sons.

Mirer, T. W. 1988. *Economic Statistics and Econometrics.* 2d ed. New York: Macmillan.

Rao, C. R. 1973. *Linear Statistical Inference and Its Applications.* 2d ed. New York: John Wiley & Sons.

Theil, H. 1971. *Principles of Econometrics.* New York: John Wiley & Sons.

Wallace, T. D., and V. G. Ashar. 1972. "Sequential methods in model construction." *Review of Economics and Statistics* 54:172–178.

Wolak, F. A. 1987. "An exact test for multiple inequality and equality constraints in the linear regression model." *Journal of the American Statistical Association* 82:782–793.

Zellner, A. 1962. "An efficient method of estimating seemingly unrelated regressions and tests for aggregation bias." *Journal of the American Statistical Association* 57:348–368.

Index